New Perspectives in Magnesium Research

Yoshiki Nishizawa, Hirotoshi Morii, and
Jean Durlach, Eds.

New Perspectives in Magnesium Research

Nutrition and Health

Yoshiki Nishizawa, MD, PhD
Dean of Medicine
Professor of Internal Medicine
Department of Metabolism, Endocrinology
   and Molecular Medicine
Osaka City University Graduate School of
   Medicine
Osaka, Japan

Hirotoshi Morii, MD, PhD
Emeritus Professor
Osaka City University
Osaka, Japan

Jean Durlach, MD, PhD
President of the International Society for the
   Development of Research on Magnesium
   (SDRM)
Université Pierre et Marie Curie
Paris, France
and
Editor-in-Chief Magnesium Research journal

British Library Cataloguing in Publication Data
International Magnesium Symposium (11th : 2006 : Osaka, Japan)
   New perspectives in magnesium research : nutrition and research
   1. Magnesium – Physiological effect – Congresses   2. Minerals in human nutrition –
   Congresses   3. Metabolism – Disorders – Congresses
   I. Title   II. Nishizawa, Yoshiki, 1945–   III. Morii, H.
   (Hirotoshi)   IV. Durlach, Jean
   612.3′924

ISBN-13: 9781846283888
ISBN-10: 1846283884

Library of Congress Control Number: 2006925869

ISBN-10: 1-84628-388-4   Printed on acid-free paper
ISBN-13: 978-1-84628-388-8

© Springer-Verlag London Limited 2007

Apart from any fair dealing for the purposes of research or private study, or criticism or review, as permitted under the Copyright, Designs and Patents Act 1988, this publication may only be reproduced, stored or transmitted, in any form or by any means, with the prior permission in writing of the publishers, or in the case of reprographic reproduction in accordance with the terms of licences issued by the Copyright Licensing Agency. Enquiries concerning reproduction outside those terms should be sent to the publishers.
The use of registered names, trademarks, etc. in this publication does not imply, even in the absence of a specific statement, that such names are exempt from the relevant laws and regulations and therefore free for general use.
Product liability: The publisher can give no guarantee for information about drug dosage and application thereof contained in this book. In every individual case the respective user must check its accuracy by consulting other pharmaceutical literature.

9 8 7 6 5 4 3 2 1

Spinger Science+Business Media
springer.com

Preface

Magnesium: New Vistas

Eighteen years after the 5th International Magnesium Symposium was held in Kyoto in 1988, the Japanese Society for Magnesium Research (JSMR) is organizing the 11th International Magnesium Symposium in ISE-Osaka. The vitality of the Japanese Society has been evidenced by the regular publication of its journal (Journal of the Japanese Society of Magnesium Research), which ensures regular updating of current Japanese research on magnesium, and by its participation in the activities of the international Society for the Development of Research on Magnesium (SDRM).

Thanks to the efficient management of Yoshiki Nishizawa, Hirotoshi Morii, and Masaaki Inaba, the publication of this book of proceedings provides new vistas on magnesium research in 2006. It comprises reports presented during the plenary sessions concerning the role of magnesium not only in physiology, chronobiology, nutrition, epidemiology, and internal medicine particularly, cardiology, neurology, rheumatology, gynecology, and nephrology, but also in stomatology, pharmacognosy, and veterinary medicine.

Unlike many books that promise more than their texts deliver, the multiplicity and interest of the discussed topics can but stimulate careful reading of this book, which gives a platform to authors originating from many parts of the world. Published 35 years after the 1st International Magnesium Symposium, it demonstrates that magnesium research stands the test of time and that a wide path remains open for further investigation.

Such as it is, this book provides a valuable tool for interdisciplinary research in the future.

Jean Durlach
September 2006

Preface

New Perspectives in Magnesium Research is published as a document of the 11th International Magnesium Symposium, a joint meeting of Japanese Society for Magnesium Research. Professor Jean Durlach has been President of International Society for Magnesium Research since it was established in 1970. There are international meetings every 2 to 3 years in many parts of the world. The last meeting hosted by the Japanese Society for Magnesium Research was in Kyoto in 1988, when the Society President was Professor Yoshinori Itokawa.

Magnesium is the most abundant cation, second only to potassium in the intracellular compartment and to calcium in bone tissues. One of the big differences between calcium and magnesium is the difference between intra- and extracellular levels. Extra- and intracellular ratio is 10,000 for calcium, but 0.33 for magnesium. Such facts have significance for maintenance of life. Intracellular space rich in magnesium is just like seawater, adequate for the origin of life and maintenance of life in sea animals and plants. Bone magnesium constitutes a part of hydroxyapatite and may be important in maintaining bone integrity.

In physiological situations, magnesium is involved in metabolism of fat, amino acids, and sugar. Magnesium plays an important role in PPAR (peroxisome proliferators-activated receptor)–mediated signaling pathways as a key cofactor in the protein phosphorylation.

In bone and calcium metabolism, magnesium seems to play roles as a constituent of bone. Experimentally, magnesium deficiency induces osteoporosis but there have not been definite evidences of correlation between osteoporosis and magnesium. In calcium- and magnesium-deficient animals, calcium restores the bone quality but magnesium only partially. Another aspect of role of magnesium is the regulation of calcium metabolism through calcium-sensing receptor, which requires magnesium for its action. Another point is how magnesium participates in vitamin D action. Vitamin D is another important factor, other than calcium-sensing receptor, that requires magnesium in its action. The correlation between calcium-sensing receptor and vitamin D receptor has been discussed. Thus, magnesium plays some roles in the regulation of calcium and bone metabolism.

Clinically, diabetes mellitus, atherosclerosis, hypertension, cardiovascular diseases, and hyperlipidemia are in some other ways influenced by magnesium. More evidence is needed to have a definite conclusion how to manage clinical problems, problems of public health, and individual nutrition.

Hirotoshi Morii
September 2006

Contents

Preface v
Jean Durlach

Preface vii
Hirotoshi Morii

List of Contributors xiii

OVERVIEW

1 Overview of Magnesium Research: History and Current
   Trends 3
   Jean Durlach

2 Magnesium and Calcium in Drinking Water 11
   *Hirotoshi Morii, Ken-ichi Matsumoto, Ginji Endo,
   Mieko Kimura, and Yoshitomo Takaishi*

MAGNESIUM HOMEOSTASIS

3 Cellular Mg^{2+} Transport and Homeostasis: An
   Overview 21
   Martin Kolisek, Rudolf J. Schweyen, and Monika Schweigel

4 TRPM6 and TRPM7 Chanzymes Essential for Magnesium
   Homeostasis 34
   *Wouter M. Tiel Groenestege, Joost G.J. Hoenderop, and
   René J.M. Bindels*

5 CorA-Mrs2-Alr1 Superfamily of Mg^{2+} Channel Proteins 46
   Rudolf J. Schweyen and Elisabeth M. Froschauer

6 Practical Interest of Circulating Total and Ionized Magnesium
   Concentration Evaluation in Experimental and Clinical
   Magnesium Disorders 55
   *Nicole Pagès, Pierre Bac, Pierre Maurois, Andrée Guiet-Bara,
   Michel Bara, and Jean Durlach*

NUTRITION

7 Overview of Magnesium Nutrition 69
  Mieko Kimura

8 Magnesium Requirement and Affecting Factors 94
  *Mamoru Nishimuta, Naoko Kodama, Eiko Morikuni, Nobue
  Matsuzaki, Yayoi H. Yoshioka, Hideaki Yamada,
  Hideaki Kitajima, and Hidemaro Takeyama*

9 Experimental Data on Chronic Magnesium Deficiency 104
  *Pilar Aranda, Elena Planells, C. Sánchez, Bartolomé Quintero,
  and Juan Llopis*

10 Clinical forms of Magnesium Depletion by Photosensitization
   and Treatment with Scototherapy 117
   Jean Durlach

11 New Data on Pharmacological Properties and Indications of
   Magnesium 127
   Pedro Serrano, Maríasol Soria, and Jesús F. Escanero

EPIDEMIOLOGY

12 Magnesium Intake and the Incidence of Type 2 Diabetes 143
   Fernando Guerrero-Romero and Martha Rodríguez-Morán

13 Magnesium Intake and Hepatic Cancer 155
   Andrzej Tukiendorf

EXERCISE

14 Exercise and Magnesium 173
   Maria José Laires and Cristina Monteiro

METABOLIC SYNDROME

15 Review of Magnesium and Metabolic Syndrome 189
   Yasuro Kumeda and Masaaki Inaba

16 Diabetes Mellitus and Magnesium 197
   Masanori Emoto and Yoshiki Nishizawa

17 Magnesium Metabolism in Insulin Resistance, Metabolic
   Syndrome, and Type 2 Diabetes Mellitus 213
   *Mario Barbagallo, Ligia J. Dominguez, Virna Brucato,
   Antonio Galioto, Antonella Pineo, Anna Ferlisi,
   Ernesto Tranchina, Mario Belvedere, Ernesto Putignano,
   and Giuseppe Costanza*

CARDIOVASCULAR DISEASE

18 Cardiovascular Disease and Magnesium 227
    Naoyuki Hasebe and Kenjiro Kikuchi

19 Magnesium: Forgotten Mineral in Cardiovascular Biology and
    Atherogenesis 239
    Burton M. Altura and Bella T. Altura

SKELETAL DISEASES AND CALCIUM METABOLISM

20 Overview of Skeletal Diseases and Calcium Metabolism in
    Relation to Magnesium 263
    Hirotoshi Morii

21 Magnesium and Osteoporosis 266
    *Hirotoshi Morii, Takehisa Kawata, Nobuo Nagano,
    Takashi Shimada, Chie Motonaga, Mariko Okamori,
    Takao Nohmi, Takami Miki, Masatoshi Kobayashi, K. Hara,
    and Y. Akiyama*

22 Calcium-Sensing Receptor and Magnesium 272
    Karl Peter Schlingmann

KIDNEY

23 Magnesium and the Kidney: Overview 289
    Linda K. Massey

24 Cellular Basis of Magnesium Transport 293
    *Pulat Tursun, Michiko Tashiro, Masaru Watanabe, and
    Masato Konishi*

25 Magnesium in Chronic Renal Failure 303
    Juan F. Navarro and Carmen Mora-Fernández

26 Magnesium in Hemodialysis Patients 316
    Senji Okuno and Masaaki Inaba

NEUROLOGY

27 Neurology Overview 333
    Keiichi Torimitsu

28 Magnesium in the Central Nervous System 338
    Renee J. Turner and Robert Vink

MAGNESIUM IN DENTAL MEDICINE

29 Tooth and Magnesium 359
    Masayuki Okazaki

MAGNESIUM IN NEUROLOGY AND PSYCHIATRY

30 Magnesium in Psychoses 369
    Mihai Nechifor

VETERINARY MEDICINE

31 Significance of Magnesium in Animals 381
    Tohru Matsui

Index 393

Contributors

Y. Akiyama
Department of Pharmacology
Eisai Co., Ltd
Tokyo, Japan

Bella T. Altura, PhD
Physiology and Pharmacology
State University of New York
Downstate Medical Center
Brooklyn, NY, USA

Burton M. Altura, PhD
Physiology and Pharmacology, and
 Medicine
State University of New York
Downstate Medical Center
Brooklyn, NY, USA

Pilar Aranda, PhD
Department of Physiology
School of Pharmacy
University of Granada
Granada, Spain

Pierre Bac, PhD
Pharmacology Laboratory
Faculty of Pharmacy
F-Châtenay Malabry, France

Michel Bara, PhD
Physiology and Physiopathology
 Laboratory
Pierre and Marie Curie University
Paris, France

Mario Barbagallo, MD, PhD
Geriatric Unit
Department of Internal Medicine
 and Geriatrics
University of Palermo
Palermo, Italy

Mario Belvedere, MD
Department of Internal Medicine
 and Geriatrics
University of Palermo
Palermo, Italy

René J. Bindels, PhD
Department of Physiology
Radboud Univesity Nijmegen
Medical Centre
Nijmegen Centre for Molecular Life
 Sciences (NCMLS)
The Netherlands Nijmegen

Virna Brucato, MD
Department of Internal Medicine
 and Geriatrics
University of Palermo
Palermo, Italy

Giuseppe Costanza, MD
Department of Internal Medicine
 and Geriatrics
University of Palermo
Palermo, Italy

xiv **Contributors**

Ligia J. Dominguez, MD
Department of Internal Medicine
 and Geriatrics
University of Palermo
Palermo, Italy

Jean Durlach, MD, PhD
International Society for the
 Development of Research on
 Magnesium (SDRM)
Université Pierre et Marie Curie
Paris, France

Masanori Emoto, MD, PhD
Department of Metabolism,
 Endocrinology, and Molecular
 Medicine
Osaka City University Graduate
 School of Medicine
Osaka, Japan

Ginji Endo, MD, PhD
Department of Preventive Medicine
 and Environmental Medicine
Osaka City University
Osaka, Japan

Jesús F. Escanero, MD, PhD
Department of Pharmacology and
 Physiology
University of Zaragoza
Zaragoza, Spain

Anna Ferlisi, MD
Department of Internal Medicine
 and Geriatrics
University of Palermo
Palermo, Italy

Elisabeth M. Froschauer
Max F. Perutz Laboratories
University of Vienna
Vienna, Austria

Antonio Galioto, MD
Department of Internal Medicine
 and Geriatrics
University of Palermo
Palermo, Italy

Wouter M. Tiel Groenestege
Department of Physiology
Radboud University Nijmegen
 Medical Centre
Nijmegen Centre for Dalccular Life
 Sciences (NCMTS)
Nijmegen, the Netherlands

Fernando Guerrero-Romero, MD,
 PhD, FACP
Medical Research Unit in Clinical
 Epidemiology
Mexican Social Security Institute
and
Research Group on Diabetes and
 Chronic Illnesses
Durango, Mexico

Andrée Guiet-Bara, MD
Physiology and Physiopathology
 Laboratory
Pierre and Marie Curie University
Paris, France

K. Hara
Department of Pharmacology
Eisai Co., Ltd
Tokyo, Japan

Naoyuki Hasebe, MD, PhD
First Department of Medicine
Asahikawa Medical College
Asahikawa, Japan

Joost G.J. Hoenderop
Department of Physiology
Radboud University Nijmegen
 Redical Centre
Nijmegen Centre for Ralcular Life
 Scicuces (NCMLS)
Nijmegen, the Netherlands

Masaaki Inaba, MD, PhD
Metabolism, Endocrinology and
   Molecular Medicine
Osaka City University Graduate
   School of Medicine
Osaka, Japan

Takehisa Kawata, PhD
Pharmaceutical Division
Pharmaceutical Development
   Laboratories
Pharmacology Group, Nephrology
Kirin Brewery Co., Ltd
Takasaki, Japan

Kenjiro Kikuchi, MD, PhD
First Department of Medicine
Asahikawa Medical College
Asahikawa, Japan

Mieko Kimura, PhD
Takeda Research Institute of Life
   Science and Preventive Medicine
Kyoto, Japan

Hideaki Kitajima, PhD
Taisho Pharmaceutical Co. Ltd
Tokyo, Japan

Masatoshi Kobayashi, PhD
Department of Pharmacology
Eisai Co., Ltd
Tokyo, Japan

Naoko Kodama, RN
Laboratory of Clinical Nutrition
Tokyo Dietitian Academy
Tokyo, Japan

Martin Kolisek, PhD
Department of
   Veterinary-Physiology
Free University of Berlin
Berlin, Germany

Masato Konishi, MD, PhD
Department of Physiology
Tokyo Medical University
Tokyo, Japan

Yasuro Kumeda, MD, PhD
Department of Internal Medicine
Jyurakukai Ohno Memorial Hospital
Osaka, Japan

Maria José Laires, PhD
Biochemistry Laboratory
Faculdade de Motricidade Humana
Universidade Técnica de Lisboa
Lisbon, Portugal

Juan Llopis, PhD
Department of Physiology
School of Pharmacy
University of Granada
Granada, Spain

Linda K. Massey, PhD, RD
Human Nutrition
Food Science and Human Nutrition
Washington State University
Spokane, WA, USA

Tohru Matsui, PhD
Laboratory of Nutritional Science
Division of Applied Biosciences
Graduate School of Agriculture
Kyoto University
Kyoto, Japan

Ken-ichi Matsumoto
Sakai Institute of Public Health
Sakai, Japan

Nobue Matsuzaki, RN
Laboratory of Mineral Nutrition
Division of Human Nutrition
The Incorporated Administrative
   Agency of Health and Nutrition
Tokyo, Japan

Pierre Maurois, MD
Pharmacology Laboratory
Faculty of Pharmacy
F-Châtenay Malabry, France

Takami Miki, MD, PhD
Department of Geriatrics and Neurology
Graduate School of Osaka City University
Osaka, Japan

Tokuhiko Miki, MD
Osaka City University
Osaka, Japan

Cristina Monteiro, PhD
Biochemistry Laboratory
Faculdade de Motricidade Humana
Universidade Técnica de Lisboa
Lisbon Portugal

Carmen Mora-Fernández, MD
Research Unit
Hospital Universitario Nuestra Señora de Candelaria
and
Instituto Canario de Investigación Biomédica
Santa Cruz de Tenerife, Spain

Hirotoshi Morii, MD, PhD
Emeritus Professor Osaka City University
Osaka, Japan

Eiko Morikuni, RN
Laboratory of Mineral Nutrition
Division of Human Nutrition
The Incorporated Administrative Agency of Health and Nutrition
Tokyo, Japan

Chie Motonaga
Central Research Laboratories
Nichinichi Pharmaceutical Co., Ltd
Mie, Japan

Nobuo Nagano, PhD
Pharmaceutical Division
Pharmaceutical Development Laboratories
Pharmacology Group, Nephrology
Kirin Brewery Co., Ltd
Takasaki, Japan

Juan F. Navarro, MD
Nephrology Service and Research Unit
Hospital Universitario Nuestra Señora de Candelaria
and
Instituto Canario de Investigación Biomédica
Santa Cruz de Tenerife, Spain

Mihai Nechifor, MD, PhD
Department of Pharmacology
University of Medicine and Pharmacy "Gr. T. Popa"
Iasi, Romania

Mamoru Nishimuta, MD, PhD
Laboratory of Mineral Nutrition
Division of Human Nutrition
The Incorporated Administrative Agency of Health and Nutrition
Tokyo, Japan

Yoshiki Nishizawa, MD, PhD
Department of Metabolism, Endocrinology and Molecular Medicine
Osaka City University Graduate School of Medicine
Osaka, Japan

Takao Nohmi
Central Research Laboratories
Nichinichi Pharmaceutical Co., Ltd
Mie, Japan

Mariko Okamori
Central Research Laboratories
Nichinichi Pharmaceutical Co., Ltd
Mie, Japan

Masayuki Okazaki, PhD
Department of Biomaterials Science
Graduate School of Biomedical Sciences
Hiroshima University
Hiroshima, Japan

Senji Okuno, MD, PhD
Kidney Center
Shirasagi Clinic
Osaka, Japan

Nicole Pagès, PhD
Toxicology Laboratory
Faculty of Pharmacy
Illkirch-Graffenstaden, France

Antonella Pineo, MD
Department of Internal Medicine and Geriatrics
University of Palermo
Palermo, Italy

Elena Planells, PhD
Department of Physiology
School of Pharmacy
University of Granada
Granada, Spain

Ernesto Putignano, MD
Department of Internal Medicine and Geriatrics
University of Palermo
Palermo, Italy

Bartolomé Quintero
Department of Physical Chemistry
School of Pharmacy
University of Granada
Granada, Spain

Martha Rodríguez-Morán, MD, MSc, PhD
Medical Research Unit in Clinical Epidemiology
Mexican Social Security Institute
Durango, Mexico
Research Group on Diabetes and Chronic Illnesses
Durango, Mexico

C Sánchez
Department of Physiology
School of Pharmacy, University of Granada
Granada, Spain

Karl Peter Schlingmann, MD
Department of Pediatrics
University Children's Hospital
Philipps-University
Marburg, Germany

Monika Schweigel, PhD
Department of Nutritional Physiology "Oskar Kellner"
Research Institute for the Biology of Farm Animals – FBN
Dummerstorf, Germany

Rudolf J. Schweyen, PhD
Max F. Perutz Laboratories
University of Vienna
Vienna, Austria

Pedro Serrano, MD, PhD, FESC
Department of Pharmacology and Physiology
University of Zaragoza
Zaragoza, Spain

Takashi Shimada, PhD
Central Research Laboratories
Nichinichi Pharmaceutical Co., Ltd
Mie, Japan

Maríasol Soria, MD, PhD
Department of Pharmacology and
   Physiology
University of Zaragoza
Zaragoza, Spain

Yoshitomo Takaishi, DDS
Takaishi Dental Clinic
Himeji, Japan

Hidemaro Takeyama, PhD, RN
Graduate School of Medical Science
Nagoya City University
Nagoya City, Japan

Michiko Tashiro
Department of Physiology
Tokyo Medical University
Tokyo, Japan

Keiichi Torimitsu, PhD
Materials Science Laboratory
NTT Basic Research Laboratories
Atsugi, Japan

Ernesto Tranchina, MD
Department of Internal Medicine
   and Geriatrics
University of Palermo
Palermo, Italy

Andrzej Tukiendorf, PhD
Opole University of Technology
Opole, Poland

Renee J. Turner, BSc
Division of Pathology
School of Medical Sciences
University of Adelaide
Adelaide, SA, Australia

Pulat Tursun
Department of Physiology
Tokyo Medical University
Tokyo, Japan

Robert Vink, PhD
Neurosurgical Research
Division of Pathology
School of Medical Sciences
University of Adelaide
Adelaide, SA, Australia

Masaru Watanabe
Department of Physiology
Tokyo Medical University
Tokyo, Japan

Hideaki Yamada, RN
Mimasaka Women's University
Tsuyama City, Japan

Yayoi H. Yoshioka, RN
Laboratory of Mineral Nutrition
Division of Human Nutrition
The Incorporated Administrative
   Agency of Health and Nutrition
Tokyo, Japan

Overview

1
Overview of Magnesium Research: History and Current Trends

Jean Durlach

The lore of magnesium in medicine, starting as far back as the 17th century up to the first quarter of the 20th century, covers a large span of the chemical and pharmacological fields of knowledge.

The modern period, from 1926—when the essential character of magnesium was demonstrated—up to the 1960s, laid the basis for the present development of magnesium research by opening new vistas in the epidemiological and clinical fields where magnesium was acknowledged as being involved.

Several leading ideas deserve to be mentioned: (1) magnesium concerns all areas of medical activities; (2) the main expression of chronic marginal deficit is long-term chronic magnesium deficiency; (3) the specific etiological treatment of the responsible illness is the specific therapeutic measure meant to correct secondary magnesium deficit; (4) atoxic physiological palliative oral magnesium therapy is basically different from potentially noxious pharmacological use of magnesium; (5) the importance of distinguishing between magnesium deficiency and magnesium depletion in case of magnesium deficit; and (6) the interest of studying the chronopathological forms of magnesium deficit with hyper- or hypofunction of the biological clock.

This overview on past, present, and future bears witness to the vitality of magnesium research in health and disease.

Magnesium, the second most important intracellular cation, is found in all tissues and may affect many functions in the body. Its multiple physiological actions have been discovered thanks to the numerous convergent efforts of multinational research.

The aim of this study is to sum up the history of magnesium research and to highlight the current trends with a special stress on the leading ideas developed at the XIth International Magnesium Symposium (ISE, Japan).

This overview, which first considers the early history of the subject during the 18th and 19th centuries and first quarter of the 20th century, is mainly an account of the development of chemical and pharmacological knowledge. The modern history follows, which includes an initial period ending in the 1970s, when physiological, analytical, and epidemiological data established a firm background for the first clinical studies in the neurological and cardiovascular fields particularly.

The present period is characterized by an exponential development of magnesium research, as will be testified by the subjects of the multiple sessions at the ISE Osaka Symposium.

History[1–7]

One may consider the recognition by N. Grew, in 1695, of magnesium sulfate as one of the essential constituents of Epsom salts, as marking the entry of magnesium into medicine. N. Grew separated the solid salt in quantity from the bitter-tasting natural water of the Epsom spring. This latter was considered as an internal remedy and purifier of the blood and used by "a great store of citizens" and especially by "persons of quality," including Marie de Medicis in the 16th century. Other important springs also contained magnesium sulfate and Epsom salts, or sal anglicum, synonymous with Sedlitz, or Egra powder, or salt, to designate the first preparation of magnesium sulfate used in medicine, mainly as a purgative. It was considered as a typical saline cathartic.

In 1707, M.B. Valentini of Giessen processed "magnesia alba" from the mother liquors obtained in the manufacture of nitre. This by-product of the preparation of nitre was considered as "a panacea for all bodily ailments," but then magnesia alba and "calcareous earth" were confused. In 1755, J. Black of Edinburgh distinguished between magnesia and lime chemically.

In 1808, H. Davy of London isolated magnesium. Conducting his studies on alkali, that is, earth compounds, H. Davy succeeded in producing the amalgams of calcium, barium, strontium, and magnesium. He then isolated the metals by distilling off the mercury. As in the case of the alkali metals, he named these alkali–earth metals after their oxides: baryta, strontia, chalk, and magnesia, calling them barium, strontium, calcium, and magnium. Magnium has long been forgotten, however, and magnesium has been adopted by general usage for the element derived from magnesia.

In 1828, the French chemist A.A. Bussy, by reducing anhydrous magnesium chloride with potassium, prepared the metal in a state approximating purity.

In 1833, M. Faraday of London was the first to succeed in producing metallic magnesium by electrolysis of molten magnesium chloride. Electrolytic methods entirely superseded the older ones in the industrial production of magnesium until about 1941, when the carbothermic and ferrosilicon technique employing a thermal process came into use for a small proportion of magnesium production. In industry, magnesium was first used in photography and to make incendiary bombs; now it is in great demand for alloys and structural materials. Because of its lightness and abundance (2.1% of the earth's crust) "it has become the glamour metal of the space age." Aluminium, the nearest rival for structural purposes, is one and a half times as heavy.[1]

Two Nobel prizes are linked to work concerning the important role of magnesium in organic chemistry. V. Grignard was awarded his in 1912 for the description of organomagnesium compounds. Grignard reagents are those with the composition $RMgX$ (where R = an akyl or aryl group and X = a halogen). These compounds are primarily important as intermediates in a large scale of synthesis, in biology in the conversion of aldehyde or ketones to alcohols, as well as in organic chemistry in the production of silicones. R. Willstatter was awarded the Nobel prize in 1915 for demonstrating that the structure of chlorophyll consisted of the porphyrin system, the central magnesium atom with its complex linkage and the phytol radical.

These discoveries laid a scientific basis for the use of magnesium in fertilizers and for assessing its importance in phytophysiology.

Around 1900, the essentiality of magnesium was established by studies concerning several algae and inferior fungi.

The same notion concerning animals was demonstrated only after a further quarter of century. However, one can cite as a predecessor J. Gaube Du Gers who, more than a century ago, noticed in mice that a diet deficient in magnesium, that is, composed of bread devoid of magnesium and of distilled water, but also no doubt deficient in other nutrients, caused progressive sterility. Magnesium already appeared to Gaube Du Gers as "the metal of vital activity for what is most precious in life: reproduction and sensation."[1,3,7]

The first modern biological studies[1,2,4,6] concerning the pharmacodynamic properties of parenteral magnesium were carried out by the French.

In 1869, F. Jolyet showed that intravenous injection of magnesium in dogs induced a peripheral paralyzing action similar to that of curare. Unfortunately, these pertinent observations were considered physiological.

Thirty years later, the same misconstruction, through the assimilation of pharmacological and physiological data, was repeated by the American groups of J. Loeb, who included magnesium among the denominators of his coefficient in which the principal humoral factors controlling neuromuscular excitability were grouped, and of J. Meltzer, who analyzed the sedative pharmacological properties of magnesium. The latter also observed that the effects of intravenous administration of a solution of a calcium salt rapidly reversed the effects of magnesium given parenterally. Correspondingly, parenteral magnesium sulphate was used to treat tetanus and eclamptic, nephritic and tetanic convulsions. In cases of respiratory depression through magnesium therapeutic overload, calcium was administered as an antidote to magnesium intoxication.

Although these observations of the pharmacodynamic effects of high doses of oral or parenteral magnesium may have some interest per se, one should nevertheless not extrapolate from these pharmacodynamic effects to physiological properties. Evidence of the latter can be demonstrated only by the occurrence of symptoms due to magnesium deficiency followed by its specific control by administration of nutritional physiological oral doses of magnesium. One should however, note, the predictive character of hypotheses presented more than a century ago, and subsequently verified, on the etiological and physiopathological role of magnesium in certain tetanies, convulsions, and cardiovascular or toxic problems. We have cited here only carefully chosen examples, where modern studies have proved past extrapolations to be well founded, and not any others. The latter have, on the contrary, been the source of a number of unwarranted attributions of paternity as to the etiology of magnesium deficit. For this reason few ions have generated as much enthusiasm and as much disdain. Heated zealots, such as the French P. Delbet, have seen in magnesium a sort of panacea, the lack of which plays a major role in the development of cancer, the spread of epidemics, and even in the frequency of suicides. To the skeptics, on the other hand, magnesium remained a trace element of unclear biological physiological and pathological importance.

Between these two extremes it is today possible to find a balance.

The Early Modern Period

Beginning with the seminal experiment of J. Leroy, who proved in 1926 the essential character of the ion in mice, this era ended in the 1960s. The physiological properties of magnesium were mainly revealed in the 1930s by the remarkable studies in the rat of the American groups of E.V. MacCollum and D.M. Greenberg, who showed the multiple effects of a lack of magnesium intake on development, reproduction, the neuromuscular apparatus, and humoral balance. The specific reversibility of such defects by oral loading of magnesium provided the experimental basis for the diagnostic test for magnesium deficiency by oral loading of nutritional physiological doses of magnesium.

These physiological data constituted the background for the first clinical observations establishing the importance of abnormalities in magnesium metabolism in the period between the 1930s and 1960s. These observations concerned acute syndromes, for example, in veterinary medicine according to the work of the Dutchman B. Sjollema in 1932 on grass tetany (in cows or ponies) and the tetany of milk-fed calves and human pathology, where Americans E.B. Flink and W.E.C. Wacker in particular led the way with reports of acute neurological manifestations of primary or secondary magnesium deficit due to alcoholism or endocrinometabolic disorders.

Epidemiological data introduced two major considerations. First, in the 1960s, the American M. Seelig, as well as J. Durlach in France, emphasized the fact that dietary magnesium in many regions appeared to be insufficient to meet daily needs. Indeed, it may be said that magnesium intake in developed countries is marginal. Previously, the Japanese J.A. Kobayashi (1957) and the American H.A. Schroeder (1958) stressed an inverse relation between total drinking water hardness and cardiovascular risk.

In 1955, the Australian A. Walsh reported the first application of atomic absorption spectrophotometry to the determination of magnesium. Henceforth magnesium could be measured with ease and accuracy.

Current Trends[2-8]

The first international symposium on magnesium deficit in human pathology (1971) may be considered as evidence of the present importance of magnesium research in medicine: 350 participants from 52 countries contributed to the publication of two heavy volumes of proceedings corresponding to a synthesis of more than 25,000 references and which still offer a wide range of valuable knowledge.

The first International Symposium on Magnesium fostered the creation of the Society for the Development of Research on Magnesium (SDRM), an international coordinating structure.

The Society for the Development of Research on Magnesium promoted the publication of magnesium books: volume of proceedings[2,3,5,11-13] and monographs.[6]

The Society for the Development of Research on Magnesium organized the publication of diverse magnesium journals. First of all, *Magnesium Research*, the official organ of the international Society for the Development of Research on Magnesium (18 volumes), but also national journals such as the *Journal of the Japanese Society for Magnesium Research* (Japan), the *Buletin informatic al societatii romane de cercetare a magneziului* (Romania), and the *Journal of Elementology* (Poland).

Several leading ideas expressed in the course of the modern period of magnesium research deserve special mention:

- In the organism, all the systems and all the functions are involved. Therefore, magnesium research is not only relevant in certain fields but in all areas of medical activity.
- The main expression of primary magnesium deficit is closely linked with the consequences of long-term chronic marginal deficiency. Experimentally, it has been widely studied after the seminal report of O. Heroux in 1972. Clinical forms of chronic magnesium deficit are better and better identified, particularly in the neuromuscular system: for example, normocalcaemic latent tetany with or without mitral valve prolapse, and in the cardiovascular system, where the importance of magnesium deficit is recognized among cardiovascular risk factors.
- Specific therapeutic measures that seek to correct secondary magnesium deficit are justified only if the etiological treatment of the responsible illness is either impossible or ineffective. Such measures are then adjuvant treatments only justified by the ultimate importance of the secondary magnesium deficits in the physiopathology of the original illness.

Pharmacological effects of magnesium are observed irrespective of the magnesium status. They are established either in vitro, in situ, or in vivo when a parenteral or an oral intake is high enough to exceed any homeostatic capacity that may prevent magnesium overload. These basic differences between pharmacological and physiological magnesium actions lead us to discriminate between the two types of magnesium load tests. Clinical efficiency of parenteral magnesium administration should not be used as a diagnostic tool attesting to magnesium deficiency. Conversely, physiological oral doses of magnesium are totally devoid of the pharmacodynamic effects of parenteral magnesium and without clinical effects when magnesium status is normal. Correction of symptoms by this oral magnesium load constitutes the best proof that they were due to magnesium deficiency.

But the main consequence of the differentiation between the pharmacological properties of magnesium is to allow us to distinguish between two different types of therapy with magnesium: atoxic physiological oral therapy and pharmacological magnesium therapy that may induce toxicity. Today the main form of magnesium therapy is oral magnesium physiological nutritional supplementation of magnesium deficiency. The palliative oral doses meant to balance magnesium deficiency are obviously devoid of any toxicity because their purpose is to restore to normal the insufficiency of the magnesium intake. In order to use the pharmacological properties of magnesium, no matter what the magnesium status is, it is necessary to go beyond the mechanisms of magnesium homeostasis to induce a therapeutic magnesium overload. Large doses of magnesium given orally are advisable when there are chronic indications and the parenteral route suitable for acute applications. Both types of pharmacological magnesium treatment are capable of inducing toxicity. It is a real scientific fraud and an ethical misconduct to fail to differentiate between the absent toxic consequences of a physiological magnesium supplementation and the effects of high pharmacological magnesium doses that are potentially dangerous.

In the case of magnesium deficit it is important to differentiate *magnesium deficiency*, where the disorder corresponds to a negative balance, and which merely requires an increased intake of magnesium as treatment, from *magnesium depletion*, where the disorder is related to a dysregulation of the control mechanisms of magnesium metabolism, and which requires appropriate correction. Modern research is concerned with the analysis of the neuro–endocrino–metabolic factors that control or disturb magnesium metabolism associated with genetic factors either linked with major leucocyte complex or studied as an etiopathogenic factor of various congenital hypomagnesemia.[6,18] The nervous forms of magnesium deficit may be considered as typical examples.

The nervous form of primary chronic magnesium deficiency represents the best documented experimental and clinical aspect of chronic magnesium deficit. This neurotic, neuromuscular, and autonomic nervous form induced by magnesium deficiency merely requires nutritional oral physiological magnesium supplementation.[4,10,13]

In the case of neurodegenerative diseases linked with various types of magnesium depletion, simple nutritional magnesium supplementation is inefficient. Magnesium depletion requires more or less specific correction of its causal disregulation. Amyotrophic Lateral Sclerosis/Parkinsonism/Dementia complex (ALSPDC) may be considered as the type of a neurodegenerative disease linked with magnesium depletion due to the sum of magnesium- (and calcium-) deficient intake plus slow neurotoxins (either inorganic, i.e., Al, or organic, particularly cycad neurotoxins). Cycad seeds can be directly eaten or indirectly through traditional feasting on flying foxes (with possible biomagnification of neurotoxins).[6,12–15]

Both in clinical forms and in animal experiments, the dysregulation mechanism of magnesium depletion associates a reduced magnesium intake with various types of stress. Among these are the biological clock dysrhythmias with two opposite groups: (1) with a biological clock hyperfunction and (2) with a biological clock hypofunction. These symmetrical physiopathologies lead to opposed therapies: phototherapy in cases of a biological clock hyperfunction and scototherapy in cases of a biological clock hypofunction.[11,16–18]

Conclusion

Upon the initiative of SDRM and of Hirotoshi Morii, President of the Japanese Society for the Development of Research on Magnesium, the Eleventh International Magnesium Symposium will be held in ISE Osaka (October 24–28, 2006) under Yoshiki Nishizawa and Mieko Kimura. Once again, after the remarkable meeting organized 18 years ago in Kyoto (1988) by Yoshinori Itokawa, the Japanese branch of SDRM is organizing a meeting testifying to the exponential development of research on magnesium covering not only basic research (cellular and subcellular channels particularly), nutrition, epidemiology, and metabolic diseases, but also multifaceted clinical forms: cardiovascular, neuromuscular, psychiatric, rhumatologic, nephrologic, and even in dental and sport medicine.

References

1. Aikawa JK. *The Role of Mg in Biologic Processes.* Springfield, IL: CC Thomas; 1963.
2. Durlach J, ed. *1er Symposium International sur le déficit magnésique en pathologie humaine. I. Volume des rapports.* Vittel: SGEMV; 1971.
3. Durlach J, ed. *1er Symposium International sur le déficit magnésique en pathologie humaine, II. Volume des communications et discussions.* Vittel: SGEMV; 1973.
4. Durlach J. Neurological manifestations of magnesium imbalance. In: Vinken PJ, Bruyn GW, eds. *Handbook of Clinical Neurology.* Amsterdam: North Holland; 1976:545–579.
5. Itokawa Y, Durlach J. *Magnesium in Health and Disease.* London: John Libbey; 1989.
6. Durlach J. *Magnesium in Clinical Practice.* 1st ed. London: John Libbey; 1988. [Monograph updated: Durlach J, Bara M. *Le magnésium en biologie et en médecine.* 2nd ed. Paris: EMInter Tec et Doc; 2000.
7. Magnesium. In: *Encyclopedia Britannica,* vol. 14. Chicago: Encyclopedia Britannica; 1961;634–636.

8. Durlach J. Editorial Policy of *Magnesium Research*: general considerations on the quality criteria for biomedical papers and some complementary guidelines for the contributors of *Magnesium Research*. *Magnes Res* 1995;8:191-206.
9. Rayssiguier Y, Mazur A, Durlach J. *Advances in Magnesium Research: Nutrition and Health (MAG 2000, Vichy, France)*. London: John Libbey; 2001.
10. Durlach J, Bac P, Durlach V, Bara M, Guiet-Bara A. Neurotic, neuromuscular and autonomic nervous form of magnesium imbalance. *Magnes Res* 1997;10:169-195.
11. Escanero JF, Alda JO, Guerra M, Durlach J. *Magnesium Research: Physiology, Pathology and Pharmacology*. Prensas Universitarias de Zaragoza; 2003.
12. Nechifor M, Porr PJ. *Magnesium: Involvements in Biology and Pharmacotherapy*. Cluj-Napoca: Casa Cartii de Stiinta; 2003.
13. Porr PJ, Nechifor M, Durlach J. *Advances on Magnesium Research*. London: John Libbey; 2006.
14. Durlach J, Bac P, Durlach V, Durlach A, Bara M, Guiet-Bara A. Are age-related neurodegenerative diseases linked with various types of magnesium depletion? *Magnes Res* 1997;10:339-353.
15. Durlach J, Pagès N, Bac P, Bara M, Guiet-Bara A. Magnesium research: from the beginnings to today. *Magnes Res* 2004;17:163-168.
16. Durlach J, Pagès N, Bac P, Bara M, Guiet-Bara A, Agrapart C. Chronopathological forms of magnesium depletion with hypofunction or with hyperfunction of the biological clock. *Magnes Res* 2002;15:263-268.
17. Durlach J, Pagès N, Bac P, Bara M, Guiet-Bara A. Magnesium depletion with hypo or hyperfunction of the biological clock may be involved in chronopathological forms of asthma. *Magnes Res* 2005;18:19-34.
18. Durlach J, Pagès N, Bac P, Bara M, Guiet-Bara A. Headache due to photosensitive magnesium depletion. *Magnes Res* 2005;18:109-122.

2
Magnesium and Calcium in Drinking Water

Hirotoshi Morii, Ken-ichi Matsumoto, Ginji Endo, Mieko Kimura, and Yoshitomo Takaishi

While magnesium is one of the elements that supports life, many studies have been performed regarding physiological functions as well as correlation with diseases. Seelig and Rosanoff reviewed interesting data describing correlations between magnesium deficiency and disease.[1] Magnesium deficiency causes arrythmia, overactivity to stress hormones (adrenalin), overproduction of cholesterol, blood clotting in blood vessels, constriction of blood vessels, high sodium–potassium ratio, insulin resistance, coronary atherosclerosis, and vulnerability to oxidative stress. Thus, Seelig and Rosanoff showed that magnesium content in hearts from cadavers of those who died of heart disease were much less than controls. Cadaver hearts from people who had lived in areas with hard drinking water had higher amounts of magnesium than cadaver hearts from soft-water areas. In 1957, Kobayashi indicated that the hardness of drinking water is related to the incidence of apoplexy.[2]

Materials and Methods

Water samples were collected from various sources in the world: hotel aqueduct, river, spring, and other sources. Countries included the United States, France, Belgium, Turkey, Greece, Chili, Egypt, China, Korea, Mongol, Indonesia, and Japan. Mineral content was measured at Sakai Institute of Public Health, Sakai, Japan, and Takeda Research Institute Life Science and Preventive Medicine, Kyoto, Japan.

Results

Calcium and magnesium as well as calcium/magnesium in drinking water showed considerably different levels in various parts of the world. Generally speaking, both calcium and magnesium levels are higher in Europe compared to other areas. Mineral content in Pamukkale in Turkey was the highest among sampled water. Contrexville in France had high levels of calcium and magnesium. In Asian countries, some areas showed relatively higher levels of calcium and magnesium. Sang Sa Village has been known as a longevity village in Korea and has its own spring from which drinking water is obtained. Zhoukoudian area, located 50 kilometers southwest of Beijing, where Peking man (*Sinanthropus pekinensis*) was discovered, has wells nearby still being utilized by neighborhood populations; water from one had calcium and magnesium levels that were remarkably high. Marie Eugene Francois Thomas Dubois (1858–1940) discovered fossils of *Pithecanthropus* near Solo River in 1890. Drinking water was collected in this region and compared with that collected near Jakarta, Indonesia. Calcium and magnesium content in the former was higher compared with the latter. In Japan, calcium and magnesium content are not so high compared with other areas in the world, especially in comparison with those in Europe.

Discussion

Cardiovascular Disease and Mineral Content

Regarding the effect of both magnesium and calcium on blood pressure and incidence of cardiovascular disease, there have been many studies that are contradictory.[1] However, Seelig and Rosanoff showed that cardiovascular and overall rates were found to be lower in hard-water areas than in soft-water areas. Deaths rates from coronary heart disease are approximately 300 per 1,000,000 people in Lincoln, Nebraska, where water hardness level was 147 ppm, a little more than 600 in Washington, DC, where water hardness level was 96 ppm, and more than 800 in Savannah, Georgia, where water hardness level was 41 ppm.

Recent statistics in the United States (Table 2.1)[3] indicate that areas with high incidence of total death (as well as cardiovascular deaths) are located in the southeastern part of the country, including approximately 10 states. Among the top 10 areas with high mortality rates for both total deaths and cardiovascular deaths, six area are included for both causes of death: Mississippi, the District of Columbia, Kentucky, Alabama, Tennessee, and Oklahoma (Table 2.1). In the state of Tennessee, the water hardness is not so high in the present study.

Water hardness in France seems to be high compared with other countries outside Europe. The idea of a French paradox has been proposed for the low incidence of cardiovascular deaths in France compared with some neighboring countries in Europe.[4] One of the contributing factors was ascribed to the high consumption of polyphenol supplied from red wine, but the water hardness could be another factor. Marque collected information about all deaths of 14,311 individuals in 69 parishes of southwest France from 1990–1996. A significant relationship was observed between calcium and cardiovascular mortality with relative risk (RR) = 0.90 for noncerebrovascular causes and

2. Magnesium and Calcium in Drinking Water

TABLE 2.1. Number of total and heart disease deaths in the United States in 2002.[3]

Rank	Total deaths		Heart disease deaths	
1	Mississippi		Mississippi	
2		1037		326.9
3	District of Columbia		Oklahoma	
4		1027.4		307.1
5	Louisiana		District of Columbia	293.5
6		1001.1	Kentucky	
7	Kentucky			288.9
8		1000.6	West Virginia	
9	Alabama			288.0
10		999.9	Alabama	
11	West Virginia			286.3
12		991.7	Tennessee	
13	Tennessee			283.7
14		985.5	Arkansas	
15	Oklahoma			279.5
16		976.2	New York	
17	Arkansas			277.6
18		966.9	Missouri	
19	Georgia			270.6
20		953.4	Louisiana	
21	South Carolina			269.0
22		949.5	Michigan	
23	Nevada			265.7
24		919.1	Georgia	
25	Missouri			263.0
26		917.1	Ohio	
27	Ohio			258.0
28		907.8	Texas	
29	North Carolina			252.7
30		907.5	Pennsylvania	250.6
31	Indiana		Indiana	248.0
32		919.1	Illinois	246.4
33	Texas		Nevada	246
34		878.4	South Carolina	244.7
35	Michigan		New Jersey	244.4
36		875.8	Rhode Island	240.3
37	Wyoming		Maryland	238.7
38		868.0	Delaware	236.7
39	Pennsylvania		North Carolina	235.6
40		863.5	Virginia	226.6
41	Maryland		California	225.9
42		863.5	Florida	222.3
43	Illinois		Kansas	221.0
44		855.8	New Hampshire	220.1
45	Virginia		Iowa	
46		855.6		219.1
47	Maine		Conneicut	216.9
48		848.3	Wisconsin	216.4
49	Montana		Nebraska	213.5
50		848.3	Wyoming	211.3
51	Kansas		South Dakota	209.9
		845.1	Maine	209.0

TABLE 2.1. *Continued*

Rank	Total deaths		Heart disease deaths	
	Delaware	840.9	Vermont	208.2
	Oregon	832.6	Arizona	204.2
	New Mexico	817.7	Massachusetts	202.3
	Nebraska	814.6	Idaho	200.9
	Rhode Island	813.3	North Dakota	200.8
	New Jersey	811.2	New Mexico	194.1
	Arizona	800.5	Washington	193.1
	Wisconsin	799.3	Oregon	191.9
	Massachusetts	796.1	Montana	190.6
	Colorado	795.7	Hawaii	187.7
	Alaska	794.1	Utah	186.0
	Florida	787.8	Colorado	180.3
	Utah	785.8	Alaska	167.0
	Idaho	785.7	Minnesota	163.9
	Washington	784.4		
	New York	783.3		
	New Hampshire	781.1		
	Vermont	774.2		
	Iowa	773.3		
	South Dakota	771.7		
	Conneticut	760.3		
	California	758.1		
	North Dakota	748.3		
	Minnesota	743.8		
	Hawaii	659.6		

RR = 0.86 for cerebrovascular deaths when calcium is higher than 94 mg/L. There was a protective effect of magnesium between 4 and 11 mg/L with RR = 0.92 for noncerebrovascular and RR = 0.77 for cerebrovascular mortality at concentration lower than 4 mg/L.[5]

Cardiovascular deaths in Japan are much less compared with those in western countries: 85.8 in males and 48.5 in females per 100,000 people in 2000, although the rank is the second next to neoplastic diseases.[6] Hardness of water is less in most of the areas in Japan compared with countries in Europe and North America (Table 2.2). But there are still differences in mineral content among areas in Japan. One of areas in Kyushu Island showed the higher level of calcium and magnesium compared to other areas (Table 2.2). This area belongs to Miyazaki Prefecture, where the cardiovascular death rate was shown to be lower than average. There are so many factors other than minerals that effect the incidence of cardiovascular death.

Mineral Intake of Prehistoric Man

Mineral content in the well of Zhoukoudian area, where the Peking man fossils were discovered, showed very high levels compared with those in Beijing City water and in the Yong Ding Hu river (Table 2.2). In Indonesia, calcium and magnesium content in natural water were not so high. However, in nearby Solo River, near where Java man was discovered 1859 by Dubois, mineral content was higher than in another area near Jakarta (Table 2.2).

Mineral intake will be influenced by mineral content in drinking water, contributing to the requirement of minerals. Total amount of minerals ingested from drinking water not only supplement the dietary allowances, but also modulates mineral metabolism by affecting various factors depending on the concentration of minerals. Meunier compared serum parathyroid hormone and biochemical markers of bone remodeling between females ingesting high- (596 mg/L) and low- (10 mg/L) calcium drinking water. One hundred and eighty healthy postmenopausal women with mean age of 70.1 ± 4.0 years and with daily average intake of calcium less than 700 mg were studied, and 152 completed the 6-month trial. There was a significant 14.1% decrease of PTH, osteocalcin (−8.6%), bone alkaline phosphatase (−11.5%), serum (−16.3%), and urine (−13.0%) type-1 collagen C-telopeptide in those who ingested high-calcium drinking water compared with control group. The additive effect of vitamin D supplement at a dose of 400 IU was not significant.[7] Regarding the effect of magnesium, Marie demonstrated that in weanling male mice, magnesium supplementation in drinking water increased serum and urinary magnesium concentrations and bone magnesium content and that both calcification rate and the extent of tetracycline double-labeled osteoid surface increased progressively in magnesium-treated mice.[8]

These data indicate that calcium and magnesium content in drinking water, as well as in diet, influence calcium and bone metabolism, thus helping to adjust individuals to the environment depending on the quantity of minerals

TABLE 2.2. Calcium, magnesium, and calcium/magnesium in drinking water in some parts of the world.

			Measured minerals (mg/L)		
Continent	Country	Area	Calcium	Magnesium	Ca/Mg
Europe	Turkey	Pamukkale	427	108	3.95
	Greece	Athens city	54.9	9.9	5.55
		Kos Island	131.1	18.5	7.09
		ITI mountain	64.4	24.2	2.70
	Belgium	Mountain natural	5.3	1.90	2.79
	France	Lourdes Fountain	77.9	5.20	14.98
		Contexeville	486	84	6.14
	Norway	North pole sea	309.5	1200	0.26
The Americas	United States	Tennessee River	28.9	6.71	4.31
		Nashville city	30.1	6.70	4.49
	Chili	Santiago city	53.4	8.7	6.14
Oceania	Australia	Sydney city	11.6	4.9	2.37
		Sydney suburb	nd	nd	—
Asia	Korea	Kwang-ju city	38.6	6.0	6.43
		Sang Sa village	53.5	7.0	7.64
	China	Zhoukoudian well	97.1	19.8	4.90
		Town near Beijing	62.1	17.4	3.57
		Beijing city	60.5	20.4	2.97
		Yong Ding River near Beijing	28.9	14.5	1.99
	Indonesia	Mountain near Jakarta	7.0	2.2	3.18
		Wonogiri Mountain near Solo River	27.3	7.1	3.85
	Mongol	Well in the plain	40.9	7.20	5.68
		Ulaanbaartar City	15.1	2.25	6.71
		Tur River	nd	0.01	—
	Japan	Kyoto City w (Biwa Lake)	13.7	2.64	5.19
		Niigata Mountain area	2.1	0.97	2.16
		R river in Osaka	10.0	1.64	6.10
		Kirishima Mountain	42.0	9.70	4.33

Abbreviation: nd, not determined.

ingested. Prehistoric man may have needed strong physical activity to have better access to foods and protection from enemies. Guillemant studied the effect of exercise on calcium and bone metabolism in 12 well-trained elite male triathletes with and without ingesting high-calcium mineral water. When exercise was performed without calcium load, both serum concentrations and total amount of crosslinking telopeptide of type-1 collagen (CTX) began to increase progressively 30 min after the start of the exercise and were still significantly elevated, by 45% to 50%, 2 h after the end of the exercise. Ingestion of high-calcium mineral water completely suppressed the CTX response.[9] Thus, high-mineral drinking water and foods may have had beneficial effects in prehistoric men.

Kaifu showed the difference in mandibular structure between inhabitants of the Jomon period (ca. 10,000 BCE–300 BCE hunter/gatherer/fishers) and those in Yayoi period (ca. 300 BCE–300 CE agriculturists). The underdevelopment of major masticatory muscle attachment in the former populations may have been caused by diminished chewing stress.[10]

Summary and Conclusion

Magnesium and calcium in drinking water were measured in various sources in the world. There were considerable differences in the mineral content in drinking water. In Europe, both minerals showed higher levels compared to other areas in the world. There have been some studies indicating the importance of magnesium content in drinking water. But there are other important other constituents in drinking water, as well as other risk and protective factors for cardiovascular diseases that should be taken into account to explain the cause of cardiovascular death.[11] The lifestyle of prehistoric man is another aspect of interest from the standpoint of the role of minerals in cardiovascular disease.

Acknowledgment. I express my sincere gratitude to Mr. Katsuyuki Kashiwagi, Teijin Pharma, Tokyo, Japan, for his help in collecting water from Java Island.

References

1. Seelig MS, Rosanoff A. *Magnesium Factor Av Database*. New York: Penguin; 2003.
2. Kobayashi J. On geographical relations between the chemical nature of river water and death rate from apoplexy. *Berichte des Ohara Institut fur Landwirtschftliche Biologie* 1957;11:12.
3. US Department of Health and Human Services, Centers for Disease Control and Prevention (CDC). *Series 20*. Washington, DC: National Centers for Health Statistics (NCHS); 2004. Publication 2H-2004.
4. Renaud S, de Lorgeril M. Wine, alcohol, platelets and the French paradox for coronary heart disease. *Lancet* 1992;339:1523–1526.
5. Maruque S, Jacqmin-Gadda H, Darigues JF, Commenges D. Cardiovascular morality and calcium and magnesium in drinking water: an ecological study in elderly people. *Eur J Epidemiol* 2003;18:305–309.
6. Ministry of Health and Welfare of Japan. *Death Status in 47 Prefectures in Japan in 2000*.
7. Meunier PJ, Jenvrin C, Munoz F, de la Gueronniere V, Garnero P, Menz M. Consumption of a high calcium mineral water lowers biochemical indices of bone remodeling in postmenopausal women with low calcium intake. *Osteoporos Int* 2005;16:1203–1209.

8. Marie PJ, Travers R, Delvin EE. Influence of magnesium supplementation on bone turnover in the normal young mouse. *Calcif Tissue Int* 1983;35:755–761.
9. Guillemant J, Accare C, Peres G, Guillemant S. Acute effects of an oral calcium load on markers of bone metabolism during endocrine endurance cycling exercise in male athletes. *Calcif Tissue Int* 2004;74:407–414.
10. Kaifu Y. Changes in mandibular morphology from the Jomon to moern periods in eastern Japan. *Am J Phys Anthropol* 1997;104:227–243.
11. Rylander R. Magnesium in drinking water and cardiovascular disease—an epidemiolopgical dilemma. *Clin Calcium* 2005;15:1773–1777.

Magnesium Homeostasis

3
Cellular Mg^{2+} Transport and Homeostasis: An Overview

Martin Kolisek, Rudolf J. Schweyen, and Monika Schweigel

Magnesium plays a vital role as a cofactor for many enzymes, as a binding partner of nucleotides, and in stabilizing nucleic acids and membranes. It acts as a modulator of ion channels, and it affects many other cellular processes such as neuromuscular excitability, secretion of hormones, and it antagonizes the actions of Ca^{2+}, to name a few effects.[1-4] Mg^{2+} deficiency was found to be associated with hypertension, ischemic heart disease, inflammation, eclampsia, diabetes, cystic fibrosis, and in the establishment of human immunodeficiency virus 1 (HIV-1) reservoirs.[5-10] Several disease phenotypes have been shown to be due to inherited disorders of Mg^{2+} homeostasis.[11-15] Therefore, the regulation of extracellular and intracellular magnesium levels by transmembrane and transepithelial transport processes is critical for numerous cellular and organ functions.

Free cytosolic [Mg^{2+}] is maintained between 0.3 to 1.2 mM, with the higher levels in excitable cells. Yet the extracellular Mg^{2+} concentration can vary considerably and the inside negative membrane potential of most cells constitutes a strong driving force for influx of the ion. This suggests a low membrane permeability to Mg^{2+} in the resting state and the presence of transport proteins allowing regulated uptake and extrusion of Mg^{2+}. Although studies of Mg^{2+} fluxes in mammalian cells and in microorganisms have indicated the presence of functionally active plasma membrane Mg^{2+} transport mechanisms, most pathways lack molecular identification and/or elucidation of the mechanisms regulating transporter expression and activity.

Mechanisms of Mg^{2+} Influx

The inside negative membrane potential is thought to serve as a driving force for cellular uptake of Mg^{2+}, either mediated by ion channels or by carriers. Whether there exist such transport pathways has been unclear until recently. Mg^{2+} uptake in some gastrointestinal and renal epithelial cells,[16,17] in cardiac myocytes,[18] and human platelets[19] have been observed to be influenced by the transmembrane voltage and/or by ion channel blockers. These findings are consistent with channel-mediated Mg^{2+} influx in these cells, but the applied methods, such as fluorescence spectroscopic measurements of $[Mg^{2+}]_i$ by the aid of Magfura-2 and/or ^{28}Mg uptake measurements, are not the methods of choice unequivocally to establish the presence of channels. In some cellular models, Mg^{2+} uptake has also been shown to involve Mg^{2+}/Cl^- or Mg^{2+}/HCO_3^- cotransport.[20–22]

Progress towards the identification and function of plasma membrane Mg^{2+} channels comes from genetic studies both in microorganisms and in mammalia. Maguire and coworkers have characterized various proteins to be involved in bacterial Mg^{2+} transport.[23] Among them, the CorA protein has been recognized as the major, constitutively expressed Mg^{2+} transport protein in eubacteria as well as in archaea. Distant homologues of the CorA protein have been shown to occur in the plasma membrane of lower eukaryotes and plants and in mitochondria of yeast, plants, and mammalia, where they constitute the major Mg^{2+} uptake channels (cf. Chapter 5 of this book). Mammalian cells lack proteins of the CorA superfamily in their plasma membrane. However, recent genetic studies identified two novel ion channels of the melastatin-related, transient receptor potential (TRPM) ion channel subfamily, TRPM6 and TRPM7, as being essential for Mg^{2+} uptake and homeostasis.[24–28] Additional candidate genes for Mg^{2+} transport proteins were identified by Goytain and Quamme[29–32] based on a screen for genes showing Mg^{2+}-regulated expression in renal epithelial cells. Two of these putative Mg^{2+} transporters belong to the solute carrier transporter superfamily (SLC) and are termed SLC41 member a1 (SLC41a1) and SLC41 member a2 (SLC41a2). Another putative transporter, named ACDP2, belongs to the family of ancient conserved domain proteins (*ACDP*). MagT1, finally, represents a member of a novel protein family without similarity to any other transporter protein family known to date. Table 3.1 gives an overview of identified and putative Mg^{2+} channels in mammalian cells.

TRPM6 and TRPM7 Mg^{2+} Channel Proteins of the Plasma Membrane

Patch-clamp studies described TRPM7 as a constitutively active ion channel with a specific permeation profile of ($Zn^{2+} \approx Ni^{2+} \gg Ba^{2+} > Co^{2+} > Mg^{2+} \geq Mn^{2+} \geq Sr^{2+} \geq Cd^{2+} \geq Ca^{2+}$). Native TRPM7-mediated currents were designated as magnesium-nucleotide–regulated metal ion currents (MagNuM). Under standard physiological conditions, it is a primarily Mg^{2+}-conducting ion channel that is regulated by $[Mg^{2+}]_i$ and/or intracellular magnesium–nucleotide complexes. At potentials where divalent ions do not experience sufficient driving force to enter the cell and therefore no longer impede monovalent outward fluxes, TRPM7 channel mediates Na^+ and K^+ efflux.[24,27,33]

TRPM7 and TRPM6 share similar molecular characteristics, such as six transmembrane (TM) domains participating on channel formation and a

TABLE 3.1. Overview of known or putative mammalian Mg^{2+} channels and transporters.

Gene	Protein superfamily	Protein family	TM	Cellular localization	Similarity	Channel	Chanzyme	Influx / efflux
Trpm7	TRP	TRPM (8)[a]	6	PM (EC)		Yes	Yes	I
Trpm6	TRP	TRPM (8)[a]	6	PM (EC)		Yes	Yes	I
Slc41a1	SLC	SLC41 (3)[a]	10	PM (EC)	MgtE (p)	??	No	I / E?
Slc41a2	SLC	SLC41 (3)[a]	10	PM ???	MgtE (p)	??	No	I / E?
MagT1		MagT (2)[a]	5	PM ???		??	No	I / E?
Acdp2		ACDP (4)[a]	4	PM ???	CorC (p)	??	No	I / E?
hMrs2	CorA-Mrs2-Alr1	MRS2 (2)[a]	2	IMM (EC)	CorA (p)	Yes	No	I

Abbreviations: ?, unknown; ??, speculative experimental evidence; E, predominantly efflux; EC, experimentally confirmed; I, predominantly influx; IMM, inner mitochondrial membrane; p, prokaryotic; PM, cytoplamatic membrane; PU, putative; TM, transmembrane domain.
[a]Numbers of family members are given in parentheses.

C-terminal functional kinase domain. TRPM6 and TRPM7 proteins (together with TRPM2 protein) in fact represent unique natural chimeras between a channel-forming polypeptide and an enzyme. To emphasize uniqueness of this kind of proteins they were named *chanzymes*.[34] Schmitz and colleagues[27,35] showed that the TRPM7 kinase domain is not essentially required for TRPM7 activation, but functional coupling between kinase and channel exist such that structural alterations of the kinase modify the sensitivity of channel activation to Mg^{2+}. Direct phosphorylation of the channel by its kinase was proposed being unlikely.

Both TRPM6 and TRPM7 have overlapping expression profiles that was confirmed by real-time polymerase chain reaction (RT-PCR) and Northern blot analyses.[36,37] TRPM7 is able to form a functional homooligomeric complex as well as heterooligomers together with TRPM6. To date, it has not been shown that TRPM6 is able to form homooligomeric complexes. Chubanov and colleagues[36] elegantly demonstrated that TRPM6 without TRPM7 is unable to target the membrane, thus it is retained intracellularly.

The naturally occurring TRPM6 missense mutation S141L was identified as associated with hypomagnesemia with secondary hypocalcemia (HSH).[25,26] This mutation prevents interaction of TRPM6 with TRPM7 and thus TRPM6/TRPM7 oligomerization. Hermosura and coworkers[14] identified a TRPM7 variant containing a missense mutation (T1482I) in DNA of a group of individuals suffering from guamanian latheral sclerosis (ALS-G) and parkinsonism dementia (PD-G). For more details, see Chapter 4 in this book.

Novel Candidate Mg^{2+} Transport Proteins in Mammalia

MagT1

Human as well as mouse *MagT1* genes are X-chromosome–linked and encode proteins with a relative molecular mass of 38 kDa with five putative TM domains.[29] *MagT1* expression was found to be high in kidney, colon, liver, and heart, while weak in the intestine, brain, spleen, and lung. Mg^{2+} starvation enhances *MagT1* expression in epithelium of the renal distal tubule in mice when compared to animals fed with normal diet. When expressed in oocytes of *Xenopus laevis*, MagT1 creates currents that are induced by elevated $[Mg^{2+}]_{ext}$, but not by elevated concentrations of other divalent cations. Goytain and Quamme[29] excluded direct coupling of Mg^{2+}-evoked currents to H^+ movement but claimed their sensitivity to external pH with maximum currents at pH 7.4. MagT1-induced currents were inhibited by Ni^{2+}, Zn^{2+}, and Mn^{2+}, as well as some other cations at high external concentrations. MagT1-mediated Mg^{2+} uptake was linear for at least 20 min and did not display any time-dependent decay during repetitive stimulation with voltage steps. From these data, the authors concluded that MagT1 is a voltage-dependent magnesium transporter with channel-like properties. The authors state that many characteristics of MgT1 expressed in oocytes parallel physiological observations with kidney distal convoluted tubule cells.[29] It remains to be shown in mammalian cells in which membrane the MagT1 protein is localized and what its contribution to Mg^{2+} in the respective compartment might be.

ACDP2

The ACDP (ancient conserved domain protein) gene family encodes large proteins with four predicted TM domains. Mouse and human genomes contain four genes (*Acdp1*, *Acdp2*, *Acdp3*, and *Acdp4*) with deduced proteins of 951, 874, 713, and 771 amino acids, respectively. ACDP sequence motifs are highly conserved in various taxonomic entities.[38] Among these homologues is the bacterial CorC protein that had previously been implicated in Mg^{2+} and Co^{2+} transport[39] and the yeast MAM3 gene product involved in vacuolar Mn^{2+}, Co^{2+}, and Zn^{2+} homeostasis.[40] Expression of these ACDP proteins in mouse varied from tissue to tissue, but was generally strong in kidney. Low Mg^{2+} diet led to a increase of ACDP2 expression relative to normal diet in mouse kidney. A similar observation was made when MDCT epithelial cells were cultivated in low-Mg^{2+} medium.[30,38] Upon expression of mouse ACDP2 in oocytes of *Xenopus laevis*, Goytain and Quamme[30] attributed saturable transport of Mg^{2+} and several other cations (Co^{2+}, Mn^{2+}, Sr^{2+}, Ba^{2+}, Cu^{2+}, and Fe^{2+}). Ca^{2+}, Cd^{2+}, Zn^{2+}, and Ni^{2+} did not induce currents, but Zn^{2+} inhibited Mg^{2+} permeation significantly. These and further results suggested that ACDP2 may constitute a regulated transporter of Mg^{2+} and other divalent cations.[30]

SLC41A1

Wabbaken and colleagues[41] identified the human gene *Slc41A1* as encoding a protein with a predicted mass of 56 kDa. Based on computer prediction, SLC41A1 contains 10 TM domains and is localized in the cytoplasmatic membrane (Kolisek et al., unpublished data). It has significant sequence homology to the bacterial MgtE, which had previously been shown to be involved in Mg^{2+} transport.[42] In human and mouse *Slc41a1*, transcripts are detected in most tissues (notably in heart, muscle, testis, thyroid, kidney) and low Mg^{2+} diet in mice led to significant increase of *Slc41a1* gene expression in the kidney, colon, and heart.[31,41] The TEV clamp of *Xenopus* oocytes heterologously expressing SLC41A1 revealed Mg^{2+}-evoked currents with a large amplitude of 1.0 μM. Uptake was linear for at least 20 min and did not display a time-dependent decay during repetitive stimulation with voltage steps. Steady-state Mg^{2+} currents were saturable. Substitution of external NaCl with choline Cl^- had no effect on SLC41A1 related Mg^{2+} currents. Similarly, substitution of external NaCl and $MgCl_2$ with appropriate amounts of Na^+ gluconate and $MgSO_4$ did not alter SLC41A1-mediated Mg^{2+} conductance. It was concluded that transport neither depends on extracellular Na^+ or Cl^-, nor is it influenced by these electrolytes.[31] A number of other divalent cations was reported to be conducted by SLC41A1, namely Sr^{2+}, Fe^{2+}, Ba^{2+}, Cu^{2+}, Zn^{2+}, and Co^{2+}, but neither Ni^{2+} nor Ca^{2+} or Gd^{3+}. Gd^{3+} and La^{3+} were identified as antagonists of Mg^{2+}-induced SLC41A1 currents. SLC41A1 thus may constitute another transporter of Mg^{2+} and other divalent cations.

Taken together, the identification of SLC41, ACDP2, and MagT1 proteins is a major step forward in the characterization of mammalian Mg^{2+} transport proteins. While their role as electrogenic ion transporters rests on a single series of experiments in a heterologous system only, their function in cation homeostasis is supported by their mode of expression in mammalia and, as far as SLC41 and ACDP2 are concerned, by studies on the roles of their homologues in ion homeostasis in microorganisms. Future ion transport studies only will reveal the capacities of these proteins in cation transport, their location in plasma membrane or an intracellular membrane, and the contribution of these proteins to Mg^{2+} homeostasis of various mammalian cell types and organs.

Mechanisms of Mg^{2+} Extrusion

Extrusion of Mg^{2+} must occur against a steep electrochemical gradient. Because no evidence exists that Mg^{2+} is extruded from the cell via an outwardly oriented Mg^{2+}–adenosine triphosphatase (ATPase), other driving forces, such as the Na^+ (mammalian) or H^+ (microorganisms) gradient, must be used to transport the ion by secondary active transport mechanisms. In accordance with this, a Na$^+$/Mg^{2+} exchanger, whose presence and characteristics have been characterized mainly by functional studies, seems to be an important efflux system in many mammalian cells. The Na$^+$/Mg^{2+} exchanger is also a candidate mechanism for the extrusion of Mg^{2+} across the basolateral membrane of the Mg^{2+} absorbing and reabsorbing intestinal and renal epithelia.[43,44] Although some authors suggest that Mg^{2+} extrusion also occurs through the Na$^+$/Ca^{2+} antiporter,[45] through a Mg^{2+}/Ca^{2+} exchanger,[46] or through anion-dependent mechanisms,[47] we will concentrate here on the proposed Na$^+$/Mg^{2+} antiport which is best characterized Mg^{2+} transporter of nonepithelial cells and tissues.

Na$^+$/Mg^{2+} Exchange

Na$^+$/Mg^{2+} exchange was first described by Günther and Vormann in chicken and turkey red blood cells,[48] and they and others studied the Na$^+$-dependent Mg^{2+} transport in numerous vertebrate erythrocytes.[49-51] Over the years, the Na$^+$/Mg^{2+} exchanger has been shown to operate in the majority of examined cells pointing to an ubiquitous expression.[52-57]

With physiological transmembrane Na$^+$ ([Na$^+$]$_i$ < [Na$^+$]$_e$) and Mg^{2+} ([Mg^{2+}]$_i$ ≤ [Mg^{2+}]$_e$) gradients the Na$^+$/Mg^{2+} antiport mediates efflux of intracellular Mg^{2+} in exchange for extracellular Na$^+$ ions. The driving force for this exchange is the inwardly directed concentration gradient of Na$^+$, making the transporter indirectly dependent on the Na$^+$/K$^+$ pump and thereby on adenosine triphosphate (ATP).[19,58] The stoichiometry of the exchange seems to vary between cell types and possibly according to the functional state, but in most cases an electroneutral ratio of 2 Na$^+$ to 1 Mg^{2+} has been reported.[48] In some cell systems, the Mg^{2+} efflux via Na$^+$/Mg^{2+} antiport can be switched to Mg^{2+} influx by reversing the Na$^+$ gradient.[59] Until now, the physiological meaning of a reverse Na$^+$/Mg^{2+} is not clear.

A characteristic feature of Mg^{2+} extrusion via the Na$^+$/Mg^{2+} exchanger is its strict dependence on extracellular Na$^+$.[52,56] Substitution of extracellular Na$^+$ by K$^+$, Li$^+$, or choline$^+$ can not support Mg^{2+} efflux.[46,60] As expected for a carrier system, Mg^{2+} efflux is saturable and its activation by [Na$^+$]$_e$ follows a Michael–Menten relationship with K_m values between 11 to 30 mM.[61,62]

It is long known that a measurable Mg^{2+} efflux took place only when [Mg^{2+}]$_i$ was increased and that it stopped when the normal Mg^{2+} content is achieved,[60] showing that the rate of Na$^+$/Mg^{2+} antiport is dependent on the [Mg^{2+}]$_i$.[49] Günther[63] suggested that the transporter performs Mg^{2+}/Mg^{2+} exchange in cells with normal [Mg^{2+}]$_i$ and that binding of two Mg^{2+} to an intracellular modifier site is required for the allosteric transformation of the Mg^{2+}/Mg^{2+} Na$^+$/Mg^{2+} exchanger. In most cell systems unstimulated Mg^{2+} influx is not high enough to activate the exchanger.[49,60] However, in some cell types, a basal Na$^+$/Mg^{2+} exchanger activity has been shown to exist without manipulation of intra- or extracellular ion concentrations.[64,65] The K_m for [Mg^{2+}]$_i$ at half maximal rate has been calculated to be in the range between 1.3 to 4 mM.[48,49,66]

Differences in cell type, $[Mg^{2+}]_i$ (loaded or unloaded cells, loading concentration, use of metabolic inhibitors, intracellular buffering capacity for Mg^{2+}) and in the method used to determine Mg^{2+} efflux (flux studies with ^{28}Mg, determination of intra- or extracellular $[Mg^{2+}]$ and $[Na^+]$ by AAS or by fluorescence probes) are partly responsible for the observed differences in the transport capacity. Therefore, a comparison of the available data is difficult. Under nearly V_{max} conditions (after preloading with 6 mM Mg^{2+}), we found a Na^+/Mg^{2+} exchanger mediated Mg^{2+} efflux of 9.4 ± 4.7 mM/l cell/15 min in ruminal epithelial cells.[56] This rate of Mg^{2+} loss is even greater than the rate reported under the same or very similar conditions for most of the other cell systems.[49,55,65] We interpret the higher efflux rate in the epithelial cell system as to reflect the capability to maintain a normal $[Mg^{2+}]_i$ of 0.6 to 1.2 mM in spite of disturbances induced by marked Mg^{2+} influx (37.5–42 μM/min), which is a characteristic of this Mg^{2+}-absorbing cell.[17,22]

Regulation

In perfused heart and liver, a transient Mg^{2+} efflux can be induced by β adrenergic substances or by application of db-cyclic adenosine monophosphate (cAMP) and different cell types respond to cAMP stimulation with a 50% to 70 % increase of Mg^{2+} efflux.[67–69] Stimulation with cAMP was ineffective in Na^+-free NMDG medium, indicating that elevation of the intracellular cAMP concentration specifically triggers the Na^+-dependent Mg^{2+} efflux.[56,70] An increased affinity of the Na^+/Mg^{2+} transporter for intracellular Mg^{2+} induced by phosphorylation of the transport protein has been postulated as the underlying mechanism for cAMP activated Mg^{2+} extrusion.[53]

Besides acting on the antiport directly, cAMP and other agents could also induce the release of Mg^{2+} from intracellular stores, for example, from mitochondria, and activation of the exchanger via the increase in cytosolic free Mg^{2+} content.[71] Alternatively to the cAMP activation, Na^+-dependent Mg^{2+} extrusion is activated by changes in intracellular Ca^{2+} content.[52]

Most likely, the Na^+/Mg^{2+} exchanger is a pathway that different hormones or mediators activate to induce Mg^{2+} extrusion. Angiotensin II,[72] PGE_2,[56] catecholamines,[52] and $INF\alpha$[70] are some examples. Recently, it has been shown that the $[Mg^{2+}]_i$ decrease resulting from Na^+/Mg^{2+} exchanger activation influences directly or indirectly cellular transport mechanisms and physiological functions. He and colleagues[72] demonstrated Na^+/Mg^{2+} exchanger-mediated Mg^{2+} efflux from smooth vascular muscle cells after acute angiotensin II stimulation. The induced reduction of $[Mg^{2+}]_i$ led than to vasoconstriction. TRPM7, which facilitates Mg^{2+} influx, may also be activated by decreased $[Mg^{2+}]_i$. A TRPM7-mediated increase of the intracellular Mg^{2+} level is postulated to play a role in smooth vascular muscle cell growth.[72] In ruminal epithelial cells (REC), a cAMP-induced reduction of $[Mg^{2+}]_i$ has been demonstrated to activate a nonselective cation conductance expressed in the apical membrane of these cells and, thereby to play an important role in the regulation of ruminal Na^+

transport.⁴⁴ By use of the Ussing chamber technique, the same authors showed a stimulating effect of cAMP on the unidirectional $^{28}Mg^{2+}$ flux across the isolated ruminal epithelium, which is in line with the stimulation of the Na^+/Mg^{2+} exchanger seen in ruminal epithelial cells.⁵⁶ For years a role of the Na^+/Mg^{2+} exchanger in the transepithelial Mg^{2+} transport performed by gastrointestinal and renal epithelia has been postulated,⁴³ but is shown here for the first time. Moreover, the suggestion of a regulation of this transport by cAMP is of general importance, and there is a need to investigate what hormones or other signals are involved.

Towards Identification of the Na^+/Mg^{2+} Exchanger

Despite the extensive evidence for a functional Na^+/Mg^{2+} exchanger in the cell membrane of nearly all mammalian cells, encoding gene(s), structural properties, and the modalities of its operation and regulation remain for the most part unknown. Among other reasons this can be explained by the fact that there is no highly selective blocker available. A reversible inhibition of the Na^+/Mg^{2+} exchanger was observed after application of amiloride, quinidine, and imipramine, compounds known to inhibit other transport systems (Na^+ channels, Na^+-dependent carriers, K^+ channels) as well.⁵⁹,⁷⁰ Imipramine binds to the Na^+ site of the Na^+/Mg^{2+} exchanger and competition between these two compounds slows Mg^{2+} efflux. Using imipramine concentrations between 10 and 500 μM, most investigators achieved a 33% to 85% inhibition of Mg^{2+} efflux.⁵²,⁶⁰,⁶⁵ However, the imipramine inhibition was always incomplete and an even greater reduction of Mg^{2+} extrusion could be obtained by omitting extracellular Na^+.⁵⁶,⁵⁹ To overcome this problem, monoclonal antibodies (mabs) raised against the porcine erythrocyte Na^+/Mg^{2+} exchanger were prepared in our laboratory.⁷³ Because of the unknown protein structure of the Na^+/Mg^{2+} exchanger, the screening for effective antibodies has been done by a specific functional test using the capability of the Na^+/Mg^{2+} exchanger of porcine red cells to transport Mn^{2+} instead of Na^+.⁴⁹ Supernatants of antibody-producing hybridoma cell clones were checked for their ability to inhibit $^{54}Mn^{2+}$ influx as a measure of the Na^+/Mg^{2+} exchange activity. Compared with control conditions, effective antibody-containing supernatants decreased the rate of Mn^{2+} uptake in porcine red cells by 68% ± 2% and the inhibition effect varies from 60% to 74% between single clones. Hybridoma supernatants or purified mabs were subsequently used to inhibit Mg^{2+} transport and to detect the Na^+/Mg^{2+} exchanger in REC.⁵⁶ Transport studies were performed with unloaded REC under conditions promoting influx (high-K^+, low-Na^+ medium) or efflux (high-Na^+, low-K^+ medium) of Mg^{2+} via the Na^+/Mg^{2+} exchanger, respectively. Addition of the mab effectively inhibited Mg^{2+} transport as reflected by a 50% reduction or 23% increase of the REC $[Mg^{2+}]_i$. In addition, the antibody specifically labeled a protein with an apparent molecular mass of 70 kDa in immunoblots of proteins from REC as well as from porcine red blood cell membranes. Also, in flow cytometric measurements the anti-Na^+/Mg^{2+} exchanger mab was specifically bound to a significant proportion of REC. With these new tools, we should be able to purify and to characterize the protein by molecularbiological techniques. The considerable variation of functional properties, such as maximum transport capacity, reversibility, capability of Mn^{2+} transport, and stoichiometry of Mg^{2+} transport via Na^+/Mg^{2+} exchanger between cell types point, to the existence of various isoforms of the transport protein. Its molecular characterization is of greatest interest because disturbance or dysregulation of the Na^+/Mg^{2+} exchanger has been assumed to participate in the pathogenesis of diseases, such as primary hypertension[6] and cystic fibrosis.[9]

References

1. Hartwig A. Role of magnesium in genomic stability. *Mutation Res* 2001;475:113–121.
2. Bara M, Guiet-Bara A. Magnesium regulation of Ca^{2+} channels in smooth muscle and endothelial cells of human allantochorial placental vessels. *Magnes Res* 2001;14:11–18.
3. White RE, Hartzell HC. Magnesium ions in cardiac function. Regulator of ion channels and second messengers. *Biochem Pharmacol* 1989;38:859–867.
4. Mooren FC, Turi S, Günzel D, et al. Calcium-magnesium interactions in pancreatic acinal cells. *FASEB J* 2001;15:659–672.
5. Altura BM, Altura BT. Magnesium and cardiovascular biology: an important link between cardiovascular risk factors and atherogenesis. *Cell Mol Biol Res* 1995;41:347–359.
6. Kisters K, Krefting ER, Barenbrock M, Spieker C, Rahn KH. Na^+ and Mg^{2+} contents in smooth muscle cells in spontaneously hypertensive rats. *Am J Hypertens* 1999;12:648–652.
7. Weglicki WB, Phillips TM. Pathobiology of magnesium deficiency: a cytokine/neurogenic inflammation hypothesis. *Am J Physiol* 1992;263:R734–R737.
8. Saris NE, Mervaala E, Karppanen H, Khawaja JA, Lewenstam A. Magnesium. An update on physiological, clinical and analytical aspects. *Clin Chim Acta* 2000;294:1–26.
9. Vormann J. Mineral metabolism in erythrocytes from patients with cystic fibrosis. *Eur J Clin Chem Clin Biochem* 1992;30:193–196.
10. Goldschmidt V, Didierjean J, Ehresmann B, Ehresmann C, Isel C, Marquet R. Mg2+ dependency of HIV-1 reverse transcription, inhibition by nucleoside analogues and resistance. *Nucleic Acids Res* 2006;34:42–52.
11. Cole DE, Quamme GA. Inherited disorders of renal magnesium handling. *J Am Soc Nephrol* 2000;11:1937–1947.
12. Schlingmann KP, Konrad M, Seyberth HW. Genetics of hereditary disorders of magnesium homeostasis. *Pediatr Nephrol* 2004;19:13–25.
13. Hoenderop JG, Bindels RJ. Epithelial Ca^{2+} and Mg^{2+} channels in health and disease. *J Am Soc Nephrol* 2005;16:15–26.
14. Hermosura MC, Nayakanti H, Dorovkov MV, et al. A TRPM7 variant shows altered sensitivity to magnesium that may contribute to the pathogenesis of two Guamanian neurodegenerative disorders. *Proc Natl Acad Sci U S A* 2005;102:11510–11515.

15. Hou J, Paul DL, Goodenough DA. Paracellin-1 and the modulation of ion selectivity of tight junctions. *J Cell Sci* 2005;118:5109–5118.
16. Dai LJ, Raymond L, Friedman PA, Quamme GA. Mechanisms of amiloride stimulation of Mg^{2+} uptake in immortalized mouse distal convoluted tubule cells. *Am J Physiol* 1997;272:F249–F256.
17. Schweigel M, Lang I, Martens H. Mg^{2+} transport in sheep rumen epithelium: evidence for an electrodiffusive mechanism. *Am J Physiol* 1999;277:G976–G982.
18. Quamme GA, Rabkin SW. Cytosolic free magnesium in cardiac myocytes: identification of a Mg^{2+} influx pathway. *Biochem Biophys Res Commun* 1990;167:1406–1412.
19. Okorodudu A, Yang H, Elghetany MT. Ionized magnesium in the homeostasis of cells: intracellular threshold for Mg^{2+} in human platelets. *Clin Chim Acta* 2001;303:147–154.
20. Günther T, Vormann J, Averdunk R. Characterization of furosemide-sensitive Mg^{2+} influx in Yoshida ascites tumor cells. *FEBS Lett* 1986;197:297–300.
21. Jüttner R, Ebel H. Characterization of Mg^{2+} transport in brush border membrane vesicles of rabbit ileum studied with mag-fura-2. *Biochem Biophys Acta* 1999;1370:51–63.
22. Schweigel M, Martens H. Anion-dependent Mg^{2+} influx and a role for a vacuolar H^+-ATPase in sheep ruminal epithelial cells. *Am J Physiol Gastrointest Liver Physiol* 2003;285:G45–G53.
23. Smith RL, Maguire ME. Microbial magnesium transport: unusual transporters searching for identity. *Mol Microbiol* 1998;28:217–226.
24. Nadler MJ, Hermosura MC, Inabe K, et al. LTRPC7 is a Mg.ATP-regulated divalent cation channel required for cell viability. *Nature* 2001;411:590–595.
25. Schlingmann KP, Weber S, Peters M, et al. Hypomagnesemia with secondary hypocalcemia is caused by mutations in TRPM6, a new member of the TRPM gene family. *Nat Genet* 2002;31:166–170.
26. Walder RY, Landau D, Meyer P, et al. Mutation of TRPM6 causes familial hypomagnesemia with secondary hypocalcemia. *Nat Genet* 2002;31:171–174.
27. Schmitz C, Perraud AL, Johnson CO, et al. Regulation of vertebrate cellular Mg^{2+} homeostasis by TRPM7. *Cell* 2003;114:191–200.
28. Voets T, Nilius B, Hoefs S, et al. TRPM6 forms the Mg^{2+} influx channel involved in intestinal and renal Mg^{2+} absorption. *J Biol Chem* 2004;279:19–25.
29. Goytain A, Quamme GA. Identification and characterization of a novel mammalian Mg^{2+} transporter with channel-like properties. *BMC Genomics* 2005;6L48.
30. Goytain A, Quamme GA. Functional characterization of ACDP2 (ancient conserved domain protein), a divalent metal transporter. *Physiol Genomics* 2005;22:382–389.
31. Goytain A, Quamme GA. Functional characterization of the human solute carrier, SLC41A2. *Biochem Biophys Res Commun* 2005;330:701–705.
32. Goytain A, Quamme GA. Functional characterization of human SLC41A1, a Mg^{2+} transporter with similarity to prokaryotic MgtE Mg^{2+} transporters. *Physiol Genomics* 2005;21:337–342.
33. Monteilh-Zoller MK, Hermosura MC, Nadler MJ, Scharenberg AM, Penner R, Fleig A. TRPM7 provides an ion channel mechanism for cellular entry of trace metal ions. *J Gen Physiol.* 2003;121:49–60.
34. Montell C. Mg2+ homeostasis: the Mg2+nificent TRPM chanzymes. *Curr Biol* 2003;13:R799–R801.

35. Schmitz C, Perraud AL, Fleig A, Scharenberg AM. Dual-function ion channel/ protein kinases: novel components of vertebrate magnesium regulatory mechanisms. *Pediatr Res* 2004;55:734-737.
36. Chubanov V, Waldegger S, Mederos y Schnitzler M, et al. Disruption of TRPM6/TRPM7 complex formation by a mutation in the TRPM6 gene causes hypomagnesemia with secondary hypocalcemia. *Proc Natl Acad Sci U S A* 2004;101:2894-2899.
37. Schmitz C, Dorovkov MV, Zhao X, Davenport BJ, Ryazanov AG, Perraud AL. The channel kinases TRPM6 and TRPM7 are functionally nonredundant. *J Biol Chem* 2005;280:37763-37771.
38. Wang CY, Yang P, Shi JD, et al. Molecular cloning and characterization of the mouse Acdp gene family. *BMC Genomics* 2004;5:7.
39. Gibson MM, Bagga DA, Miller CG, Maguire ME. Magnesium transport in Salmonella typhimurium: the influence of new mutations conferring Co2+ resistance on the CorA Mg2+ transport system. *Mol Microbiol* 1991;5:2753-2762.
40. Yang M, Jensen LT, Gardner AJ, Culotta VC. Manganese toxicity and *Saccharomyces cerevisiae* Mam3p, a member of the ACDP (ancient conserved domain protein) family. *Biochem J* 2005;386:479-487.
41. Wabakken T, Rian E, Kveine M, Aasheim HC. The human solute carrier SLC41A1 belongs to a novel eukaryotic subfamily with homology to prokaryotic MgtE Mg^{2+} transporters. *Biochem Biophys Res Commun* 2003;306:718-724.
42. Smith RL, Thompson LJ, Maguire ME. Cloning and characterization of MgtE, a putative new class of Mg^{2+} transporter from *Bacillus firmus* OF4. *J Bacteriol* 1995;177:1233-1238.
43. Quamme GA. Control of magnesium transport in the thick ascending limb. *Am J Physiol Renal Fluid Electrolyte Physiol* 1989;25:F197-F210.
44. Leonhard-Marek S, Stumpff F, Brinkmann I, Breves G, Martens H. Basolateral Mg^{2+}/Na^+ exchange regulates apical nonselective cation channel in sheep rumen epithelium via cytosolic Mg^{2+}. *Am J Physiol Gastrontest Liver Physiol* 2005;288: G630-G645.
45. Tashiro M, Konishi M, Iwamoto T, Shigekawa M, Kurihara S. Transport of magnesium by two isoforms of the Na^+-Ca^{2+} exchanger expressed in CCL39 fibroblasts. *Plügers Arch* 2000;440:819-827.
46. Cefaratti C, Romani A, Scarpa A. Characterization of two Mg^{2+} transporters in sealed plasma membrane vesicles from rat liver. *Am J Physiol* 1998;275:C2995-C1008.
47. Ödblom MP, Handy RD. A novel DIDS-sensitive, anion-dependent Mg^{2+} efflux pathway in rat ventricular myocytes. *Biochem Biophys Res Commun* 1999;264: 334-337.
48. Günther T, Vormann J, Förster R. Regulation of intracellular magnesium by Mg^{2+} efflux. *Biochem Biophys Res Commun* 1984;119:124-131.
49. Büttner S, Günther T, Schäfer A, Vormann J. Magnesium metabolism in erythrocytes of various species. *Magnes Bull* 1998;101-109.
50. Feray JC, Garay R An Na^+-stimulated Mg^{2+} transport system in human red blood cells. *Biochim Biophys Acta* 1986;856:76-84.
51. Flatman P, Smith LM. Magnesium transport in magnesium-loaded ferret red blood cells. *Pflügers Arch* 1996;432:995-1002.
52. Fagan T, Romani A. α1-adrenoreceptor-induced Mg^{2+} extrusion from rat hepatocytes occurs via Na^+-dependent transport mechanism. *Am J Physiol Gastrointest Liver Physiol* 2001;280:G1145-G1156.

53. Günther T, Vormann J. Activation of Na^+/Mg^{2+} antiport in thymocytes by cAMP. *FEBS Lett* 1992;297:132–134.
54. Hintz K, Günzel D, Schlue W-R. Na^+-dependent regulation of the free Mg^{2+} concentration in neuropile glial cells and P neurones of the leech Hirudo medicinalis. *Pflügers Arch* 1999;437:354–362.
55. Zhang GH, Melvin JE. Regulation by extracellular Na^+ of cytosolic Mg^{2+} concentration in Mg^{2+}-loaded rat sublingual acini. *FEBS Lett* 1995;371:52–56.
56. Schweigel M, Park HS, Etschmann B, Martens H. Characterization of the Na^+-dependent Mg^{2+} transport in sheep ruminal epithelial cells. *Am J Physiol Gastrointest Liver Physiol* 2006;290:G56–G65.
57. Yoshimura M, Oshima T, Matsuura H, et al. Effect of transmembrane gradient of magnesium and sodium on the regulation of cytosolic free magnesium concentration in human platelets. *Clin Sci* 1995;89:293–298.
58. Frenkel EJ, Graziani M, Schatzmann HJ. ATP requirement of the sodium-dependent magnesium extrusion from human red blood cells. *J Physiol* 1989;414:385–397.
59. Schweigel M, Vormann J, Martens H. Mechanisms of Mg^{2+} transport in cultured ruminal epithelial cells. *Am J Physiol Gastrointest Liver Physiol* 2000;278: G400–G408.
60. Kubota T, Tokuno K, Nakagawa J, et al. Na^+/Mg^{2+} transporter acts as a Mg^{2+} buffering mechanism in PC12 cells. *Biochem Biophys Res Commun* 2003;303:332–336.
61. Cefaratti C, Romani A, Scarpa A. Differential localization and operation of distinct Mg^{2+} transporters in apical and basolateral sides of rat liver plasma membrane. *J Biol Chem* 2000;275:3772–3780.
62. Günther T, Vormann J. Mg^{2+} efflux is accomplished by an amiloride-sensitive Na^+/Mg^{2+} antiport. *Biochem Biophys Res Commun* 1985;130:540–545.
63. Günther T. Putative mechanism of Mg^{2+}/Mg^{2+} exchange and Na^+/Mg^{2+} antiport. *Magnes Bull* 1996;18:2–6.
64. Handy RD, Gow IF, Ellis D, Flatman PW. Na-dependent regulation of intracellular free magnesium concentration in isolated rat ventricular myocytes. *J Mol Cell Cardiol* 1996;28:1641–1651.
65. Tashiro M, Konishi M. Na^+ gradient-dependent Mg^{2+} transport in smooth muscle cells of guinea pig tenia cecum. *Biophys J* 1997;73:3371–3384.
66. Willis JS, Xu W, Zhao Z. Diversities of transport of sodium in rodent red cells. *Comp Biochem Physiol* 1992;102:609–614.
67. Fatholahi M, LaNoue K, Romani A, Scarpa A. Relationship between total and free cellular Mg^{2+} during metabolic stimulation of rat heart myocytes and perfused hearts. *Arch Biochem Biophys* 2000;374:395–401.
68. Romani A, Scarpa A. Norepinephrine evokes a marked Mg^{2+} efflux from liver cells. *FEBS Lett* 1990;269:37–40.
69. Matsuura T, Kanayama Y, Inoue T, Takeda T, Morishima I. cAMP-induced changes of intracellular free Mg^{2+} levels in human erythrocytes. *Biochim Biophys Acta* 1993;1220:31–36.
70. Wolf FI, Di Francesco A, Covacci V, Cittadani, A. Regulation of Na-dependent magnesium efflux from intact tumor cells. *Magnes Res* 1995;(Suppl 1):490–496.
71. Romani A, Dowell E, Scarpa A. Cyclic AMP induced Mg^{2+} release from rat liver hepatocytes, permeabilized hepatocytes and isolated mitochondria. *J Biol Chem* 1991;266:24376–24384.

72. He Y, Yao G, Savoia C, Touyz RM. Transient receptor potential melastatin 7 ion channels regulate magnesium homeostasis in vascular smooth muscle cells. Role of angiotensin II. *Circ Res* 2005;96:207–215.
73. Schweigel M, Buschmann F, Etschmann B, Vormann J. Development of monoclonal antibodies directed against the Na^+/Mg^{2+}-antiporter and their use in ruminal epithelial cells. *Proc Soc Nutr Physiol* 2005;13:94.

4
TRPM6 and TRPM7 Chanzymes Essential for Magnesium Homeostasis

Wouter M. Tiel Groenestege, Joost G. J. Hoenderop, and René J. M. Bindels

Mg^{2+} is the second most abundant intracellular cation and plays an essential role as cofactor in many enzymatic reactions. Regulation of the total body Mg^{2+} balance principally resides within the kidney that tightly matches the intestinal absorption of Mg^{2+}. The identification of epithelial Mg^{2+} transporters in the kidney has been greatly facilitated by studying hereditary disorders with primary hypomagnesemia. Identification of the gene defect in hypomagnesemia with secondary hypocalcemia has recently elucidated the TRPM6 protein, a member of the transient receptor potential melastatin (TRPM) family. TRPM6 shows the highest homology with TRPM7, which has been identified as a Mg^{2+}-permeable ion channel primarily required for cellular Mg^{2+} homeostasis. TRPM6 and TRPM7 are distinct from all other ion channels because they are composed of a channel linked to a protein kinase domain and therefore referred to as chanzymes. These chanzymes are essential for Mg^{2+} homeostasis, which is critical for human health and cell viability. This chapter describes the characteristics of epithelial Mg^{2+} transport in general and highlights the distinctive features and the physiological relevance of these new chanzymes in (patho)physiological situations.

Mg^{2+} (Re)absorption in Kidney and Intestine

The kidney is the principal organ responsible for the regulation of the body Mg^{2+} balance. About 80% of the total plasma Mg^{2+} is filtered in the glomeruli,[1,2] of which the majority is subsequently reabsorbed along the nephron.[3] Approximately 10% to 20% of Mg^{2+} is reabsorbed by the proximal tubule. However, the bulk amount of Mg^{2+} (50%–70%) is reabsorbed by the thick ascending limb of Henle (TAL), which mediates Mg^{2+} reabsorption via paracellular transport. The distal convoluted tubule (DCT) reabsorbs 5% to 10% of the filtered Mg^{2+} and the reabsorption rate in this segment defines the final urinary Mg^{2+} concentration, because virtually no reabsorption takes places beyond this segment.[3] Mg^{2+} transport in DCT is transcellular in nature and influenced by

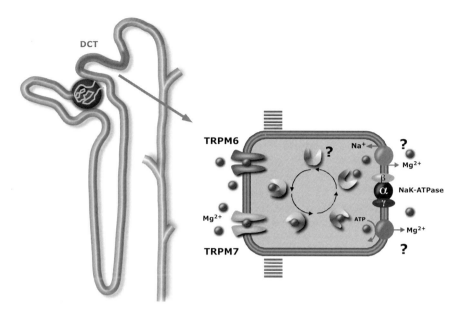

FIGURE 4.1. Transcellular Mg^{2+} reabsorption. Active Mg^{2+} reabsorption is carried out as a three-step process in the distal convoluted tubule (DCT). Following entry of Mg^{2+} in DCT through the epithelial Mg^{2+} channel, TRPM6 (and TRPM7), Mg^{2+} diffuses to the basolateral membrane. As extrusion mechanisms are postulated a basolateral Na^+/Mg^{2+} exchanger and/or ATP-dependent Mg^{2+} pump. The Na^+,K^+-ATPase complex including the γ-subunit controls this transepithelial Mg^{2+} transport. In this way, there is net Mg^{2+} reabsorption from the luminal space to the extracellular compartment.

dietary Mg^{2+} restriction and various hormones.[4,5] However, the molecular details and regulation of this pathway remain largely unknown.[3,4,6,7]

Hypothetically, the process of transcellular Mg^{2+} transport is envisaged by the following sequential steps (Figure 4.1). Driven by a favorable transmembrane potential, Mg^{2+} enters the epithelial cell through the apical epithelial Mg^{2+} channel TRPM6. Next, Mg^{2+} will diffuse through the cytosol to be extruded actively against an electrochemical negative gradient across the basolateral membrane.[8] The molecular identity of these latter basolateral Mg^{2+} transporters is not known. Most physiological studies favor a Na^+-dependent exchange mechanism.[9] Other candidate mechanisms include an ATP-dependent Mg^{2+} pump. The Mg^{2+} entry appears to be the rate-limiting step and the site of regulation.

Genes Involved in Primary Inherited Hypomagnesemia

In the last decade, several genes encoding proteins that are directly involved in renal Mg^{2+} handling have been identified following a positional cloning strategy in families with primary hereditary hypomagnesemia. First, the gene

PCLN-1 (or CLDN16) encoding the protein paracellin-1 (or claudin-16) was found to be mutated in patients who have hypomagnesemia, hypercalciuria, and nephrocalcinosis (HHN; MIM 248250).[10] The tight junction protein paracellin-1 is specifically expressed in TAL and is important for paracellular Mg^{2+} and Ca^{2+} reabsorption in this segment.[10] Second, the gene FXYD2, encoding the γ-subunit of the Na^+,K^+-adenosine triphosphatase (ATPase) pump, is mutated in patients with autosomal dominant renal hypomagnesemia associated with hypocalciuria (IDH; MIM 154020). The γ-subunit of the Na^+,K^+-ATPase pump is predominantly expressed in the kidney and shows the highest expression levels in DCT and medullary TAL.[11] Although the γ-subunit of the Na^+,K^+-ATPase pump is critical for active transcellular Mg^{2+} reabsorption in DCT, the molecular mechanism for renal Mg^{2+} loss in this autosomal dominant type of primary hypomagnesemia remains to be elucidated.[11] The key molecules that represent the basolateral Mg^{2+} extrusion mechanism in the process of transcellular Mg^{2+} transport are still elusive. Until recently the molecular identity of the protein that facilitates the apical influx of Mg^{2+} in active Mg^{2+} transport was unknown. A promising candidate was found by the elucidation of the genetic basis of hypomagnesemia with secondary hypocalcemia (HSH; MIM 602014). HSH is an autosomal recessive disease of which the gene locus was mapped on chromosome 9q22.[12] Affected individuals show neurological symptoms of hypomagnesemic hypocalcemia, including seizures and muscle spasms during infancy.[13-15] Physiological studies indicate that the pathophysiology of HSH is primarily caused by a primary defect in intestinal Mg^{2+} transport.[16,17] In most patients, renal Mg^{2+} conservation has been reported to be normal. However, for some patients, inappropriately high fractional Mg^{2+} excretion rates with respect to their low serum Mg^{2+} levels were reported.[14,18] This renal Mg^{2+} leak suggests an impaired renal Mg^{2+} reabsorption. HSH can be treated by high dietary Mg^{2+} intake because passive Mg^{2+} absorption is not affected.[19] When untreated, the disease may be fatal or may lead to severe neurologic damage. The observed hypocalcemia is a secondary effect possibly caused by parathyroid failure resulting from Mg^{2+} deficiency.[20] Using a positional candidate gene-cloning approach, mutations in TRPM6 were found to be the cause of autosomal recessive HSH.[13,14,21] The TRPM6 protein shows 52% homology with TRPM7, which has been identified as a Mg^{2+}-permeable ion channel particularly required for cellular Mg^{2+} homeostasis.[8,22,23] In summary, a genetic screen in patients with primary hypomagnesemia revealed the identification of the transient receptor potential (TRP) cation channel TRPM6 as potential gatekeeper in the maintenance of the Mg^{2+} balance.

TRPM6 and TRPM7

The TRP superfamily is a newly discovered family of cation-permeable ion channels.[24] There are at least three previously recognized subfamilies of proteins; TRPC (conical), TRPV (vanilloid), and TRPM (metastatin), that are

4. TRPM6 and TRPM7 Chanzymes Essential for Magnesium Homeostasis

expressed throughout the animal kingdom (http://clapham.tch.harvard.edu/trps/). Recently, the polycystins were also included in the TRP superfamily abbreviated as TRPP (polycystin).[25] Each of the proteins is a cation channel composed of six transmembrane-spanning domains and a conserved pore-forming region that assemble in a tetrameric configuration (Figure 4.2).[24,26,27] TRPM6 is one of the eight members of the identified TRPM cation channel subfamily and is composed of 2022 amino acids encoded by a large gene that contains 39 exons.[13,14,24]

TRPM6 and its closest homologue TRPM7 are unique bifunctional proteins combining Mg^{2+}-permeable cation channel properties with protein kinase activity.[24,28] However, the precise function of this kinase domain remains to be established. To date, TRPM7 regulation has received most of the attention. TRPM7 is ubiquitously expressed and implicated in cellular Mg^{2+} homeostasis. In contrast to TRPM7, TRPM6 has a more restricted expression pattern and is predominantly present in (re)absorbing epithelia.[8,13,14] In mice, TRPM6 is predominantly present in kidney, lung, cecum, and colon, that therefore likely constitute the main sites of active Mg^{2+} (re)absorption.[29] TRPM6 and TRPM7 are not functionally redundant in humans because the concurrent co-expression of TRPM7 cannot rescue the severe phenotype of HSH patients.[30] In kidney, TRPM6 is localized along the apical membrane of DCT, known as the main site of active transcellular Mg^{2+} reabsorption along the nephron.[8] In line with the expected function of being the gatekeeper of Mg^{2+} influx, TRPM6 was predominantly localized along the apical membrane of these immunopositive tubules. Immunohistochemical studies of TRPM6 and the Na^+, Cl^- cotrans-

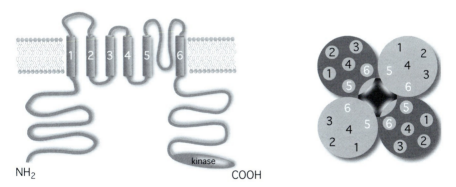

FIGURE 4.2. Structural organization of TRPM6 and TRPM7. TRPM6 and TRPM7 belong to the largest TRP channels, consisting of around 2000 amino acids including very large cytosolic amino- and carboxyl-termini including an atypical protein kinase domain. The six transmembrane unit is one of four identical or homologous subunits presumed to surround the central pore. The gate and selectivity filter are formed by the four 2 transmembrane domains (TM5 pore loop TM6) facing the center of the channel. Cations are selected for permeation by the extracellular-facing pore loop, held in place by the TM5 and TM6 α-helices.

porter (NCC), which were used as specific markers for DCT, indicated a complete colocalization of these transport proteins in kidney.[8] Until now, specific Mg^{2+}-binding proteins have not been identified, but it is interesting to mention that the Ca^{2+}-binding proteins parvalbumin and calbindins also bind Mg^{2+}.[31] Importantly, TRPM6 colocalized with parvalbumin in the first part of the DCT and with calbindin-D_{28K} in the last part of the DCT.[8] In the small intestine, absorptive epithelial cells stained positively for TRPM6 detected by in situ hybridization and immunohistochemistry.[8,14] Voets and colleagues demonstrated that TRPM6 was localized along the brush-border membrane of these intestinal cells.[8]

Interestingly, Schmitz and coworkers[23] demonstrated that Mg^{2+} supplementation of cells that lack TRPM7 expression rescued growth arrest and cell lethality that was caused by TRPM7 inactivation. Although TRPM7 is permeable for Ca^{2+}, as well as trace divalents such as Zn^{2+}, Ni^{2+}, Ba^{2+}, and Co^{2+}, supplementation with these cations was ineffective, indicating the specific effect of Mg^{2+} on these cellular processes. Thus, the ubiquitously expressed TRMP7 is important for cellular viability and cellular Mg^{2+} homeostasis rather than the extracellular Mg^{2+} homeostasis. TRPM7 has also been implicated to play a role in anoxic neuronal death[22,32,33] and regulation of cell adhesion as a result of a combined effect of kinase-dependent and -independent pathways on actomyosin contractility.[34] This suggests that TRPM7 is also important for various cell biological processes, including cytokinesis, adhesion, and migration, because these processes are regulated by actomyosin contractility.

Regulation of TRPM6 and TRPM7

Although in HSH the defect was originally established at the level of the intestine, there is also evidence for impaired renal Mg^{2+} reabsorption.[14,18,35] It was demonstrated that individuals with HSH, with respect to their low serum Mg^{2+} levels, show inappropriately high fractional Mg^{2+} excretion rates. This indicates an additional role of impaired renal Mg^{2+} reabsorption in HSH.[18,35] This is confirmed by Walder and colleagues, who characterized a considerable renal leak of Mg^{2+} in HSH patients.[13] The renal expression of TRPM6, in addition to the renal Mg^{2+} leak in patients with HSH, emphasizes the important role of TRPM6 in renal Mg^{2+} reabsorption. The gatekeeper function of TRPM6 in transepithelial Mg^{2+} transport is further supported by studies investigating the regulation of TRPM6 by dietary Mg^{2+} and hormones. Dietary Mg^{2+} restriction in mice resulted in hypomagnesemia and renal Mg^{2+} conservation, and significantly upregulated renal TRPM6 mRNA levels. The role of TRPM7 in cellular and not total body Mg^{2+} homeostasis is supported by the fact that dietary Mg^{2+} did not alter TRPM7 mRNA expression in mouse kidney and colon. Furthermore, it was shown that the hormone 17β-estradiol, but not 1,25-dihydroxyvitamin D_3 and parathyroid hormone, regulates renal TRPM6 mRNA levels.[29] Renal TRPM7 mRNA abundance remained unaltered under these conditions. This study demonstrated that the renal TRPM6 mRNA level in ovariectomized rats is significantly reduced, whereas 17β-estradiol treatment normalized TRPM6 mRNA levels.[29] Future studies should point out if these hormones regulate the expression of TRPM6 in the intestine.

Chronic metabolic acidosis results in renal Mg^{2+} wasting, whereas chronic metabolic alkalosis is known to exert the reverse effect. It was hypothesized that these adaptations are mediated at least in part by the renal Mg^{2+} transport protein TRPM6. Chronic metabolic acidosis decreased renal TRPM6 expression, increased Mg^{2+} excretion, and decreased serum Mg^{2+} concentration, whereas chronic metabolic alkalosis resulted in the exact opposite effects.[36] Thus, these data suggest that regulation of TRPM6 contributes importantly to the effects of acid–base status on renal Mg^{2+} handling. In addition, TRPM6 is regulated in NCC-knockout mice and upon NCC inhibition.[37] Thiazide diuretics enhance renal Na^+ excretion by blocking NCC, and mutations in NCC result in Gitelman syndrome. The mechanism underlying the accompanying hypomagnesemia remains debated. In NCC-knockout mice, an animal model of Gitelman syndrome and during chronic HCTZ administration, the hypomagnesemia is accompanied by downregulation of TRPM6. Thus, TRPM6 downregulation may represent a general mechanism involved in the pathogenesis of hypomagnesemia accompanying NCC inhibition or inactivation.

Functional Characteristics of TRPM6 and TRPM7

To functionally characterize TRPM6, the protein was heterogeneously expressed in human embryonic kidney 293 (HEK293) cells. TRPM6-transfected HEK293 cells perfused with an extracellular solution that contained 1 mM Mg^{2+} or Ca^{2+} exhibited characteristic outwardly rectifying currents upon establishment of the whole-cell configuration as demonstrated for TRPM7.[8,22,28] It is intriguing that at physiological membrane potentials of the DCT cell (−80 mV), small but significant inward currents were observed in TRPM6-expressing HEK293 cells with all tested divalent cations as the sole charge carrier.[8] However, mutations in TRPM6 are linked directly to HSH, emphasizing that this channel is an essential component of the epithelial Mg^{2+} uptake machinery. Interestingly, HEK293 cells transfected with the TRPM6 mutants identified in HSH patients (TRPM6^{Ser590X} and TRPM6^{Arg736fsX737}) displayed currents with similar amplitude and activation kinetics as nontranfected HEK293 cells, indicating that these mutant proteins are nonfunctional, in line with the postulated function of TRPM6 being Mg^{2+} influx step in epithelial Mg^{2+} transport.[8] It is possible that the TRPM6-mediated Mg^{2+} inward current is more pronounced in native DCT and intestinal cells as a result of specific cofactors, such as intracellular Mg^{2+} buffers, that are missing in overexpression systems, for example, HEK293 cells. The unique permeation rank order determined from the inward current amplitude at −80 mV was comparable to TRPM7 (Ba^{2+} ⩾ Ni^{2+} > Mg^{2+} > Ca^{2+}).[8,28] Micropuncture studies have demonstrated that the luminal concentration of Mg^{2+} in DCT ranges from 0.2 to 0.7 mM.[3] Because the Ca^{2+} concentration is in the millimolar range, the apical Mg^{2+} influx pathway should exhibit a higher affinity for Mg^{2+} than for Ca^{2+}. It is interesting that dose-response curves for the Na^+ current block at −80 mV indicated four times higher K_D values for Ca^{2+} compared with Mg^{2+}.[8] These data suggest that the pore of TRPM6 has a higher affinity for Mg^{2+} than for Ca^{2+}. In this way, TRPM6 comprises a unique channel because all known Ca^{2+}-permeable channels, including members of the TRP superfamily, generally display a 10 to 1000 times lower affinity for Mg^{2+} than for Ca^{2+}.

Voets and colleages demonstrated a coherent relationship between the applied extracellular Mg^{2+} concentration and the measured intracellular Mg^{2+} level in TRPM6-expressing cells by using the Mg^{2+}-sensitive radiometric fluorescent dye Magfura-2.[8] To study the effect of intracellular Mg^{2+} on TRPM6 activity, the Mg^{2+} concentration was altered directly in a spatially uniform manner using flash photolysis of the photolabile Mg^{2+} chelator DM-nitrophen.[8] Elevation of the intracellular Mg^{2+} concentration reduced the TRPM6-induced current, indicating that the channel is regulated by intracellular Mg^{2+} or Mg^{2+}·ATP. Previous studies reported that TRPM7 channel activity is strongly suppressed by Mg^{2+}·ATP concentrations in the millimolar range.[22,38] Kozak and Cahalan demonstrated that intracellular Mg^{2+} rather than ATP inhibits TRPM7 channel activity.[39] Runnels and coworkers demonstrated that in addition to Mg^{2+} inhibition, TRPM7 is also inactivated by hydrolysis of the phosphoinositide lipid PIP_2.[40] It is interesting that TRPM7 was discovered on the basis of binding to phospholipase C (PLC), which cleaves PIP_2.[28] This raises the question whether there is a link between the two apparently dissimilar modes of channel inhibition: an increase in Mg^{2+} and hydrolysis of PIP_2. It was postulated that Mg^{2+} could possibly activate PLC that associates with TRPM7.[41] This possibility is strengthened by a study in B lymphocytes that demonstrates the presence of phosphoinositide-specific PLC, which is activated as the Mg^{2+} concentration is raised from 30 to 1000 μM.[42] Although, a Mg^{2+}-activated PLC has not been described in other cell types, it may be present in various cell types including other lymphoid-derived cell lines, for example, Jurkat T cells and rat basophilic leukemia cells. The latter mentioned cell lines contain an endogenous TRPM7-like current, referred to as a Mg^{2+}-inhibited cation channel (MIC), which makes them attractive candidates for the presence a Mg^{2+}-activated PLC.[43,44]

Recently, it was suggested that at physiological pH (7.4), Ca^{2+} and Mg^{2+} bind to TRPM7 and inhibit the monovalent cationic currents, whereas at high H^+ concentrations, the affinity of TRPM7 for Ca^{2+} and Mg^{2+} is decreased, thereby allowing monovalent cations to pass through TRPM7. In addition, these investigators demonstrated that the endogenous TRPM7-like current, which is known as Mg^{2+}-inhibitable cation current (MIC) or Mg^{2+} nucleotide-regulated metal ion current (MagNuM) in rat basophilic leukemia (RBL) cells was also significantly potentiated by acidic pH, suggesting that MIC/MagNuM is encoded by TRPM7. In conclusion, the pH sensitivity represents a novel feature of TRPM7 and implies that TRPM7 may play a role under acidic pathological conditions.[45]

It was postulated that TRPM6 requires assembly with TRPM7 to form functional channel complexes in the plasma membrane and that disruption of multimer formation by a mutated TRPM6 variant, TRPM6[S141L], results in HSH.[46] In this study, TRPM6[S141L] was not directed to the cell surface by TRPM7 and failed to interact with the co-expressed TRPM7. Remarkably, in contrast to TRPM7, Gudermann and coworkers found that TRPM6 expression in *Xenopus* oocytes and HEK293 cells did not entail significant ion currents.[46] In addition, Schmitz and colleagues also supported the idea that TRPM6 and TRPM7 associate and that trafficking of TRPM6 to the cell surface is strongly dependent on TRPM7 co-expression.[30] In variance, Voets and colleagues measured significantly larger currents in TRPM6-transfected HEK293 cells compared to mock-transfected cells.[8] An explanation for this discrepancy might be the existence of specific TRPM6 splice variants with different functional properties. Chubanov and colleagues. demonstrated that 5′ rapid amplification of cDNA ends revealed three short alternative 5′ exons, called 1A, 1B, and 1C, that were found to be individually spliced onto exon 2, suggesting that the TRPM6 gene harbors a promoter with alternative transcription start sites.[46] These transcripts have been named, accordingly, TRPM6a, TRPM6b, and TRPM6c, and additional functional measurements are needed to explain possible biophysical differences.

α-Kinase Domain

Unlike other members of the TRP channel family, TRPM6 and TRPM7 contain long carboxy-terminal domains including an active threonine/serine kinase, which belongs to the atypical family of eukaryotic α-kinases.[28] Genomic studies identified six α-kinases in mammals, including the ones fused to TRPM6 and TRPM7. The α-kinases share no sequence homology with the conventional kinases.[47] A key question concerns the nature of the mechanisms underlying the activation and regulation of TRPM6 and TRPM7. In particular, what is the function of the atypical protein α-kinase domain located in the carboxyl terminus? To characterize the TRPM7 kinase activity in vitro, the catalytic domain was expressed in bacteria.[48] This kinase is able to undergo autophosphorylation and to phosphorylate substrates such as myelin basic protein and histone H3 on serine and threonine residues. The kinase is specific for ATP and Mg^{2+} or Mn^{2+} is required for optimal activity. Schmitz and coworkers demonstrated that, using a phosphothreonine-specific antibody, TRPM6 can phosphorylate TRPM7, but not vice versa.[30] Recently, annexin 1 and myosin II A have been identified as endogenous substrates of the TRPM7 kinase.[34,49] Although, the biological role of annexin 1 phosphorylation via

TRPM7 is currently unknown, both proteins have been linked to processes of cell survival and growth.[33,50] Future studies should elucidate whether the kinase domain of TRPM6 and TRPM7 has specific cellular phosphorylation targets that modulate ion channel activity and, therefore, the Mg^{2+} balance. It has been suggested that TRPM7 could potentially serve both as a Mg^{2+} uptake mechanism and a Mg^{2+} sensor when the ion channel domain could modulate the phosphotransferase activity by increasing intracellular Mg^{2+} or directly via conformational changes induced by gating of the channel.[51] This modification could result in the phosphorylation of yet unidentified substrates providing real time information on channel activity or cellular Mg^{2+} status. While there is general consensus that TRPM7 is inhibited by free intracellular Mg^{2+}, the functional role of intracellular levels of Mg^{2+} ATP and the kinase domain in regulating TRPM7 channel activity remain controversial. Several groups suggested that the kinase domain is essential for channel activity,[28,52] whereas others indicated that it is not involved.[23,53] Furthermore, the phosphotransferase activity of the TRPM7 kinase domain affected channel activity by regulating the sensitivity of the channel to inhibition by Mg^{2+} and $Mg^{2+}\cdot ATP$,[23,54,55] but this finding was not generally confirmed.[39,52] Further experiments are needed to elucidate these inconsistent results and firmly establish the functional role of the kinase domain and intracellular levels of $Mg^{2+}\cdot ATP$ in regulating TRPM6 and TRPM7 channel activity.

Prospectives

This chapter focused on the identification, function, and regulation of the epithelial Mg^{2+} channels TRPM6 and TRPM7. The discovery of these channels was a significant advance in our knowledge about cellular Mg^{2+} transport and hormonal regulation of the Mg^{2+} balance. Although, several studies have investigated the regulation of TRPM6 and TRPM7, many questions remain to be investigated. Future research is needed to unravel the nature of the mechanisms underlying the activation and regulation of TRPM6 and TRPM7 and to elucidate the reported controversy in the functioning and necessity of the kinase domain in regulating TRPM7 channel activity. In addition, the phosphorylation targets of these chanzymes that can modulate channel activity and, therefore, the Mg^{2+} balance remain to be identified. Taken together, the discovery of TRPM6 and TRPM7 provides a unique opportunity to investigate the molecular mechanisms determining the specificity, regulation, and activation of these chanzymes in detail with respect to their distinct roles in Mg^{2+} homeostasis.

Acknowledgment. The authors were supported by grants of the Dutch Kidney Foundation (C02.2030 and C03.6017) and the Dutch Organization of Scientific Research (Zon-Mw 016.006.001 and 9120.6110).

References

1. Brunette M, Crochet M. Fluormetric method for the determination of magnesium in renal tubular fluid. *Anal Biochem* 1975;65:79–88.
2. Grimellec C, Poujeol P, Rouffignia C. ^{3}H-inulin and electrolyte concentrations in Bowman's capsule in rat kidney. Comparison with artificial ultrafiltration. *Pflugers Arch* 1975;354:117–131.
3. Dai L, Ritchie G, Kerstan D, et al. Magnesium transport in the renal distal convoluted tubule. *Physiol Rev* 2001;81:51–84.
4. de Rouffignac C, Quamme G. Renal magnesium handling and its hormonal control. *Physiol Rev* 1994;74:305–322.
5. Shafik I, Quamme G. Early adaptation of renal magnesium reabsorption in response to magnesium restriction. *Am J Physiol* 1989;257:F974–F977.
6. Quamme G, Dirks J. Magnesium transport in the nephron. *Am J Physiol* 1980;239:F393–F401.
7. Quamme G, Dirks J. Intraluminal and contraluminal magnesium on magnesium and calcium transfer in the rat nephron. *Am J Physiol* 1980;238:F187–F198.
8. Voets T, Nilius B, Hoefs S, et al. TRPM6 forms the Mg^{2+} influx channel involved in intestinal and renal Mg^{2+} absorption. *J Biol Chem* 2004;279:19–25.
9. Flatman P. Mechanisms of magnesium transport *Annu Rev Physiol* 1991;53:259–271.
10. Simon D, Lu Y, Choate K, et al. Paracellin-1, a renal tight junction protein required for paracellular Mg^{2+} resorption. *Science* 1999;285:103–106.
11. Meij I, Koenderink J, van Bokhoven H, et al. Dominant isolated renal magnesium loss is caused by misrouting of the Na^+,K^+-ATPase gamma-subunit. *Nat Genet* 2000;26:265–266.
12. Walder R, Shalev H, Brennan T, et al. Familial hypomagnesemia maps to chromosome 9q, not to the X chromosome: genetic linkage mapping and analysis of a balanced translocation breakpoint. *Hum Mol Genet* 1997;6:1491–1497.
13. Walder R, Landau D, Meyer P, et al. Mutation of TRPM6 causes familial hypomagnesemia with secondary hypocalcemia. *Nat Genet* 2002;31:171–174.
14. Schlingmann K, Weber S, Peters M, et al. Hypomagnesemia with secondary hypocalcemia is caused by mutations in TRPM6, a new member of the TRPM gene family. *Nat Genet* 2002;31:166–170.
15. Paunier L, Radde I, Kooh S, et al. Primary hypomagnesemia with secondary hypocalcemia in an infant. *Pediatrics* 1968;41:385–402.
16. Milla P, Aggett P, Wolff O, et al. Studies in primary hypomagnesaemia: evidence for defective carrier-mediated small intestinal transport of magnesium. *Gut* 1979;20:1028–1033.
17. Yamamoto T, Kabata H, Yagi R, et al. Primary hypomagnesemia with secondary hypocalcemia. Report of a case and review of the world literature. *Magnesium* 1985;4:15364.
18. Matzkin H, Lotan D, Boichis H. Primary hypomagnesemia with a probable double magnesium transport defect. *Nephron* 1989;52:83–86.
19. Shalev H, Phillip M, Galil A, et al. Clinical presentation and outcome in primary familial hypomagnesaemia. *Arch Dis Child* 1998;78:127–130.
20. Anast C, Mohs J, Kaplan S, et al. Evidence for parathyroid failure in magnesium deficiency. *Science* 1972;177:606–608.

21. Schlingmann K, Sassen M, Weber S, et al. Novel TRPM6 mutations in 21 families with primary hypomagnesemia and secondary hypocalcemia. *J Am Soc Nephrol* 2005;16:3061–3069.
22. Nadler M, Hermosura M, Inabe K, et al. LTRPC7 is a Mg.ATP-regulated divalent cation channel required for cell viability. *Nature* 2001;411:590–595.
23. Schmitz C, Perraud A, Johnson C, et al. Regulation of vertebrate cellular Mg^{2+} homeostasis by TRPM7. *Cell* 2003;114:191–200.
24. Clapham D, Runnels L, Strubing C. The TRP ion channel family. *Nat Rev Neurosci* 2001;2:387–96.
25. Montell C. Physiology, phylogeny, and functions of the TRP superfamily of cation channels. *Sci STKE* 2001;2001:RE1.
26. Hoenderop J, Nilius B, Bindels R. Molecular mechanism of active Ca^{2+} reabsorption in the distal nephron. *Annu Rev Physiol* 2002;64:529–549.
27. Clapham D. TRP channels as cellular sensors. *Nature* 2003;426:517–524.
28. Runnels L, Yue L, Clapham DE. TRP-PLIK, a bifunctional protein with kinase and ion channel activities. *Science* 2001;291:1043–1047.
29. Groenestege W, Hoenderop J, van den Heuvel L, et al. The epithelial Mg^{2+} channel transient receptor potential melastatin 6 is regulated by dietary Mg^{2+} content and estrogens. *J Am Soc Nephrol* 2006;17:1035–1043.
30. Schmitz C, Dorovkov M, Zhao X, et al. The channel kinases TRPM6 and TRPM7 are functionally nonredundant. *J Biol Chem* 2005;280:37763–37771.
31. Yang W, Lee H, Hellinga H, et al. Structural analysis, identification, and design of calcium-binding sites in proteins. *Proteins* 2002;47:344–356.
32. Nicotera P, Bano D. The enemy at the gates. Ca^{2+} entry through TRPM7 channels and anoxic neuronal death. *Cell* 2003;115:768–770.
33. Aarts M, Iihara K, Wei WL, et al. A key role for TRPM7 channels in anoxic neuronal death. *Cell* 2003;115:863–877.
34. Clark K, Langeslag M, van Leeuwen B, et al. TRPM7, a novel regulator of actomyosin contractility and cell adhesion. *EMBO J* 2006;25:290–301.
35. Cole D, Quamme G. Inherited disorders of renal magnesium handling. *J Am Soc Nephrol* 2000;11:1937–1947.
36. Nijenhuis T, Renkema K, Hoenderop J, et al. Acid-base status determines the renal expression of Ca^{2+} and Mg^{2+} transport proteins. *J Am Soc Nephrol* 2006;17:617–626.
37. Nijenhuis T, Vallon V, van der Kemp A, et al. Enhanced passive Ca^{2+} reabsorption and reduced Mg^{2+} channel abundance explains thiazide-induced hypocalciuria and hypomagnesemia. *J Clin Invest* 2005;115:1651–1658.
38. Hermosura M, Monteilh-Zoller M, Scharenberg A, et al. Dissociation of the store-operated calcium current I(CRAC) and the Mg-nucleotide-regulated metal ion current MagNuM. *J Physiol* 2002;539:445–458.
39. Kozak JA, Cahalan M. MIC channels are inhibited by internal divalent cations but not ATP. *Biophys J* 2003;84:922–927.
40. Runnels L, Yue L, Clapham D. The TRPM7 channel is inactivated by PIP(2) hydrolysis. *Nat Cell Biol* 2002;4:329–336.
41. Montell C. Mg^{2+} homeostasis: the Mg^{2+}nificent TRPM chanzymes. *Curr Biol* 2003;13:R799–R801.
42. Chien M, Cambier J. Divalent cation regulation of phosphoinositide metabolism. Naturally occurring B lymphoblasts contain a Mg^{2+}-regulated phosphatidylinositol-specific phospholipase C. *J Biol Chem* 1990;265:9201–9207.

43. Kozak J, Kerschbaum H, Cahalan M. Distinct properties of CRAC and MIC channels in RBL cells. *J Gen Physiol* 2002;120:221–235.
44. Prakriya M, Lewis R. Separation and characterization of currents through store-operated CRAC channels and Mg^{2+}-inhibited cation (MIC) channels. *J Gen Physiol* 2002;119:487–507.
45. Jiang J, Li M, Yue L. Potentiation of TRPM7 inward currents by protons. *J Gen Physiol* 2005;126:137–150.
46. Chubanov V, Waldegger S, Mederos y Schnitzler M, et al. Disruption of TRPM6/TRPM7 complex formation by a mutation in the TRPM6 gene causes hypomagnesemia with secondary hypocalcemia. *Proc Natl Acad Sci U S A* 2004;101:2894–2899.
47. Ryazanov A, Ward M, Mendola C, et al. Identification of a new class of protein kinases represented by eukaryotic elongation factor-2 kinase. *Proc Natl Acad Sci U S A* 1997;94:4884–4889.
48. Ryazanova L, Dorovkov M, Ansari A, et al. Characterization of the protein kinase activity of TRPM7/ChaK1, a protein kinase fused to the transient receptor potential ion channel. *J Biol Chem* 2004;279:3708–3716.
49. Dorovkov M, Ryazanov A. Phosphorylation of annexin I by TRPM7 channel-kinase. *J Biol Chem* 2004;279:50643–50646.
50. Perretti M, Solito E. Annexin 1 and neutrophil apoptosis. *Biochem Soc Trans* 2004;32:507–510.
51. Schmitz C, Perraud A, Fleig A, Scharenberg AM. Dual-function ion channel/protein kinases: novel components of vertebrate magnesium regulatory mechanisms. *Pediatr Res* 2004;55:734–737.
52. Matsushita M, Kozak JA, Shimizu Y, et al. Channel function is dissociated from the intrinsic kinase activity and autophosphorylation of TRPM7/ChaK1. *J Biol Chem* 2005;280:20793–20803.
53. Yamaguchi H, Matsushita M, Nairn A, et al. Crystal structure of the atypical protein kinase domain of a TRP channel with phosphotransferase activity. *Mol Cell* 2001;7:1047–1057.
54. Takezawa R, Schmitz C, Demeuse P, et al. Receptor-mediated regulation of the TRPM7 channel through its endogenous protein kinase domain. *Proc Natl Acad Sci U S A* 2004;101:6009–6014.
55. Demeuse P, Penner R, Fleig A. TRPM7 channel is regulated by magnesium nucleotides via its kinase domain. *J Gen Physiol* 2006;127:421–434.

5
CorA-Mrs2-Alr1 Superfamily of Mg^{2+} Channel Proteins

Rudolf J. Schweyen and Elisabeth M. Froschauer

CorA proteins are ubiquitously expressed in eubacteria and archaea. They constitute the first ever described proteins for cellular Mg^{2+} uptake. Nearly all eukaryotic genomes encode distant relatives of CorA, named Alr1p and Mrs2p. Members of each of these subfamilies have been shown to form oligomeric membrane complexes mediating electrogenic uptake of Mg^{2+}. The recently published crystal structure of a CorA-type protein and electrophysiological data of a Mrs2-type protein consistently show that this superfamily of proteins forms Mg^{2+} permeable channels.

Uptake of Mg^{2+} into bacteria, mitochondria, or eukaryotic cells has been described as a diffusive process, driven by the inside negative membrane potential.[1-6] If these processes were unlimited, free ionized Mg^{2+} ($[Mg^{2+}]$) would reach unphysiologically high values, particularly in compartments with high inside negative membrane potentials like bacterial cells or eukaryotic mitochondria. Yet concentrations of $[Mg^{2+}]$ in cells or mitochondria were found to be small, ranging according to most determinations from 0.3 to 1.2. Accordingly, cells and organelles can be assumed to have developed mechanisms to control intracellular and intraorganellar $[Mg^{2+}]$. This may either be mediated by proteins involved in Mg^{2+} uptake processes or via activation of Mg^{2+} extrusion systems, which may involve Na^+/Mg^{2+} or H^+/Mg^{2+} exchange or anion/Mg^{2+} cotransport.[1] The latter processes occur at the expense of other, actively generated ion gradients (Na^+, H^+; cf. Chapter 3, this book). Therefore, it will be energetically favorable to limit influx to the needs of cells or organelles via regulation of Mg^{2+} influx systems.

Early literature essentially agreed on the presence of very slow turnover of Mg^{2+} across cell or mitochondrial membranes under quiescent conditions. Yet more recent reports provided a number of experimental observations suggesting that large fluxes of Mg^{2+} can cross these membranes following a variety of stimuli.[5-7]

Prokaryotic Mg^{2+} Transport Proteins

Genetic screens for bacteria resistant to toxic levels of cobalt led to the identification of corA, mgtA, and mgtBC as the first known loci encoding Mg^{2+} transport proteins. A screen for suppressors of CorA-deficient bacteria resulted in the identification of an additional transporter named MgtE.[3,8,9]

CorA proteins are constitutively expressed proteins of the plasma membrane without any similarity to other proteins families. They constitute the major Mg^{2+} transporters in *Eubacteria* and *Archaea*. Co^{2+} and Ni^{2+} also serve as CorA transport substrates, but with lower affinities and velocities than Mg^{2+}. CorA of the bacterium *Salmonella enterica serovar Typhimurium* has been reported to transport Mg^{2+} with an apparent Km of 15 to 20 μM, allowing them to grow in media with very low $[Mg^{2+}]$.[3]

While earlier studies on CorA mediated cation transport involved radioactively labeled isotopes, Froschauer and colleagues[6] introduced measurements of free intracellular $[Mg^{2+}]$ by use of the Mg^{2+} sensitive dye Magfura-2. Bacterial cells were found to exhibit a rapid change of internal free $[Mg^{2+}]$ upon increase of extracellular $[Mg^{2+}]$, consistent with a high capacity of Mg^{2+} uptake (Figure 5.1). Driving force of this Mg^{2+} uptake is the high inside-negative membrane potential of bacterial cells. Efflux from bacterial cells was comparatively small, and only visible upon dissipation of the membrane potential.[6] CorA-mediated efflux, as invoked previously,[10] is at variance with the findings that this protein forms a channel (see below) mediating Mg^{2+} influx driven by the inside negative membrane potential.

MgtA and its homologue MgtB are polytopic membrane proteins belonging to the large group of P-type ATPases in pro- and eukaryotes. MgtA is present in widely divergent bacteria, but it is not as common as CorA. Expression of MgtA and MgtB is under the control of a two-component signal-transduction

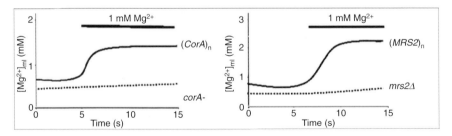

FIGURE 5.1. Influx of Mg^{2+} into bacterial cells (left) and isolated yeast mitochondria (right). Bacterial cells or isolated yeast mitochondria were loaded with the Mg^{2+}-sensitive fluorescent dye Magfura-2 and Intramitochondrial $[Mg^{2+}]$ was determined by ratiometric measurements[5,6] at resting conditions (nominally Mg^{2+}-free buffer) and upon addition of Mg^{2+} to the buffer to final concentrations of 1 mM. (Left) $(CorA)_n$ and corA−: *S. typhimurium* corA−, mgtA−, and mgtB− cells transformed with multicopy plasmid expressing CorA or lacking this gene, respectively. (Right) $(Mrs2)_n$ and mrs2Δ: mitochondria isolated from cells expressing *MRS2* from a high copy number vector or lacking the *MRS2* gene, respectively.

system regulated by Mg^{2+} and Ca^{2+}. Their expression is thus restricted to conditions when bacteria are starved for Mg^{2+}.[3]

Triple mutants lacking MgtA, MgtB, and CorA have a Mg^{2+}-dependent growth phenotype. Compared to CorA-expressing cells, mutant mgtA, mgtB, and corA cells had about twofold reduced steady state [Mg^{2+}] and exhibited about tenfold reduced rates of Mg^{2+} uptake when provided with 1 mM external Mg^{2+} (Figure 5.1).[6]

Bactrial MgtE proteins have been poorly studied, but the recent finding of putative homologues in mammalian cells (SLC41) has raised renewed interest in this protein family (cf. Chapter 3, this book). This polytopic membrane protein has been shown to be present in various *Eubacteria* as well as *Archaea*, but not as ubiquitously as CorA. MgtE has been found to transport Mg^{2+} and Co^{2+}, but (unlike CorA) not Ni^{2+}. It remains to be shown if Mg^{2+} transport is its physiological role in bacteria.[11]

Structure and Function of CorA Proteins

CorA proteins constitute a gene family of their own.[11,12] Comparisons of proteins from distantly related prokaryotes reveal very little primary sequence conservation. As a common denominator all CorA proteins have two predicted transmembrane domains, separated by a short loop-oriented to the outside of the membrane, and a YGMN/F motif at the end of the N-terminal of one of these two transmembrane (TM) domains, frequently followed by a MPEL motif in the loop sequence. Mutational studies identified this motif as critical for CorA function in *S. typhimurium* and recognized further residues in each of the two conserved TM domains as important for ion transport and/or for ion selectivity.[13,14] These studies and the finding of a CorA homotetrameric complex were consistent with the formation of Mg^{2+}- (as well as Ni- and Co^{2+}-) permeable channels.[15]

Expressing different tagged isomers of partially deleted CorA proteins in *S. typhimurium*, suggested the presence of the large N-terminal protein sequence in the periplasm, and the short C-terminus in the cytoplasm. A third TM segment, N-terminal to the two predicted TM domains near the C-terminus, was postulated according to Smith and colleagues.[16] Yet, this additional TM segment frequently contained charged residues and thus remained questionable. Other CorA-related proteins have been reported to have N-in, C-in orientation, much like the eukaryotic homologues Mrs2p and Alr1p (see below).

A crystal structure of the CorA protein from the bacterium *Thermotoga maritima* has recently been published.[17] It forms a channel, composed of five CorA monomers each of which spans the membrane twice. These TM domains form two concentric rings in the membrane. The inner ring is made up of the five N-terminal domains and forms the pore. The five C-terminal TM domains envelope this inner ring. The short loop connecting the TM domains is oriented towards the outside and presumably contains an ion selective filter. The short C-terminal as well as the long N-terminal sequence are oriented towards the cytoplasm.

The structure published appears to represent the closed form of the channel. While it is not informative with respect to predict ion conductance values, it reveals details about amino acid residues in the pore, their putative interaction with Mg^{2+} and their role in gating of the pore. The five large N-terminal domains fold into a funnel, opening from the membrane pore towards the cytoplasm. Clusters of negatively charged or hydroxyl-bearing amino acids inside this funnel may constitute an electrostatic sink for incoming ions. A magnesium binding site has been observed in the N-teminal sequences, involving two neighboring monomers. Metal binding at this site might locally alter helix conformation with possible long-range effects to the membrane pore affecting its conductance.[17]

Alr1-Type Proteins

MacDiarmid and Gardner[18] screened for genes of which overexpression rendered cells of the yeast *Saccharomyces cerevisiae* tolerant to growth on toxic levels of Al^{3+}. Two genes, *ALR1* and *ALR2*, were shown to encode closely related proteins with some similarity to the bacterial CorA proteins, particularly two adjacent TM domains near the proteins C-terminus, and a GMN motif in the short loop connecting them. Studies on the cellular topology of Alr1p revealed its location in the plasma membrane.[19] Based on splitubiquitin assays both N- and C-termini were shown to be in the cytoplasm and, accordingly, the TM-connecting loop on the outer side of the plasma membrane.[20]

Disruption of the *ALR1* gene resulted in a dependence of yeast cells for $[Mg^{2+}]$ less than 20 mM in growth media and led to about twofold reduced levels in intracellular $[Mg^{2+}]$ and thus suggested a major role of the Alr1 protein in cellular Mg^{2+} homeostasis. Patch-clamp studies confirmed that Alr1p mediates Mg^{2+} transport into yeast cells.[9,19,21] Interestingly, turnover of Alr1p via endocytosis and vacuolar degradation were shown to become active when $[Mg^{2+}]$ in growth media was less than 0.1 mM.[19] Besides transcriptional regulation of *ALR1* expression, Mg^{2+}-dependent degradation of Alr1p appears to regulate the Mg^{2+} transport capacity of yeast cells.

Disruption of *ALR2* did not affect viability of yeast cells, but overexpression of Alr2p partially suppressed the growth phenotype of *alr1Δ* mutant cells. This gene is poorly expressed in yeast and, additionally, it appears to have mutations rendering it more inefficient in Mg^{2+} transport than Alr1p.[18,20]

Alr1-related proteins are ubiquitous in fungi and occasionally found in other lower eukaryotes. They appear to be absent in metazoa and plants.[12] Compared to CorA and to Mrs2 proteins, Alr1-related proteins frequently have extended N-terminal sequences, exceeding those of CorA or Mrs2 proteins by up to 300 residues. These extensions show little sequence conservation within the Alr1-type proteins, except when phylogenetically related fungi are compared, and partial deletions had little effect on Mg^{2+} transport activity of Alr1p (Graschopf, unpublished observations).

Mrs2

A genetic screen for proteins affecting splicing of group II introns in *Saccharomyces cerevisiae* mitochondria led to the identification of Mrs2p, an integral protein of the mitochondrial inner membrane. While Mrs2p had little similarity to CorA proteins, the presence to two adjacent TM domains and the GMN motif suggested it might be a distant relative of this bacterial Mg^{2+} transporter. Homologues of the yeast Mrs2 protein were found to be encoded by most eukaryotic genomes. The single functional *MRS2* gene in the human genome codes for a mitochondrial protein. The genome of the plant *Arabidopsis thaliana*, in contrast, contains 15 genes encoding CorA-related proteins. Most of them are Mrs2-like and appear to occur in mitochondrial as well as in plasma membranes.[12]

Evidence for the contribution of Mrs2p in mitochondrial Mg^{2+} uptake came from the findings that (1) knockout of the *MRS2* gene (*mrs2Δ* mutant) resulted in functional defects of mitochondria associated with RNA splice defects and reduced $[Mg^{2+}]$ and (2) that expression of a bacterial CorA protein in yeast could partially compensate for these mutant effects.[22] Moreover, RNA splicing was restored in isolated *mrs2Δ* mutant mitochondria when provided with high $[Mg^{2+}]$ in the buffer, consistent with low intramitochondrial $[Mg^{2+}]$ inhibiting RNA processing.[23]

The use of the fluorescent, Mg^{2+}-sensitive dye Magfura-2 in isolated mitochondria revealed the presence of a high-capacity influx system into wild-type yeast as well as mammalian mitochondria, resulting in the increase of free $[Mg^{2+}]$ from $0.6\,mM$ to about $2\,mM$ within seconds [Fig. 1(B)]. Flux rates were dependent on the mitochondrial membrane potential $\Delta\psi$ and on the expression level of Mrs2p. The rapid, high-capacity influx was absent in *mrs2Δ* knockout yeast cells as well as in knockdown cultured mammalian cells. A slow Mg^{2+} uptake remained to be seen in these mutant mitochondria, which led to steady state $[Mg^{2+}]$ of about half of those seen in wild-type cells over extended periods of time. The strict correlation between Mg^{2+}-influx rates and expression levels of Mrs2p, and the restoration of this influx by expression of related proteins (human or plant Mrs2p, bacterial CorA) in *mrs2Δ* mutant mitochondria led to the conclusion that this protein constitutes the Mg^{2+} transporter.[5,22,24] In addition, a plant Mrs2-type protein was shown to complement a bacterial CorA mutant.[26]

Previous studies on mammalian heart mitochondria also involved the Magfura-2 potentiometric method and determined similar steady state $[Mg^{2+}]$ values and revealed a slow Mg^{2+} uptake dependent on the extramitochondrial $[Mg^{2+}]$.[2] Stimulation of $\Delta\psi$ by nigericin addition led to an increased Mg^{2+} uptake. Based on their results, Jung and colleagues[2] suggested that mitochondria take up Mg^{2+} by electrophoretic flux through membrane leak pathways, rather than via a specific Mg^{2+} transporter.

In contrast, Kolisek and colleagues[5] and Piskacek[27] and coworkers observed high-capacity Mg^{2+} influx into yeast and human mitochondria, mediated by the Mrs2 protein and driven by $\Delta\psi$. Both yeast and human Mrs proteins were found to form homooligomeric complexes in mitochondrial membranes.[5,27] Accordingly, they concluded that the Mrs2 protein might form a high-conductance ion channel. Patch-clamp studies involving liposomes fused with yeast mitochondrial membranes now have fully confirmed this notion.[28] Mg^{2+} conductance of this mitochondrial channel has been found to be surprisingly high (155 pS), exceeding those of most cation channels in plasma membranes. Features of this channel varied with the mutational state of Mrs2p, leaving no doubt that currents measured were mediated by the Mrs2p homooligomer.[28]

The requirement for a high-conductance Mg^{2+} channel may be specific for bacterial cells and mitochondria with their rapid shifts of adenosine diphosphate/adenosine triphosphate (ADP/ATP) ratios, for example, when switching from anaerobiosis to aerobiosis or upon ischemic reperfusion. The affinity of ATP^{4-} (K_{Mg-ATP} 1×10^{-4}) is close to tenfold greater than that of ADP^{3-}, and mitochondrial adenine nucleotide concentrations are in the same molar range as free $[Mg^{2+}]$. A rapid conversion of ADP to ATP thus is expected to require influx of Mg^{2+} into mitochondria, and this is supported by experimental data.[29]

Common Features of CorA, Mrs2, and Alr1 Channel Proteins

Sequence conservation among the many members of the superfamily of proteins is restricted to the very C-terminal part with its two adjacent TM domains of which the N-terminal one ends with a GMN signature motif. All studies place the short loop connecting these two TM domains towards the outer side of the membrane. The long N-terminal and short C-terminal parts are highly variable in length and show very little sequence similarity. The crystal structure of the *T. maritima* CorA protein[17] is oriented towards the cytoplasm (N-in, C-in orientation), and this orientation has also been reported for yeast Alr1p and Mrs2p.[20,22] At variance with these data is the report of an N-out, C-in orientation of the CorA protein of *S. typhimurium*[16] and further experiments will be required to determine if membrane topologies of bacterial CorA proteins can differ.

Mutations in the signature sequence of this superfamily of proteins, GMN, almost invariably abolish transport function.[5,14] It appears to be critical for the positioning of the external loop and to form part of the entrance to the pore. Most proteins exhibit a surplus of negatively charged residues in the external loop, notably a conserved glutamic acid residue at position +6 of those proteins shown to transport Mg^{2+}. Mutational studies revealed a key role of this residue for Mg^{2+} transport by yeast Mrs2 and Alr1 channels.[20,28,30] Forming the mouth of the channels, these charged residues may electrostatically attract Mg^{2+} and may constitute a part of the selectivity filter. Various mutations in the TM domains of *S. typhimurium* have been shown to affect Mg^{2+} transport function and ion selectivity. A series of mutations in the soluble part of yeast

Mrs2p N-terminal to the TM domains are known to result increased Mg^{2+} flux, consistent with a role of these sequences in homeostasis control. A full interpretation of their effects will only be possible when three-dimensional (3D) structures are modeled.

Patch clamping provided physiological evidence for the presence of Mg^{2+} permeable channels[28] and will be the method of choice for further characterization of these channel proteins and mutations therein. Conductance of yeast Mrs2p was found to be exceptionally high (155 ps). It may reflect that the physiology of mitochondria (and probably also in bacterial cells) requires rapid Mg^{2+} influx to meet fast changes in components with highly different Mg^{2+} affinities, for exmaple, shifts of ADP to ATP.

Control of intracellular or mitochondrial Mg^{2+} homeostasis is likely to be mediated by the channel proteins themselves.[5] Steady-state $[Mg^{2+}]$ in mitochondria is kept rather independent of the level of Mrs2p expression, but mutations in this protein can result in a considerable upshift of free $[Mg^{2+}]$ and in Mg^{2+} conductance.[28,30] The finding of a putative Mg^{2+} binding site in the *T. maritima* structure is consistent with the notion that this cation directly may interfere with the channel to adjust internal $[Mg^{2+}]$ to physiological needs of cells and organelles.

Perspectives

The crystal structure of the *T. maritima* CorA protein forms a basis to model respective folding structures of Alr1 and Mrs2 proteins, although sequence conservation is very low. In combination with already available or future data on effects of mutations in CorA, Alr1, and/or Mrs2, a more detailed picture of structure/function relationships of this interesting protein superfamily can be expected in the near future.

A major focus of future work will be on sequences defining ion selectivity and on Mg^{2+} homeostasis control via opening/closing of the channel. Mutations in CorA, Alr1, or Mrs2 proteins with effects on Mg^{2+} homeostasis will help to evaluate the contribution of these proteins themselves and of other factors in Mg^{2+} homeostasis control. Yet a full understanding of Mg^{2+} homeostasis will require the identification of other genes and proteins contributing to the maintenance of physiological $[Mg^{2+}]$ and for Mg^{2+} extrusion from cells or organelles (presumably H^+/Mg^{2+} or Na^+/Mg^{2+} exchangers).

Finally, mutations in genes encoding the Mg^{2+} transport systems will experimentally allow obtaining unphysiologically high or low $[Mg^{2+}]$ in cells and organelles and thus will help to recognize and evaluate possible pathological effects of Mg^{2+} homeostasis disorders. Mutant CorA bacterial cells arrest growth while the absence of Mrs2p in yeast causes mitochondrial dysfunction, but no growth arrest. In contrast, mammalian cells die when expression of Mrs2p in their mitochondria is severely reduced. Low steady-state $[Mg^{2+}]$ obtained with yeast Mrs2 and bacterial CorA mutants indicated that RNA processing is a Mg^{2+}-sensitive function, possibly because Mg^{2+} has a key function in RNA structure formation.[23,31,32] Yet rapid influx of Mg^{2+} in response to sudden increase of Mg^{2+} binding components (e.g., ATP) may be equally or even more important for cellular function and a lack of this function may explain lethality observed in mammalian mMrs2 knockdown cells.

References

1. Beyenbach KW. Transport of magnesium across biological membranes. *Magnes Trace Elem* 1990;9:233-254.
2. Jung DW, Panzeter E, Baysal K, Brierley GP. *Biochim Biophys Acta* 1997;1320: 310-320.
3. Smith RL, Maguire ME. Microbial magnesium transport: unusual transporters searching for identity. *Mol Microbiol* 1998;28:217-226.
4. Schweigel M, Vormann J, Martens H. Mechanisms of Mg($^{2+}$) transport in cultured ruminal epithelial cells. *Am J Physiol Gastrointest Liver Physiol* 2000;278: G400-G408.
5. Kolisek M, Zsurka G, Samaj J, Weghuber J, Schweyen RJ, Schweigel M. Mrs2p is an essential component of the major electrophoretic Mg^{2+} influx system in mitochondria. *EMBO J* 2003;22:1235-1244.
6. Froschauer EM, Kolisek M, Dieterich F, Schweigel M, Schweyen RJ. Fluorescence measurements of free [Mg^{2+}] by use of mag-fura 2 in *Salmonella enterica*. *FEMS Microbiol Lett* 2004;237:49-55
7. Romani AM, Scarpa A. Regulation of cellular magnesium. *Front Biosci* 2000;5: D720-D734.
8. Kehres DG, Maguire ME. Structure, properties and regulation of magnesium transport proteins. *Biometals* 2002;15:261-270.
9. Gardner RC. Genes for magnesium transport. *Curr Opin Plant Biol* 2003;6: 263-267.
10. Snavely MD, Florer JB, Miller CG, Maguire ME. Magnesium transport in Salmonella typhimurium: expression of cloned genes for three distinct Mg^{2+} transport systems. *J Bacteriol* 1989;171:4752-4760.
11. Smith RL, Thompson LJ, Maguire ME. Cloning and characterization of MgtE, a putative new class of Mg^{2+} transporter from *Bacillus firmus* OF4. *J Bacteriol* 1995;177:1233-1238.
12. Knoop V, Groth-Malonek M, Gebert M, Eifler K, Weyand K. Transport of magnesium and other divalent cations: evolution of the 2-TM-GxN proteins in the MIT superfamily. *Mol Genet Genomics* 2005;274:205-216.
13. Smith RL, Szegedy MA, Kucharski LM, et al. The CorA Mg^{2+} transport protein of *Salmonella typhimurium*. Mutagenesis of conserved residues in the third membrane domain identifies a Mg^{2+} pore. *J Biol Chem* 1998;273:28663-28669.
14. Szegedy MA, Maguire ME. The CorA Mg^{2+} transport protein of *Salmonella typhimurium*—mutagenesis of conserved residues in the second membrane domain. *J Biol Chem* 1999;274:36973-36979.
15. Warren MA, Kucharski LM, Veenstra A, Shi L, Grulich PF, Maguire ME. The CorA Mg^{2+} transporter is a homotetramer. *J Bacteriol* 2004;186:4605-4612.
16. Smith RL, Banks JL, Snavely MD, Maguire ME. Sequence and topology of the CorA magnesium transport systems of *Salmonella typhimurium* and *Escherichia coli*.

Identification of a new class of transport protein. *J Biol Chem* 1993;268: 14071–14080.
17. Lunin VV, Dobrovetsky E, Khutoreskays G, et al. Crystal structure of the CorA Mg^{2+} transporter. *Nature* 2006 (in press).
18. MacDiarmid CW, Gardner RC. Overexpression of the *Saccharomyces cerevisiae* magnesium transport system confers resistance to aluminum ion. *J Biol Chem* 1998;273:1727–1732.
19. Graschopf A, Stadler JA, Hoellerer MK, et al. The yeast plasma membrane protein Alr1 controls Mg^{2+} homeostasis and is subject to Mg^{2+}-dependent control of its synthesis and degradation. *J Biol Chem* 2001;276:16216–16222.
20. Wachek M, Schweyen RJ, Graschopf A. Expression and oligomerization of Mg^{2+} transport proteins Alr1 and Alr2 in yeast plasma membrane. 2006 (submitted for publication).
21. Liu GJ, Martin DK, Gardner RC, Ryan PR. Large Mg^{2+}-dependent currents are associated with the increased expression of *ALR1* in *Saccharomyces cerevisiae*. *FEMS Microbiol Lett* 2002;213:231–237.
22. Bui DM, Gregan J, Jarosch E, Ragnini A, Schweyen RJ. The bacterial magnesium transporter CorA can functionally substitute for its putative homologue Mrs2p in the yeast inner mitochondrial membrane. *J Biol Chem* 1999;274:20438–20443.
23. Gregan J, Kolisek M, Schweyen RJ. Mitochondrial Mg($^{2+}$) homeostasis is critical for group II intron splicing in vivo. *Genes Dev* 2001;15:2229–2237.
24. Zsurka G, Gregan J, Schweyen RJ. The human mitochondrial Mrs2 protein functionally substitutes for its yeast homologue, a candidate magnesium transporter. *Genomics* 2001;72:158–168.
25. Schock I, Gregan J, Steinhauser S, Schweyen R, Brennicke A, Knoop V. A member of a novel *Arabidopsis thaliana* gene family of candidate Mg^{2+} ion transporters complements a yeast mitochondrial group II intron-splicing mutant. *Plant J* 2000;24:489–501.
26. Li L, Tutone AF, Drummond RS, Gardner RC, Luan S. A novel family of magnesium transport genes in *Arabidopsis*. *Plant Cell* 2001;13:2761–2775.
27. Piskacek M, Szurka M, Weghuber J, Schweyen RJ. (2006) Conditional knock-down of the mitochondrial magnesium channel protein, Mrs2p, in HEK-293 cells. 2006 (in press).
28. Weghuber J, Schindl R, Romanin C, Schweyen RJ. Mrs2p forms a high conductance Mg^{2+}-selective channel in mitochondria. 2006 (submitted for publication).
29. Jung DW, Apel L, Brierley GP. Matrix free Mg^{2+} changes with metabolic state in isolated heart mitochondria. *Biochemistry* 1990;29:4121–4128.
30. Weghuber J, Schindl R, Romanin C, Schweyen RJ. Mutational analysis of functional domains in Mrs2p, the mitochondrial Mg^{2+} channel protein of *Saccharomyces cerevisiae* FEBS J 2006 (in press).
31. Wiesenberger G, Waldherr M, Schweyen RJ. The nuclear gene *MRS2* is essential for the excision of group II introns from yeast mitochondrial transcripts in vivo. *J Biol Chem* 1992;267:6963–6969.
32. Dieterich F, Froschauer EM, Schweyen RJ. Sensitivity of RNA processing to intracellular Mg^{2+} concentrations in bacteria. 2006 (in press).
33. Hmiel SP, Snavely MD, Florer JB, Maguire ME, Miller CG. Magnesium transport in *Salmonella typhimurium*: genetic characterization and cloning of three magnesium transport loci. *J Bacteriol* 1989;171:4742–4751.

6
Practical Interest of Circulating Total and Ionized Magnesium Concentration Evaluation in Experimental and Clinical Magnesium Disorders

Nicole Pagès, Pierre Bac, Pierre Maurois, Andrée Guiet-Bara, Michel Bara, and Jean Durlach

Magnesium (Mg) is essentially an intracellular ion, which makes it difficult to evaluate Mg status. Both circulating total (MgT) and ionized Mg^{2+} (MgI) are used in clinical practice but their respective interest is still debated. In the present review, we list various studies comparing MgT and MgI in either Mg imbalances or Mg status dysregulations. In simple Mg imbalances (either therapeutic Mg overload or Mg deficiency), the evaluation of MgT appears a better marker than MgI because it seems that a subtle physiological homeostatic compensatory reaction modifies the proportion of MgI, the most biologically active fraction of blood Mg, in order to reduce the effects of Mg imbalance. In contrast, in Mg dysregulations (either Mg pathological overload or Mg depletion) both fractions may vary independently, depending mainly on the renal status and on the equilibrium between MgI and Mg complexed with proteins and anions. The choice of the more appropriate marker is discussed.

Magnesium is an essential cofactor in numerous cellular functions.[1] In addition to a balance between intestinal absorption and renal excretion, the bone is involved in magnesium homeostasis.[2] Exchanges between the circulating Mg forms also occur.[1,3]

Although MgI is considered as the bioactive fraction and generally presented as the best marker in Mg imbalances, most laboratories still measure only MgT. In healthy humans, MgT concentration is maintained within narrow limits (0.88 ± 0.05 mmol/L^{-1}).[1] MgT consists of three different circulating Mg forms: (1) ionized (65%); (2) complexed to anions (15%) phosphate, bicarbonate, lactate, citrate; and (3) bound to proteins, mainly albumin (20%).

MgI concentration, measured by using Mg-selective electrodes, is about 0.54 ± 0.05 mmol/L^{-1}.[1,3]

The less severe disorders are Mg imbalances. Hypermagnesaemia is unusual, observed after inadequate parenteral Mg treatment and requires only a posologic adaptation. Mg deficiency, resulting from low Mg dietary intake, is more common and may be corrected by dietary supplementation.[3,4]

More worrying, Mg status may be dysregulated.[3,5] Dysregulated hypermagnesemia appear mainly in renal failure. Dysregulated hypomagnesemia (Mg depletion) appears when both Mg deficiency and any type of stress, including clinical or iatrogenic disorders, coexist.

In the present review, the practical interest of both markers were compared in various clinical or experimental situations of Mg disorders.

Magnesium Imbalances

In magnesium imbalances, MgT varies more importantly than the finely regulated MgI form. Magnesium balance and the ratio R (MgI/MgT) in serum or plasma are inversely correlated: in Mg excess, R decreases; in Mg deficiency, R increases.[3] This might reflect a buffer function of the bound serum fraction, allowing the concentration of the functionally active MgI fraction to stay constant as long as possible to reduce the effects of Mg imbalances.[6] This is the case in magnesium excess[1,7,8] or in experimental and clinical situations of dietary-induced Mg deficiency,[1-3,6,9-13] leading most authors to the conclusion that this mechanism erodes MgI diagnostic usefulness and that MgT is the best clinical marker.[12]

The mechanisms implicated in this adaptative phenomenon are poorly documented but include various factors involved in Mg homeostasis, such as neuroendocrine factors or compensatory inter- or intracell or tissue exchanges. For example, (1) the parathyroid hormone–calcitonin couple and perhaps some digestive polypeptide hormones mainly controlling exchanges between the extra cellular compartment and hard tissues; (2) catecholamines and insuline, especially presiding over exchanges between extracellular components and the soft tissues; (3) taurine and L-glutamyl taurine, which may oppose the harmful intracellular effects of hypersecretion of epinephrine; and (4) a central regulation of magnesemia accounting for the biological links between photoperiods and Mg status.[1,3,14]

Magnesium Dysregulations

Magnesium dysregulations may occur during failure of the Mg homeostasis controlling mechanisms or after intervention of Mg status disturbing factors. Generally, the compensatory exchanges between blood fractions are insufficient to face the major disturbances induced by pathological or iatrogenic agents.

Dysregulated Hypermagnaesemia

Chronic renal failure is known to increase both MgI and MgT, their levels raising progressively with deterioration in renal function.[15-19] Although some studies reported similar R in patients and controls,[20-22] most studies showed a mild tendency towards a reduced R in patients because of a relative decrease in MgI compared to MgT.[16-20,23,24] It has been postulated that, before dialysis, the decreased MgI would result from Mg complexation to anions (inorganic phosphate) or from Mg binding to anionic toxins that accumulate in the setting of uremia[16,18,23] but, even after dialysis that removes these anions, MgI complexation persisted.[22] It may be assumed that in the postdialysis period, the tendency to metabolic alkalosis leads to a higher affinity of albumin for MgI, and consequently to a decrease in MgI, as observed in vitro.[25] These complexations might reflect a trend for the homeostatic adaptation proposed above, even in this critical situation, in order to preserve as long as possible MgI levels.

Dysregulated Hypomagnaesemia

Magnesium depletion cannot be corrected by a simple oral physiological Mg supplementation, but requires the more specific correction of its causal dysregulation.[1,3] It is often linked with renal loss: when the renal Mg handling is impaired, hypomagnesaemia ensues because, unlike calcium, equilibration with cellular stores (mainly the bones) does not occur for several weeks.[26]

Studies on various diseased human subjects indicate that plasma MgI often exhibit slight significant alterations from normality, despite no change in MgT.[27] Consequently, in most of the following studies taken as examples, circulating MgI is regarded as the best marker for an accurate estimate of Mg status, but this result is sometimes controversial.

Postoperative, Injured, and Critically Ill Patients

This heterogeneous population often suffers hypomagnesemia, but the fractions concerned vary according to the clinical feature and do not reflect the Mg body stores.[28] Even though some studies in adult intensive care unit (ICU) patients showed a decrease in both MgT and MgI,[29,30] most studies revealed a decrease in MgI (associated with a worse prognosis) with normal MgT concentration.[31-33] The same pattern was also observed after the administration of chelating agents such as citrate with blood transfusions.[34] Consequently, these patients would not be recognized as Mg-deficient based on routine MgT.

Conversely, in the following examples taken among other ICU patients, MgI is unnecessary to detect low Mg levels.[34] For example, after cardiopulmonary bypass and in agreement with our hypothesis, the protein-bound fraction of Mg decreased consecutively to the dilutional hypoalbuminemia following surgery, without affecting the MgI levels.[35,36] Finally, in abdominal surgery, without massive transfusion and no cardiopulmonary bypass, MgI and MgT both decreased. The MgT decrease correlated with the degree of hypoalbuminemia. The MgI decrease may be linked to the intraoperative administration of fluids.[34]

Cardiovascular Diseases and Risk Factors

Magnesium deficits have been shown to be associated with fatal cardiovascular diseases, such as cardiac arrythmias and coronary heart diseases, as well as with risk factors for these diseases, such as essential hypertension, hypercholesterolemia, and diabetes mellitus.[37] In all these patient groups, MgT concentration was similar to that of controls, whereas patients with arrythmia and diabetes mellitus presented a decrease in MgI. In those patients, MgT measurement obscured the diagnosis of an abnormality in Mg metabolism.[37]

Low MgT level has been reported in 25% to 38% of patients with poorly controlled type 1 and type 2 diabetes.[38,39] It is associated with late complications (cardiovascular disease, retinopathy) and might contribute to insulin resistance.[40] Magnesium depletion is related to enhanced urinary loss, associated with the diabetic state.[41] Most studies have reported decreases in MgT, whereas only sparse information is available regarding MgI.[42] Well-controlled type 1 or 2 diabetes patients have MgT generally comparable to controls,[43,44] whereas MgI are significantly lower[44–46] except in one study.[47]

Hypomagnesemia appears during gestation, mainly in diabetics; similar decreases in both MgT and MgI levels were reported in control and diet-controlled diabetic gestational women, whereas the intracellular free Mg decreased only in diabetics, indicative of a true Mg depletion.[48] Infants of either insulin-dependent diabetic mothers[49,50] or gestational-diabetic mothers[51] showed that Mg deficiency plays a role in neonatal hypocalcemia. Hypocalcemia was more accurately correlated through MgI that decreased significantly than using MgT, which remained similar to controls.[49–51]

Finally, in chronic renal failure, one of the major complication of diabetes, both MgI and MgT levels were lower in diabetic than in nondiabetic patients (with a mild tendency towards a reduced R in diabetics because of a relative decrease in MgI).[52]

Kidney Diseases

Primary Renal Tubular Disorders

They represent unusual and heterogeneous disorders marked by renal hypokalemic alkalosis and various alterations of calcium and Mg homeostasis.[53] According to the authors, MgI determination is of little value in the diagnostic workup of most of these primary disorders because MgI levels generally correlates with either normal or decreased MgT. In contrast, low MgI with normal MgT levels might perhaps disclose latent hypomagnaesemia in nephrocalcinosis of unknown cause.[53]

Renal Transplants Patients

Hypomagnesemia is a well-documented metabolic derangement of renal transplantation, mainly occurring after surgery, in close correlation with acute tubular necrosis and exposure to high levels of immunosuppressive drugs.[54-56] However, it frequently persists even in the long term, probably as a consequence of metabolic acidosis, hypercalcemia, hypophosphatemia, increased urinary excretion fraction, diuretics, and cyclosporin A (CyA).[56-58] In feline renal transplants, MgT concentrations were within the normal range, whereas MgI were below normal in the postoperative period.[59] In human transplant patients, the significantly lower MgI was confirmed but only a minority (7%) had really pathologically low values.[60] In other studies taking into account the urinary Mg excretion and the possible role of confounding factors (CyA, acid–base balance, other divalent cations), MgT values were decreased consecutively to a slight hypoalbuminemia or an altered plasma Mg binding (with 26% of patients having pathological levels) and MgI concentrations were also lower (with only 10% of patients having pathologically low values), indicating that the prevalence of a true ionized hypomagnesemia is low, especially in the absence of diuretics.[56,61] In conclusion, the diagnostic of hypomagnesaemia in renal transplant cannot lie only on the determination of plasma MgT or MgI and requires a more complete workup, including evaluation of urinary excretion and search for clinical signs of Mg deficit.

Acid–Base Imbalances

MgI concentrations decrease with increasing pH, indicating the stronger binding of MgI with protein, mainly albumin, in an alkaline environment.[62] In addition, evidence from micropuncture studies indicates that acid–base changes alter active Mg absorption within the distal convoluted tubule (DCT), leading to alterations in urinary Mg excretion: alkalosis stimulates Mg^{2+} uptake, whereas acidosis diminishes entry rates.[63]

Alkalosis

Respiratory alkalosis induced by 30-min hyperventilation, in healthy subjects, induced no change in MgT and a significant decrease in MgI levels.[64]

Similarly, chemical alkalosis induced by addition of bicarbonate in vitro to neonatal serum decreased MgI.[65]

Acidosis

Despite major advances in the care of diabetes, ketoacidosis (DKA) remains a leading cause of morbidity in children and adolescents with type 1 diabetes. In 62% of children with DKA, MgT was significantly lower than in diabetic and nondiabetic control groups but returned to normal value with correction of DKA,[66] whereas MgI was not measured. But in dogs with DKA, MgI was significantly higher than in diabetic and nondiabetic control groups, in agreement with the homeostatic adaptative mechanism.[67]

Iatrogenic Dysregulations

Many therapeutics drugs (diuretics, chemotherapeutics, immunosuppressive agents, some antibiotics) cause hypomagnesaemia due to increased urinary loss. Thiazide diuretics or tacrolimus chronic administrations result in a similar defect in active Mg^{2+} reabsorption, that is, downregulation of the epithelial Mg^{2+}-channel transient receptor potential melastatin 6 (TRPM6) localized along the apical membrane of DCT to which active Mg^{2+} reabsorption is restricted.[68,69]

Immunosuppressants

Many studies showed that in renal transplant patients on CyA with allografts functioning stably for more than 6 months, both MgT and MgI were significantly lower than in control or renal transplant patients not on cyclosporin.[61,70–74] The MgI decrease was more pronounced and inversely correlated with urinary Mg and directly correlated with platelet Mg.[71] The MgT decrease was inversely correlated with blood CyA concentration.[74]

Chemotherapeutics

Patients with transient severe hypomagnesemia due to cisplatin or interleukin-2 therapies had MgT profoundly decreased, whereas R rapidly increased to values up to 93% to 128% of the total, indicative of the previously evoked adaptive mechanism.[3] In that case, MgI measurement failed to detect a severe hypomagnesemia.[75]

Antibiotics: Aminoglycosids

Renal Mg wasting is a rare complication of treatment with aminoglycosides, which is mostly associated with acute renal failure.[76,77] However, recent observations in animals[78–81] and humans[26] indicate the frequent occurrence of renal magnesium wasting even in the absence of renal failure. For example, in cystic fibrosis patients both MgT and MgI significantly decreased in a rather similar way.[26]

Antiviral Agent: Foscarnet

As a pyrophosphate analogue, foscarnet is a potent chelator of ionized Ca^{2+} and Mg^{2+} and acutely reduces their corresponding plasma ionized concentrations, whereas the decrease MgT is less severe[82] or absent.[76]

Conclusion

In Mg imbalances or dysregulated hypermagnesemia (renal failure), MgT measurement should be preferred in clinical practice because, in addition to its clinical value, it is available and cheap. In most of the pathological situations leading to Mg depletion, generally by Mg^{2+} renal loss, MgI often exhibits slight significant decreases despite no change in MgT. Consequently, circulating MgI is sometimes regarded as the best marker for an accurate estimate of Mg status or considered as having low added value, because the variations are always discrete and do not reflect the intracellular stores.[5,83] It reflects sometimes only changes in either blood pH or anion concentrations. More frequently, this ultrafiltrable Mg fraction decreases after pathological or iatrogenic renal Mg wasting that cannot be completely reversed by homeostatic corrections involving bone or blood exchanges. However, this review pointed out that in some severe situations, the exchanges between the different blood fractions still occur, thus complicating the MgI measurement accuracy. In addition, technical and chronobiological factors must be considered. Some evidences have shown that MgI may be quite variable and thus not a reliable measure[84]: the MgI assay could be improved with respect to the sensitivity, selectivity, and nonspecific interferences.[83] Finally, because of a circadian rhythm for both markers, the collection of specimens at appropriate times is important.[83]

References

1. Durlach J, Bara M. *Le magnésium en biologie et en médecine*. Paris: EMInter; 2000.
2. Dimai HP, Porta S, Wirnsberger G, et al. Magnesium supplementation for 30 days leads to correlative changes in circulating ionized magnesium and parathormone. *Magnes Bull* 1994;16:113–118.
3. Durlach J, Pagès N, Bac P, et al. Importance of the ratio between ionized and total Mg in serum or plasma: new data on the regulation of Mg status and practical importance of total Mg concentration in the investigation of Mg imbalance. *Magnes Res* 2002;15:203–205.
4. Hoshino K, Ogawa K, Kitazawa R, et al. Ionized magnesium level in whole blood of healthy Japanese children. *Acta Paediatr Jpn* 1998;40:116–121.
5. Sanders GT, Huijgen HJ, Sanders R. Magnesium in disease: a review with special emphasis on the serum ionized magnesium. *Clin Chem Lab Med* 1999;37: 1011–1033.
6. Zimmermann P, Weiss U, Classen HG, et al. The impact of diets with different Mg contents on Mg and Ca in serum and tissues of the rat. *Life Sci* 2000;67:949–958.
7. Standley CA, Cotton DB. Brain ionized magnesium and calcium levels during magnesium supplementation and deficiency in female Long-Evans rats. *Obstet Gynecol* 1996;88:184–188.
8. Mesinger V, Jäger J, Smetana R. Laboratory analysis of magnesium: total and ionized magnesium. *Magnes Res* 1997;10:85–86.
9. Altura BT, Wilimzig C, Trnovec T, et al. Comparative effects of a Mg-enriched diet and different orally administered magnesium oxide preparations on ionized Mg, Mg metabolism and electrolytes in serum of human volunteers. *J Am Coll Nutr* 1994;13:447–454.
10. Zimmermann P, Weiss U, Classen HG, et al. The impact of diets with different Mg contents on Mg and Ca in serum and tissues of the rat. *Life Sci* 2000;67:949–958.
11. Mullis PE, Bianchetti MG. High-fiber diets may be responsible for hypomagnesaemia in diabetic patients. *Nephron* 1998;78:238–239.
12. Norris CR, Christopher MM, Howard KA, et al. Effect of magnesium-deficient diet on serum and urine magnesium concentrations in healthy cats. *Am J Vet Res* 1999; 60:1159–1163.

13. Riond JL, Hartmann P, Steiner P, et al. Long term excessive magnesium supplementation is deleterious whereas suboptimal supply is beneficial for bones in rats. *Magnes Res* 2000;13:249-264.
14. Durlach J, Pagès N, Bac P, et al. Biorythms and possible central regulation of magnesium status, phototherapy, darkness therapy and chronopathological forms of magnesium depletion. *Magnes Res* 2000;15:49-66.
15. Garcia-Ortiz R, Milovic I, Uribe, C, et al. Levels erythrocyte and plasma magnesium in patients in chronic dialysis. Rev Med Chil 1989;117:544-548.
16. Saha H, Harmoinen A, Pietlä K, et al. Measurement of serum ionized versus total levels of magnesium and calcium in hemodialysis patients. *Clin Nephrol* 1996;46: 326-331.
17. Saha HH, Harmoinen AP, Pasternack AI. Measurement of serum ionized magnesium in CAPD patients. *Perit Dial Int* 1997;17:347-352.
18. Pedrozzi NE, Truttmann AC, Faraone R, et al. Circulating ionized and total Mg in end-stage disease. *Nephron* 1998;79:288-292.
19. Ariceta G, Rodriguez-Soriano J, Vallo A. Renal magnesium handling in infants and children.*Acta Paediatr* 1996;85:1019-1023.
20. Saha H, Harmoinen A, Nissula M, et al. Serum ionized versus total magnesium in patients with chronic renal disease. *Nephron* 1998;80:149-152.
21. Dewitte K, Dhondt A, Lameire N, et al. The ionized fraction of serum total magnesium in hemodialysis patients: is it really lower than in healthy subjects. *Clin Nephrol* 2002;58:205-210.
22. Huijgen HJ, Van Ingen HE, Kok WT, et al. Magnesium fractions in serum of healthy individuals and CPAD patients, measures by an ion-selective electrode and ultra-filtration. *Clin Biochem* 1996;29:261-266.
23. Markell MS, Altura BT, Sarn Y, et al. Deficiency of serum ionized magnesium in patients receiving hemodialysis or peritoneal dialysis. *ASAIO J* 1993;39:801-804.
24. Truttmann AC, Faraone R, Von Vigier RO, et al. Maintenance hemodialysis and circulating ionized magnesium. *Nephron* 2002;92:616-621.
25. Zaidenberg G, Mimouni FB, Dollberg S. Effect of bicarbonate on neonatal serum ionized magnesium *in vitro*. *Magnes Res* 2004;17:90-93.
26. Von Vigier RO, Truttmann AC, Zindler-Schmocker K, et al. Aminoglycosides and renal magnesium homeostasis in humans. *Nephrol Dial Transplant* 2000;15: 822-826.
27. Altura BT, Altura BM. Measurement of ionized magnesium in whole blood, plasma and serum with a new ion-selective electrode in healthy and diseased human subjects. *Magnes Trace Elem* 1991;10:90-98.
28. Hebert P, Mehta N, Wang J, et al. Functional magnesium deficiency in critically ill patients identified using a magnesium-loading test. *Crit Care Med* 1997;25:749-755.
29. Koch SM, Warters RD, Mehlhorn U. The simultaneous measurement of ionized and total calcium and ionized and tota magnesium in intensive care unit patients. *J Crit Care* 2002;17:203-205.
30. Mallon A, Brockmann C, Fijalkowska-Morawska J, et al. Ionized magnesium in erythrocytes- the best magnesium parameter to observe hypo or hypermagnesemia. *Clin Chim Acta* 2004;349:67-73.
31. Soliman HM, Mercan D, Lobo SS, et al. Development of ionized hypomagnesemia is associated with higher mortality rates. *Crit Care Med* 2003;31:1082-1087.

32. Kasaoka S, Tsuruta R, Nakashima K, et al. Effect of intravenous magnesium sulfate on cardiac arrhythmias in critically ill patients with low serum ionized magnesium. *Jpn Circ J* 1996;60:871–875.
33. Wilkes NJ, Mallett SV, Peachey T, et al. Correction of ionized plasma magnesium during cardiopulmonary bypass reduces the risk of postoperative cardiac arrhythmia. *Anesth Analg* 2002;95:828–834.
34. Lanzinger MJ, Moretti EW, Wildermann RF, et al. The relationship between ionized and total serum magnesium concentrations during abdominal surgery. *J Clin Anesth* 2003;15:245–249.
35. Aziz S, Haigh WG, Van Norman GA, et al. Blood ionized magnesium concentrations during cardiopulmonary bypass and their correlation with other circulating cations. *J Card Surg* 1996;11:341–347.
36. Scott VL, De Wolf AM, Kang Y, et al. Ionized hypomagnesemia in patients undergoing orthotopic liver transplantation: a complication of citrate intoxication. *Liver Transpl Surg* 1996;2:343–347.
37. Sasaki S, Oshima T, Teragawa H, et al. Magnesium (Mg) status in patients with cardiovascular diseases. *Rinsho Byori* 1999;47:396–401.
38. Tosiello L. Hypomagnesemia and diabetes mellitus. A review of clinical implications. *Arch Intern Med* 1996;156:1143–1148.
39. Lima M de L, Cruz T, Pousada JC, et al. The effect of magnesium supplementation in increasing doses on the control of type 2 diabetes. *Diabetes Care* 1998;21:682–686.
40. Eibi N, Schnack C, Schernthaner G. Magnesium supplementation in type 2 diabetes. *Diabetes Care* 1998;21:2031–2032.
41. Gurlek A, Bayraktar M, Ozaltin N. Intracellular magnesium depletion relates to increased urinary magnesium loss in type I diabetes. *Horm Metab Res* 1998;30:99–102.
42. Matthiesen G, Olofsson K, Rudnicki M. Ionized magnesium in Danish children with type 1 diabetes. *Diabetes Care* 2004;27:1216–1217.
43. Roffi M, Kanaka C, Mullis PE, et al. Hypermagnesiuria in children with newly diagnosed insulin-dependent diabetes mellitus. *Am J Nephrol* 1994;14:201–206.
44. Husmann MJ, Fuchs P, Truttmann AC, et al. Extracellular magnesium depletion in pediatric patients with insulin-dependent diabetes mellitus. *Miner Electrolyte Metab* 1997;23:121–124.
45. Resnick LM, Barbagallo M, Bardicef M, et al. Cellular-free magnesium depletion in brain and muscle of normal and preeclamptic pregnancy: a nuclear magnetic resonance spectroscopic study. *Hypertension* 2004;44:322–326.
46. Corsonello A, Ientile R, Buemi M, et al. Serum ionized magnesium levels in type 2 diabetic patients with microalbuminuria or clinical proteinuria. *Am J Nephrol* 2000;20:187–192.
47. Matthiesen G, Olofsson K, Rudnicki M. Ionized magnesium in Danish children with type 1 diabetes. *Diabetes Care* 2004;27:1216–1217.
48. Bardicef M, Bardicef O, Sorokin Y, et al. Extracellular and intracellular magnesium depletion in pregnancy and gestational diabetes. *Am J Obstet Gynecol* 1995;172:1009–1013.
49. Tsang RC, Strub R, Steichen JJ, et al. Hypomagnesemia in infants of diabetic mothers. Perinatal studies. *J Pediatr* 1976;80:115–119.
50. Jones CW. Gestational diabetes and its impact on the neonate. *Neonatal Netw* 2001;20:17–23.

51. Banerjee S, Mimouni FB, Mehta R, et al. Lower whole blood ionized magnesium concentrations in hypocalcemic infants of gestational diabetic mothers. *Magnes Res* 2003;16:127–130.
52. Dewitte K, Dhondt A, Lameire N, et al. The ionized fraction of serum total magnesium in hemodialysis patients: is it really lower than in healthy subjects. *Clin Nephrol* 2002;58:205–210.
53. Morger ID, Truttmann AC, Von Vigier RO, et al. Plasma ionized magnesium in tubular disorders with and without total hypomagnesemia. *Pediatr Nephrol* 1999;13:50–53.
54. Kim HJ, Ahn YH, Kee CS, et al. Early short-term profile of serum magnesium concentration in living donor renal transplant recipients on cyclosporine. *Transplant Proc* 1994;26:2178–2180.
55. Haag-Weber M, Schollmeyer P, Horl WH. Failure to detect remarkable hypomagnaesemia in renal transplant recipients receiving cyclosporine. *Miner Electrolyte Metab* 1990;16:66–68.
56. Mazzaferro S, Barberi S, Scarda A, et al. Ionized and total serum magnesium in renal transplant patients. *J Nephrol* 2002;15:275–280.
57. Ramos EL, Barri YM, Kubilis P, et al. Hypomagnaesemia in renal transplant patients: Improvements over time and association with hypertension and cyclosporine levels. *Clin Transplant* 1995;9:185–189.
58. Quamme GA, Dai LJ. Presence of a novel influx pathway for Mg2+ in MDCK cells. *Am J Physiol* 1990;259:C521.
59. Wooldridge JD, Gregory CR. Ionized and total serum magnesium concentrations in feline renal transplants recipients. *Vet Surg* 1999;28:31–37.
60. Saha HH, Harmoinen AP, Nisula M, et al. Serum ionized versus total magnesium in patients with chronic renal disease. *Nephron* 1998;80:149–152.
61. Vanini SD, Mazzola BL, Rodoni L, et al. Permanently reduced plasma ionized magnesium among plasma recipients on cyclosporine. *Transpl Int* 1999;12:244–249.
62. Wang S, McDonnnel EH, Sedor FA, et al. pH effects on measurements of ionized calcium and ionized magnesium in blood. *Arch Pathol Lab Med* 2002;126:947–950.
63. Dai LJ, Friedman PA, Quamme GA. Acid-base changes alter Mg2+ uptake in mouse distal convoluted tubule cells. *Am Physiol Soc* 1997;272:F759–F766.
64. Hafen G, Laux-End R, Truttmann AC, et al. Plasma ionized magnesium during hyperventilation in humans. *Clin Sci* 1996;91:347–351.
65. Zaidenberg G, Mimouni FB, Dollberg S. Effect of bicarbonate on neonatal serum ionized magnesium *in vitro*. *Magnes Res* 2004;17:90–93.
66. Bauza J, Ortiz J, Dahan M, et al. Reliability of serum magnesium values during diabetic ketoacidosis in children. *Bol Assoc Med P R* 1998;90:108–112.
67. Fincham SC, Drobatz KJ, Gillespie TN, et al. Evaluation of plasma ionized magnesium concentration in 122 dogs with diabetes mellitus: a retrospective study. *J Vet Intern Med* 2004;18:612–617.
68. Nijenhuis T, Hoenderop JG, Bindels RJ. Downregulation of Ca(2+) and Mg(2+) transport proteins in the kidney explains tacrolimus (FK506)-induced hypercalciuria and hypomagnesemia. *J Am Soc Nephrol* 2004;15:549–557.
69. Nijenhuis T, Vallon V, van der Kemp AW, et al. Enhanced passive Ca2+ reabsorption and reduced Mg^{2+} channel abundance explains thiazide-induced hypocalciuria and hypomagnesemia. *J Clin Invest* 2005;115:1651–1658.

70. Markell MS, Altura BT, Sarn Y, Barbour R, Friedman EA, Altura BM. Relationship of ionised magnesium and cyclosporine level in renal transplant recipients. *Ann N Y Acad Sci* 1993;696:408-411.
71. Allegra A, Corica F, Ientile R, et al. Plasma (total and ionized), erythrocyte and platelet magnesium levels in renal transplant recipients during cyclosporine and/or azathioprine treatment. *Magnes Res* 1998;11:11-18.
72. Mazzola BL, Vannini SDP, Truttmann AC, et al. Long-term calcineurin inhibition and magnesium balance after renal transplantation. *Transpl Int* 2003;16:76-81.
73. June CH, Thompson CB, Kennedy MS, Nims JE, Donnall T. Profound hypomagnesiaemia and renal magnesium wasting associated with the use of cyclosporine for marrow transplantation. *Transplantation* 1985;39:620-624.
74. Quamme GA. Renal magnesium handling: new insights in understanding old problems.*Kidney Int* 1997;52:1180-1195.
75. Csako G, Rehak NN, Elin RJ. Falsely high ionized magnesium results by an ion-selective electrode method in severe hypomagnesmia. *Eur J Clin Chem Biochem* 1997;35:701-709.
76. Huycke MM, Naguib MT, Stroemmel MM, et al. A double-blind placebo-controlled crossover trial of intravenous magnesium sulfate for foscarnet-induced ionized hypocalcemia and hypomagnesemia in patients with AIDS and cytomegalovirus infection. *Antimicrob Agents Chemother* 2000;44:2143-2148.
77. Wilkinson R, Lucas GL, Heath DA, et al. Hypomagnesaemic tetany associated with prolonged treatment with aminoglycosids. *Br Med J* 1986;292;818-819.
78. Foster JE, Harpur ES, Garland HO. A investigation of the acute effect of gentamicin on the renal handling of electrolytes in the rat. *J Pharmacol Exp Ther* 1992;261:38-43.
79. Garland HO, Birdsey TJ, Davidge CG, et al. Effects of gentamicin, neomycin and tobramycin on renal calcium and magnesium handling in two rat strains. *Clin Exp Pharmacol Physiol* 1994;21:109-115.
80. Parson PP, Garland HO, Harpur ES, et al. Acute gentamicin-induced hypercalciuria and hypermagnesiuria in the rat : dose-response relationship and role of tubular injury. *Br J Pharmacol* 1997;122:570-576.
81. Adams JP, Conway SP, Wilson C. Hypomagnesaemic tetany associated with repeated courses of intravenous tobramicyn in a patient with cystic fibrosis. *Respir Med* 1988;92:602-604.
82. Noormohamed FH, Youle MS, Tang B, et al. Foscarnet-induced changes in plasma concentrations of total and ionized calcium and magnesium in HIV-positive patients. *Antivir Ther* 1996;1:172-179.
83. Saris NE, Mervaala E, Karppanen H, et al. An update on physiological, clinical and analytical aspects. *Clin Chim Acta* 2000;294:1-26.
84. Newhouse IJ, Johnson KP, Montelpare WJ, et al. Variability within individuals of plasma ionic magnesium concentrations. *BMC Physiol* 2002;26:2-6.

Nutrition

7
Overview of Magnesium Nutrition

Mieko Kimura

Magnesium plays a very important role in wide-range biochemical metabolism, especially fundamental cellular reactions. It is explained that magnesium is essential ion for cofactors of over 300 enzyme systems, for example, as a part of the magnesium–adenosine triphosphate (Mg-ATP) complex or as an enzyme activator in experimental and clinical observations.

In 1932, McCollum and colleagues[1] observed the first symptomatic results from magnesium deprivation in rats and dogs. In a small human population of patients with various diseases, the first clinical depletion was reported by Hirschfelder and Hauary in 1934,[2] and in 1942 Hauary and Cantarow found magnesium deficiency in normal and pathological patients assessed for magnesium serum concentration.[3] Flink and coworkers discovered the clinical depletion of magnesium under prolonged parenteral fluid administration and chronic alcoholism in 1954.[4] Since then, many experimental and clinical studies for magnesium deficiency appeared in cases in unhealthy subjects as clinical disorders with magnesium depletion.

Chemistry

The atomic number of magnesium is 12; its atomic weight is 24.305. Magnesium is an alkaline earth metal and has an s orbital that can be filled with two electrons, that is, ionized magnesium (Mg^{2+}) and has potential to bind six ligands, which is characteristic for the octahedral structure of magnesium complexes. In nature, magnesium is existent magnesium cabonate ($MgCO3$), calcium magnesium carbonate [$CaMg(CO3)2$], calcium magnesium silicate [$CaMg(SiO3)2$], calcium magnesium hydroxide [$Ca2Mg5(OH)2$], magnesium sulfate ($MgSO4$) and magnesium chloride ($MgCl2$) etc for magnesium complexes.

Body Composition

Magnesium is the second most common cation found in the body, and the total body content in healthy adults is about 25 g. It is distributed with 50% to 60% of the total found in the bone and 40% to 50% of the total in the soft tissues,

TABLE 7.1. Distribution and concentrations of magnesium in a healthy adult. (Total Body: 833–1170 Mmol[a] or 20–28 g).[2]

Distribution Percentage	Concentration
Bone	0.5% of bone ash
Muscle	3.5–5 mmol/kg wet weight
Other cells	3.5–5 mmol/kg wet weight
Extracellular	
Erythrocytes	1.65–2.73[d] mmol/L[b]
Serum	0.65–0.88[d] mmol/L[c]
55% free Mg^{2+}, 13% complexed	Ultrafilterable: 0.48–0.66 mmol/L[d]
(citrate, phosphate, etc)	Ion electrode 0.53–0.66 mmol/L[e]
32% bound primarily to albumin	0.1–0.3 mmol/L
Mononuclear blood cells[f]	2.91 ± 0.6 fmol/cell[c]
	2.79 ± 0.6 fmol/cell[h]
	3.00 ± 0.04 fmol/cell[i]
Carebrospinal fluid	1.25 mmol/L
55% free Mg^{2+}	
45% complexed	
Sweat	0.3 mmol/L (in hot environment)
Secretions (saleva, gastric, bile)	0.3–0.7 mmol/L

1 mmol = 2 mEq = 24 mg. [†]Magnesium falls slowly with aging. [‡]Similar at various ages. [§]Data from Hosseini, E.: Trace Elem. Med., 5:47–51, 1988. [¶]Data from Altura: Magnesium Trace Elem., 9:3111, 1990. [**]Monocytes and lymphocytes in venous blood. [††]Data from Elin, Hoseini: Clin. Chem., 31:377–380, 1985. 1 fmol = 24.3 fg. [‡‡]Data from Reinhart, et al.: Clin. Clim. Acta, 167:187–195, 1987. [§§]Data from Yang, et al.: J. Am. Coll. Nutr., 9:328, 1990.

as shown in Table 7.1. Although there are limited data for human bone, magnesium content is approximately 200 mmol/kg of bone ash.[5] Magnesium is the most cation mineral in cells, second to potassium. Most intracellular magnesium is in the bound form and in frog muscle cells magnesium is bound to adeonsine triphosphate (ATP) at 5.8 mmol/L, to phosphocreatine at 1.7 mmol/L, to myosin at 0.3 mmol/L, and to free Mg^{2+} at 0.6 mmol/L.[6] In the skeleton, one third of intracellular magnesium is exchangeable and may be reserved to maintain normality in the intracellular concentration of magnesium.[5] The extracellular magnesium in humans is only 1% of total body magnesium, that is, 0.7 to 0.9 mmol/L in serum, with 55% being in the free ionized form, 15% in complexes to anions, and 30% bound to proteins (mostly albumin).[5] Magnesium is distributed in all of the parts within cells. That content is 5 to 20 mmol/L and is associated with metabolic activity of cells, for example, 75% in erythrocytes, fundamentally combined with anions with 80% to 90% being bound to ATP.[7] Intracellular free magnesium is an important metabolic regular factor, but only 1% to 5% of total intracellular magnesium—0.3 to 0.6 mmol/L.[8]

Biochemical and Physiological Roles

The chemical character of ionized magnesium (Mg^{2+}) binds to anions, and it becomes stable, such as in enzymes, or neutral, such as for substrate use. Magnesium has many biochemically important roles, mainly required as a cofactor for over 300 vital enzymatic reaction steps.[9] For example, there are the synthesis of protein synthesis, fatty acids, activation of amino acids, phosphorylation of glucose and its derivatives in the glycolytic pathway, oxidative decaboxylation of citrate, and transletolase reactions.[10] For aerobic and anaerobic energy production, for glycolysis that is as a part of the magnesium-adenopsine triphosphate (Mg-ATP) complex, and as an enzyme activator, magnesium is demanded.[9] In the glycolytic cycle to convert glucose to pyruvate, seven key enzymes are requested: Mg^{2+} alone or associated with ATP or adenosine monophosphate (AMP). The role of magnesium as Mg-ATP, magnesium adenosine diphosphate (Mg–ADP), or Mg-GTP, etc., is evident in that intracellular magnesium plays an essential role for cofactor of enzymes in energy storage, transfer, use, and ion transport.[11] Magnesium is also required for a substrate in all enzymes employing ATP or ADP functions, such as pyruvate kinase and hexokinase from glycolysisi, adenylate, and guanylate cyclase, and ATP-dependent pumps [sodium–potassium adenosine triphosphatase (Na-K ATPase), calcium–adenosine triphosphatase (Ca-ATPase)].[12]

In other biological functions, magnesium is important, too. Mg^{2+} has led that regulated changes in its level are compatible with its functions as a physiological modulator affecting cardiac physiology, with coupling of neurotransmitters and enzymes, to receptors, with activation proteins and with modulation of various types of ion channels.[13]

Absorption, Metabolism, and Excretion

Magnesium homeostasis is preserved by control of intestinal absorption and losses through the urine as other minerals. The excretion process from urine is the most powerful regulator for magnesium control mechanism.

Intestinal magnesium absorption is barely replied on the opposite amount of magnesium ingested in adults or children. Although magnesium is absorbed from the entire intestinal tract, the maximal magnesium absorption is found in the distal jejunum and ileum.[14] In rat studies, magnesium is absorbed in the proximal small intestine and in the lower bowel.[15] An in vitro study using isolated segments of intestine, the capability for magnesium absorption was shown in following order: colon = ileum > jejunum > duodenum. But in rats with low magnesium intakes, quantitative colonic absorption may be very important.

It is reported that magnesium absorption is throughout the small intestine in humans, that is, in the experiment using stable ^{26}Mg, ^{26}Mg is found in the plasma at 1 h after an oral dose and its absorption is 80% within 6 h.[16] In balance studies of healthy men under controlled diet, 380 mg/day magnesium was absorbed on average in net absorption of 40% to 60%, with true absorption 51% to 60% under various foodstuffs. Under higher magnesium levels, daily intakes of 550 to 570 mg or with dietary bran and oxalate, net absorption is estimated at 15% to 36%.[17]

Magnesium goes through the intestinal epithelium following one of three types of mechanisms: passive diffusion; solvent drag, for example, water movement; and active transport. Intestinal absorption of nutrients is based generally in the active transport system. The higher rate magnesium absorption under low dietary magnesium intakes is done in both an unsaturable passive and saturable active system.[18] In the experiments for humans given an oral ^{26}Mg and magnesium retention 5 days later,[19] for perfused human jejunum and ileum,[20] for perfused rat small intestine,[21] and for isolated colonic segments of rat intestine,[22] active transport is distinct according to a linear relationship between the rate of nutrient transport and the amount present in the intestine.

The mechanism for intestinal magnesium absorption is not yet made clear. The regulation of magnesium absorption by vitamin D status is adopted based the chemical similarity of magnesium to calcium. Intestinal magnesium absorption is enhanced a little by vitamin D and its metabolites: 25-hydroxyvitamin D and 1,25-dihydroxyvitamin D.[23,24] Magnesium absorption in vitamin D–deficient rats was reduced and by vitamin D repletion reversed,[25] and also promoted magnesium absorption by elevated parathyroid hormone (PTH) levels.

Magnesium absorption is prevented by high calcium concentrations in rats. This interactive effect may be indirect, depending on the reduction of tight junction permeability with calcium and reducing of the paracellular magnesium movement. Magnesium absorption in humans has been not influenced by increased calcium intake (for example, augmenting from 300 mg to 2000 mg),[26] rather, calcium and magnesium have different sites for maximal absorption, so that any effects may be a segment specialty.[27] It is sometimes reported that magnesium absorption was inhibited by phosphate. That is, free phosphate may shape insoluble salt complexes with magnesium or phytate inhibit magnesium absorption.[28] Fiber-rich foods induce lower magnesium absorption, possibly because there is an independent effect of fiber or an effect from the phyate content in food.[29]

The kidney plays an important role as the principal organ in magnesium homeostasis, which is controlled in renal excretion.[30] Renal excretion of magnesium is reduced in a low-magnesium diet, but output of other minerals is not affected by low magnesium intake and the effect will rapidly adapt.[31]

Renal magnesium re-absorption is carried out in the thick leading loop of Henle and the distal convoluted tubule. Renal magnesium handling in humans is done in a filtration–re-absorption process to be no tubular secretion of magnesium. The filtrated magnesium is re-absorbed about 65% in the loop of Henle by active transport, namely a decrease of magnesium re-absorption is independent of sodium chloride transport in either hypermagnesia or hypercalcemia.[32] The re-absorption of magnesium (20%–30%) in the proximal convoluted tubule is passive and mainly paracellular, associated with the changes in salt and water re-absorption, and with the rate of fluid flow. So the tubule in humans has a positive charged rather than a serosal cell side movement; the electrochemical gradient is down. This paracellular pathway has especially high permeability for magnesium and some calcium. High magnesium on the peritubular site of cell is induced on increasing excretion, so that it inhibits reabsorption of magnesium. High calcium in plasma also inhibits magnesium reabsorption. Two means of magnesium reabsorption in paracellular are recognized, because magnesium can alter reabsorption and calcium alters reabsorption of calcium, magnesium, and water.

The reabsorption of magnesium in the loop of Henle is an active transport; namely, a decrease in magnesium reabsorption is independent of sodium chloride transport in either hypermagnesium or hypercalcemia.[33] Magnesium reabsorption in the proximal convoluted tubule is less important than in the loop of Henle, but it is very efficiently routed. That is, magnesium that is divided in the proximal convoluted tubule has a reabsorbtion rate of 80%. This result suggests that for magnesium absorption, both reabsorption routes are important, but there are remarkable differences in the transport mechanism in these sites. The specific pathway controlled by the receptor is sensitive to magnesium and calcium may not be in the peritubular side of the loop of Henke.[33] The genetic disease for the autosomal recessive renal hypomagnesium with hypercalciura may induce the specific magnesium reabsorption. This disease is caused by a mutation in the gene encoding the protein paracellin-1.[34] Paracellin-1 is in the tight junction of the loop of the Henle. That is, it has to do with the claudin family of tight junction proteins. An affinity for Mg2+ is bestowed by the negatively charged extracellular sphere. Numerous hormones, such as PTH, calciatonin, glucagons, and vasopressin, achieve simultaneously reabsorption of paracellular magnesium in the loop of Henle by way of their cell surface receptors, which increase paracellular flux. But it is not clear if the effect on paracellin-1 function come from serum magnesium or hormones. Magnesium reabsorption in the proximal convoluted tubule also is found against an electrochemical gradient, where the transport may be active and energy dependent. And so magnesium transport in the proximal convoluted tubule is stimulated by hormones; PTH, etc., as same as in the loop of Henle.[35]

Magnesium in urine decreases to less than 20 mg/day within 3 to 4 days in experimental humans magnesium deficiency.[36] Parathyroid hormone (PTH) may be not important for regulation of magnesium homeostasis in patients, for usually with primary hyper- or hypoparathyroidism, normal serum magnesium concentrations and a normal tubular maximum for magnesium are shown.[37] Glucagons, calcitonin, and ADH affect magnesium transport in the loop of Henle in a similar system to PTH, with no clear evidence of physiological action mechanisms.[38]

Excess alcohol intake may cause renal magnesium loss. All chronic alcoholics have been shown to have symptoms of magnesium depletion. But the evidence is not clear.[39] Hypermagnesuria may be caused by medical treatment using diuretics (commonly used for treatment of hypertension, heart disease, and other edematous states).[40]

Magnesium Deficiency

Experimental Animal Magnesium Deficiency

Magnesium depletion leads to specific biochemical abnormalities and clinical symptoms. McCollum and coworkers in 1931 reported at first the characteristic syndrome of magnesium deficiency in rats eating a diet of only 2 mg percentage magnesium.[1] After, the Johns Hopkins investigators clarified some changes in animal organism under dietary magnesium insufficiency, and also magnesium deficiency in rabbits,[41] dogs,[42] and calves,[43] etc., were found. In mice eating the same diet as rats hyperemia did not appear, but hypocalcemia followed by hypomagnesemia did, and the mice frequently died under sudden and strong convulsion.[44–46]

The dilatation of cutaneous vessels, hyperirritability, and convulsive seizures are shown in the specific syndrome of magnesium depletion developed. The convulsive seizures in almost every animal appeared for first attack. The following syndrome in magnesium depletion appears: the excitable animals startle by sound and rush at rapid speed, and usual acutely fatal, tonic–clonic conversion. The entire body of animal stiffens as head stretched back, forelimbs extend to upper joints and flex to the metacarpophalangeal joint, and hind limbs extend backward. Often the teeth clench and perforate the tongue. Next, waxy skin appears. The respiratory movements suspend during the attack and come back with the relaxation of the musculature. Additionally, other changes in animals under a magnesium-deficient diet are reduced growth rate, alopecia, and more edema, hypertrophic gums, leukocytosis, and splenomegaly. Thymic abnormalities also appear, that is, malignant lymphosarcoma, disseminated lumphoblastic leukemia, or atrophy.[47] Usually serum in magnesium deficiency is with high calcium level and low PTH.

Experimental Human Magnesium Deficiency

The experimental diet included 9.6 mg magnesium per day, that is, following a baseline period with complete diet including adequate magnesium can induce signs and symptoms in magnesium depletion. Plasma magnesium fell gradually to 10% to 30% of those during control periods. Erythrocyte magnesium decreased more slowly and to lesser levels. Urine and fecal magnesium went down to very low level within 7 days. In all patients with these symptoms, hypomagnesium, hypocalcemia, and hypokalemia were shown. Positive calcium balance was found in good intestinal absorption of calcium and low urinary excretion. For increase of urinary losses, hypokalemia and negative potassium balance resulted in almost subjects. These subjects keep in positive sodium balance. These symptoms and signs reversed with re-addition of magnesium. In repletion of magnesium, serum magnesium concentration quickly return to normal range but serum calcium and potassium do not rapidly rise, and to return to baseline levels, a week or more is needed. Potassium balances are positive and sodium balances are negative.

Magnesium Deficiency in Humans

Clinical conditions contributing to magnesium depletion are shown in Table 7.2. The symptoms of magnesium deficiency in human may relate to disease states. These will depend on declined intake and intestinal or renal malabsorptive conditions.[48,49]

In clinical magnesium deficiency, the symptoms and signs are the same as in above experimental deficient cases of hypomagnesemia. The clinical signs for magnesium depletion are, for example, tremor, muscle spasm, muscle fasciculations, normal or depressed deep tendon reflexes, personality changes,

TABLE 7.2. Clinical conditions contributing to magnesium depletion.[2]

Malabsorption syndromes
   Inflammatory bowel disease
   Gluten enteropathy; sprue
   Intestinal fistulas, bypass, or resection
   Bile insufficiency states, e.g., ileal dysfunction with steatorrhea
   Immune diseases with villous atrophy
   Radiation enteritis
   Lymphangiectasia; other fat absorptive defects
   Primary idiopathic hypomagnesemia
   Gastrointestinal infections

Renal dysfunction with excessive losses
   Tubular diseases
   Metabolic disorders
   Hormonal effects
   Nephrotoxic drugs

Endocrine disorders
   Hyperaldosteronism
   Hyperparathyroidism with hypercalcemia
   Postparathyroidectomy ("hungry bone" syndrome)
   Hyperthyroidism

Pediatric genetic and familial disorders
   Primary idiopathic hypomagnesemia
   Renal wasting syndrome
   Bartter's syndrome
   Infants born of diabetic or hyperparathyroid mothers
   Transient neonatal hypomagnesemic hypocalcemia

Inadequate intake, provision, and/or retention of magnesium
   Alcoholism
   Protein-calorie malnutrition (usually with infection)
   Prolonged infusion or ingestion of low-magnesium nutrient solutions or diets
   Hypercatabolic states (burns, trauma), usually associated with above entry
   Excessive lactation

anorexia, nausea, vomiting, Frank tetany, myoclonic jerks, athetoid movements, convulsions, etc.[50] Magnesium supply was recovered tetany or convulsions with hypomagnesemia, hypomagunesuria, hypocalcemia, and hypokalemia. Only calcium and vitamin D supplies were not affected on improvement of calcemia without magnesium supply.

On the other hand, magnesium depletion in clinical practices can be found. A hypomagnesemia is generally led by surgery, and more severe cardiac surgery. This reason is not clear now. Mass blood transfusions, administration of large volumes of magnesium-free fluids, binding to free fatty acids, and catecholamine-induced intracellular magnesium shifts may be some of the causes.

Magnesium imbalance based clinical treatments may also lead magnesium depletion. For example, cellulose phosphate is used to limit calcium absorption, also bind to magnesium and limit to its absorption or re-absorption. Many common procedures—infusion of normal saline, the use of osmotic diuretics, treatment loop diuretics (furosemide, bumetanide, ethacrynic acid, etc.)—inhibit liminal Na-Cl-K channels in the loop of Henle, causing increasing magnesium excretion. More pharmacological agents may also lead to renal magnesium excretion, for example, cisplatin causes hypokalemic metabolic alkalosis, aminoglycosides promote renal magnesium losses in consequence of secondary hyperaldosteronism, cyclosporine causes a mild and asymptomatic hypomagnesemia. The patients under severe burns also have magnesium losses through skin.

A number of genetic disease are associated with magnesium deficiency: (1) hypomagnesemia with secondary hypocalcemia causes reduced magnesium absorption from intestine; (2) primary hypomagnesemia with hypercalciuria causes increased renal magnesium loss and paracellin-1 mutation; (3) renal hypomagnesemia 2 causes excess renal magnesium loss; (4) Barter syndrome causes mutations in K channel ROMK or Na-K-Cl cotransporter SLC12A1; (5) Gitelman syndrome causes mutations in thiazide-sensitive Na-Cl cotransporter.[51] Metabolic acidosis increases renal magnesium excretion. Both primary and secondary hyperaldosteronism bring augmentation of the extracellular fluid volume that increase renal magnesium loss. Inappropriate antidiuretic hormone secretion also brings high renal magnesium excretion. Parathyroidectomy and diffuse osteoblastic metastasis show hungry bone syndrome, that is, excess bone formation was done drawing magnesium from serum. For example, under Crohn's disease, Whipple's disease, ulcerative colitis, celiac disease, and short bowel syndrome, magnesium absorption is reduced by diarrhea or steatorrhea, etc.[52]

Hypomagnesemia is found in alcoholics.[53] The reasons may depend on generally reduced food intake and poor nutritional status in alcoholics. Malabsorption of magnesium in alcoholics is often found under diarrhea, vomiting, and steatorrhea caused by liver disease and chronic pancreattis, and renal magnesium excretion may be led by alcoholic ketoacidosis, phpophosphatemia, and nyperaldosteronism.

Marginal Deficiency with Various Diseases

Cardiovascular Disease

Many reports for possible biochemical roles of magnesium in cardiovascular function are shown.[54-56] Kobayashi and colleagues[57] reported that coronary artery disease was frequent in populations living in hard water (i.e., calcium- and magnesium-rich) areas, compared with those in soft water areas, based epidemiological survey.[58-61]

It is reported that after an acute myocardial infarction serum, magnesium level is quick to decline. But these facts may occur under marginal deficiency by low magnesium intake. And as these subjects have also been given many clinical treatments in which hypomagnesemia may be caused, its cause and result are not clear. Seelig and coworkers reviewed the results of a large number of studies for magnesium in acute myocardial infraction, which still show this as an open question.[62-64]

Magnesium deficiency has been associated with significantly increased serum/plasma triglycerides, phospholipids, variable total cholesterol, low-density cholesterol, and oleic and linoleic acids, and decreased stearic and aeachidonic acids. The effects of magnesium deficiency on tissue levels of total lipids and fatty acids were evaluated. Serum cholesterol and total phospholipids were significantly higher in the magnesium-deficient rats than in the controls. Edema and polycystic degeneration of the kidneys were present in the magnesium-deficient rats. The main change in tissue fatty acid composition in magnesium deficiency was the higher docosahexaenoic acid in serum, liver, and aorta than in controls.[65]

In the magnesium-deficient rats, the percentage composition of triglycerides in very-low-density lipoprotein (VLDL), low-density lipoprotein (LDL), and high-density lipoprotein (HDL) was elevated and that of protein was reduced. Although the proportion of cholesterol was reduced in LDL and HDL, that of phospholipid was decreased only in HDL. Magnesium deficiency induced a decrease in the percentage composition of apolipoprotein E (apo E) and a relative increase in the apo C for VLDL. In HDL from magnesium-deficient rats, the percentage composition of oleic and linoleic acids was increased.[66]

The mean prostanoids levels in plasma and tissue were significantly higher in magnesium-deficient rats than in controls The increased synthesis of prostanoids is apparently linked to enhanced influx and translocation of Ca^{2+} into the cells. It is possible that the changes in prostaglandin synthesis in magnesium deficiency are linked to the development of different diseases.[67] Deficiency of magnesium with cardiovascular effects tend to be related to alterations in the biosynthesis of prostaglandins in the vasculature.[68,69]

Extensive experimental and clinical studies in animals and humans during the past 25 years have yielded considerable data regarding the availability of internal magnesium stores for the maintenance of extracellular and critical tissue concentrations of magnesium during periods of magnesium deprivation. The bulk of body magnesium is present in bone and skeletal muscle, and it is these two sites that provide essentially all the available magnesium. In the rat, approximately 15% of bone magnesium (equivalent to 1.5 mmol/kg body weight) can be lost, whereas less than one tenth of that amount is available from skeletal muscle. In humans, up to 35% of bone magnesium can be lost, but the mean loss in extant studies is 18% (equivalent to 1.2 mmol/kg body weight). In contrast to the rat, up to 40% human skeletal-muscle magnesium can be lost with an average loss of 15% (equivalent to 0.45 mmol/kg). In the human, the bone and skeletal-muscle magnesium pools can provide an average of 1.7 mmol/kg body weight, which is equivalent to 15% of total body magnesium. Release of magnesium from these stores appears to depend on the presence of hypomagnesemia, which may also result in small, significant, and potentially adverse magnesium losses from certain vital organs, such as the heart, kidney, and brain. The liver and other organs appear not to lose magnesium despite magnesium deprivation although intracellular magnesium shifts of importance cannot be ruled out. These data indicate that the body reserve to combat magnesium depletion is not designed to protect the extracellular magnesium pool and certain critical organs from magnesium deficiency. A continuous optimal intake of magnesium is needed for good nutrition and health.

Blood Pressure

The relationship between blood pressure and magnesium intake was reported.[70,71] Serum concentrations of magnesium were similar between the hypertensive group and the control group, but urinary excretion of magnesium was significantly lower in hypertensive subjects and there was an inverse correlation between magnesium excretion and blood pressure.[72]

About 3 h after 200 mg Mg was infused into normal humans, serum magnesium level went on average from 0.83 mmol/L to 1.75 mmol/L, and systolic and diastolic blood pressure were in a drop. After 200 mg or Mg magnesium was infused into a normal human for about 3 hours, the serum magnesium level increased on average from 0.83 mmol/L to 1.75 mmol/L. Systolic and diastolic blood pressure dropped. The magnesium infusion also increased renal blood flow and urinary 6-keto-PGF1 alpha excretion. These data suggest that prostacyclin release via changes in Ca+2 flux may be the mechanism of magnesium vasodilatory action. Magnesium loading blunted the rise in BP and the aldosterone-stimulating effect of AII, whereas magnesium depletion significantly enhanced these AII effects. These results support the hypothesis that magnesium may be an antagonist of the pressor and steroidogenic effects of AII.[73,74] Also, as a result of infusing MgCl2 in healthy subjects for 1.5 hours, PTH concentration declined. Plasma rennin activity and rennin increased.[75] For patients with uncomplicated mild to moderate primary hypertension receiving MgO 3 times a day at a daily dose of 1.0 g, oral magnesium reduced significantly the systolic, diastolic, and mean blood pressure. After magnesium supplementation intraerythrocyte sodium concentration was reduced and intraerythrocyte magnesium concentration was increased. The diminution of the blood pressure correlated positively with the reduction in intraerythrocyte sodium after magnesium. However, our results have shown that blood pressure response to oral magnesium was not homogeneous. Intraerythrocyte potassium and calcium, serum aldosterone, plasma renin activity, and urinary sodium excretion remained unchanged after magnesium supplementation.[76] Touyz et al. studied the relationships between serum and erythrocyte magnesium and calcium concentrations and the plasma renin activity in black and white hypertensive patient. The relationship among the plasma renin activity, magnesium, and calcium may be more important in white than in black hypertensive patients.[77] Dyckner et al. suggested that the effect of magnesium on blood pressure may be direct or through influences on the internal balance of potassium, sodium, and calcium.[78] There is a report too that in magnesium-repleted hypertensive subjects, magnesium supplementation does not affect blood pressure or lipids, probably because magnesium has no effect on cellular cation metabolism in magnesium-replete individuals.[79] Dietary magnesium intake does not appear to play an important role in long-term regulation of blood pressure in rats; normotensive Wistar-Kyoto rats and spontaneously hypertensive rats.[80,81]

In a survey of 3,318 men and women who lived in rural China, it was reported that supplementation with a multivitamin/mineral combination may have reduced mortality from cerebrovascular disease and the prevalence of hypertension in this rural population with a micronutrient-poor diet.[82] A "combination" diet that emphasized fruits, vegetables, and magnesium-rich and low-fat dairy products provides significant round-the-clock reduction in BP, especially in 354 hypertensive participants.[83]

Diabetes

For many years magnesium loss in diabetes patients was well known. Type 2 diabetes patients are 25% to 38% hypomagnesemic,[84] and have associated with low serum magnesium and low magnesium intake.[85] A low magnesium concentration in nondiabetic subjects was associated with relative insulin resistance, glucose intolerance, and hyperinsulinemia.[86] Long-term (4 weeks) studies are needed to determine whether magnesium supplementation (15.8 mmol/day) is useful in the management of type 2 diabetes.[87] Insulin-dependent diabetic pregnant women are at risk for magnesium deficiency, predominantly because of increased urinary magnesium losses.[88] A magnesium-supplemented diet did not reverse diabetes once already established in rats. An increased dietary magnesium intake in male obese Zucker diabetic fatty rats prevents deterioration of glucose tolerance, thus delaying the development of spontaneous NIDDM.[89]

Osteoporosis and Bone

Recently, there is evidence that magnesium depletion plays a important role as a cause of osteoporosis in humans, and magnesium supplementation is beneficial in preventing and treatment of osteoporosis. It is evident that magnesium plays a major role in bone and mineral homeostasis, especially in its direct effects on bone-cell function and its influence on hydorxyapatite crystal formation and growth.[90] Magnesium effects mineral metabolism in hard and soft tissues indirectly through hormonal an other modulating factors, and directly effects the processes of bone formation and resorption and of crystallization (mineralization). Its causative and therapeutic relationships to calcium urolithiasis are controversial, despite an association between low urinary magnesium and urolithiasia. A tendency to low serum and/or lymphocyte magnesium levels exists in urolithiasis. There is an inhibitory effect of magnesium supplementation on experimental urolithiasis in animals and in spontaneous urolithiasis in humans. Small decreases in serum and/or erythrocyte magnesium in osteoporotic patients and improved bone mineral density with a multinutrient supplement rich in magnesium have been reported.[91] Magnesium depletion adversely affects many phases of skeletal metabolism and has been implicated as a risk factor in several forms of osteoporosis in the survey study for 60 patients. A highly significant decline in serum magnesium concentrations, associated with continued urinary magnesium losses, occurs after cardiac transplantation. Patients with low serum magnesium levels had significantly lower rates of bone loss, lower serum PTH concentrations, and lower bone turnover.[92]

The mean bone density on magnesium-deficient postmenopausal patients who received two to six tablets daily of 125 mg each of magnesium hydroxide for 6 months and two tablets for another 18 months increased significantly after one year and remained unchanged after two years. The mean bone density of the responders increased significantly both after one year and two years, while in untreated controls, the mean bone density decreased significantly. The disparity between the initial mean bone density and bone density after one year in all osteoporotic patients and in the responders differed significantly from that of the controls.[93] In a study of healthy premenopausal women with higher intakes of zinc, magnesium, potassium, and fiber, bone mineral density was significantly higher, and a significant difference in bone mineral density was also found between the lowest and highest quartiles for these nutrients and vitamin C intake. High, long-term intake of these nutrients may be important to bone health, possibly because of their beneficial effect on acid-base balance.[94] Osteo-porosis, a major health problem in all Western countries, is a condition with many dietary factors. No significant correlation was found between current calcium intake and bone mass at any site, and iron, zinc, and magnesium intake were positively correlated with forearm bone mineral content in Caucasian premenopausal women. Iron and magnesium were significant predictors of forearm bone mineral content in premenopausal and postmenopausal women, respectively. Bone mass is influenced by dietary factors other than calcium.[95] It is supporting that alkaline-producing dietary components, specifically potassium and magnesium, fruit, and vegetables, contribute to maintenance of bone mineral density.[96] Increasing the mean magnesium intake in healthy young adult females above the usual dietary intake, which is currently above the U.S. estimated average requirement (EAR), but below the U.S. recommended dietary allowance (RDA) for magnesium, does not affect blood pressure or the rate of bone turnover.[97]

Parathyroid hormone

PTH release and end-organ responsiveness to parathyroid extract were evaluated in women with magnesium deficiency associated with hypocalcemia and inappropriately low levels of serum immunoreactive PTH, which increase very rapidly with magnesium infusion. The evidence provided that the release of parathyroid hormone is impaired in magnesium deficiency and that the level of circulating calcium required for the suppression of parathyroid hormone secretion is lower than that in normal subjects.[98] Shils said that magnesium depletion was associated with a failure of parathyroid gland to either mainufacture or mecrete the hormone, so that both immunoreactive PTH secretion and bone reactivity to parathyroid extract were involved.[99] A response to PTH in magnesium deficiency are shown also in many reports. The hypercalcemic reaction in magnesium deficient animals has been noted, so that parathyroid hormone levels in rats responded to the rising calcium levels induced by the magnesium deficiency. In magnesium depleted rats, plasma magnesium levels were significantly decreased. A fall in plasma phosphate paralleled the decrease in plasma magnesium, and plasma calcium levels were significantly increased after 14 days of magnesium deficiency. A significant rise in plasma parathyroid hormone was observed on days 7 and 14 after magnesium deficiency in Rayssiguier's study.[100] The results over a 30-day period in rats maintained on a magnesium-deficient diet indicate that: (1) in the rat, an increase in parathyroid hormone secretion occurs early in the genesis of magnesium deficiency in the presence of a modest increase in serum calcium; however, the subsequent further increase in serum calcium counteracts the stimulatory effect of hypomagnesemia on parathyroid hormone secretion; (2) unlike the human parathyroid gland, the rat parathyroid gland responds appropriately to both hypo- and hypercalcemia in magnesium deficiency; and (3) the hypercalcemia that occurs in the magnesium-deficient rat is not due to increased parathyroid hormone secretion and must be accounted for by another mechanism.[101–103]

The very rapid increase in serum immunoreactive parathyroid hormone produced by magnesium infusion suggests an effect of magnesium on hormone secretion rather than an effect on hormone synthesis, and the release of parathyroid hormone is impaired in magnesium deficiency and that the level of circulating calcium required for the suppression of parathyroid hormone secretion is lower than that in normal subjects.[104] Impairment of receptor responsiveness to parathyroid hormone of the osteoclasts follows with reduction of active bone resorption.[105] Despite increased levels of circulating parathyroid hormone, hypocalcemia progresses. An effect may play a role in the hypocalcemia associated with magnesium depletion in vivo.[106] Under progressed magnesium depletion, secretion of parathyroid hormone goes down to a very low level, but there are adequate intraparashyroid grand hormonal reserves.[107] Very low circulating parathyroid hormone, unresponsive bone, hypocalcemia, hypocalciuria, hypokaleemia, sodium retention, neuromuscular, and other clinical symptoms and signs are indications of severe magnesium depletion.[108]

The calcemic effect of vitamin D—even in high dose—is blunted in the presence of magnesium depletion in human rickets, in human malabsorption, in vitamin-D–deficient animals, and in idiopathic or surgically induced hypoparathyroidism.[109]

Other Diseases with Magnesium Depletion

Several other diseases with reduced serum magnesium level have been found, such as cerebral palsy, migraine, headache, asthma, and intrauterine fetal growth retardation. These situations will be improved mostly by supplementation with magnesium in at-risk populations. But this treatment has not been completed for all of them, and the link between cerebral palsy and magnesium status is shown on very low magnesium status, as little as 20% of normal body accrual.[110]

Magnesium was neuroprotective in many but not all of a variety of experimental studies and has a variety of biological effects that might account for its benefit.[111] Administration of magnesium sulfate during labor may protect against the development of neonatal brain lesions and cerebral palsy in low birth weight infants. Magnesium exposure may be associated with a reduction in the risk of cerebral palsy in low birth weight infants who have late-onset brain lesions, but this unpredicted observation requires confirmation in another data set.[112] The importance of magnesium in the pathogenesis of migraine headaches has been clearly established by a large number of clinical and experimental studies. Magnesium concentration has an effect on serotonin receptors, nitric oxide synthesis and release, NMDA receptors, and a variety of other migraine related receptors and neurotransmitters. Refractory patients can sometimes benefit from intravenous infusions of magnesium sulfate.[113] Infusion of magnesium into patients with documented low-ionized magnesium results in rapid and sustained relief of acute migranes in human.[114] There is some evidence that magnesium, when infused into asthmatic patients, can produce bronchodilation in addition to that obtained from standard treatments. Current evidence does not clearly support routine use of intravenous magnesium sulfate in all patients with acute asthma. Magnesium sulfate appears to be safe and beneficial for patients who present with severe acute asthma.[115]

There were 12 interventions, including protein-energy, vitamin, mineral, and fish oil supplementation, as well as the prevention and treatment of anemia and hypertensive disorders. A primary concern is the limited data supporting the effectiveness of recommended nutritional interventions during pregnancy, some of which are widely used even in women without nutritional deficiencies. Magnesium is one of many nutrients.[116] The patients with malabsorption and magnesium depletion had subnormal levels of serum citrate and magnesium, decreasing 24-hour levels of urinary citrate, magnesium, and calcium, and excessive levels of urinary oxalate. Daily citrate excretion averaged only 15% normal. The hypocitraturia in the patients resulted from a subnormal filtered load of citrate and abnormally high net tubular reabsorption of the anion. An oral citrate supplement raised both the serum concentration and the filtered load of citrate to normal fasting values, but net tubular reabsorption remained abnormally high and urinary excretion abnormally low.[117]

Assessment of Magnesium Nutritional Status

Serum Magnesium

Magnesium plays an essential role in both extracellular and intracellular metabolism. Some clinicians recommend routine measurement of magnesium when other electrolyte analyses are required for the care of patients. The assessment of magnesium status is difficult because of the distribution of magnesium within the intracellular compartment. The serum magnesium level may not reflect intracellular magnesium availability. For assessment of magnesium status, measurement of serum magnesium is most easy and commonly adopted. The serum magnesium concentration may be influenced by changes in serum pH, serum albumin, other anionic ligands. Serum magnesium level is held within a tight range (1.7–2.2 mg/dL) by homeostasis mechanism. A serum magnesium concentration under 1.7 mg/dL indicates usually a limit of magnesium depletion.[118]

The magnesium in adult human body is mostly in bone and soft tissue. Only about 0.3% of the total body magnesium is present in serum, yet the majority of analytical data obtained is from this body fluid. Assessing the magnesium status of an individual is difficult, as present, no simple, rapid, and accurate test determines intracellular magnesium, but determination of total and free magnesium in tissues and physiological tests provide some information. Changes in magnesium status have been linked to cardiac arrhythmias, coronary heart disease, hypertension, and premenstrual syndrome. A better understanding of magnesium transport and of factors controlling magnesium metabolism is needed to elucidate the role of magnesium in disease processes.[119] Dietary magnesium depletion induced experimentally leads to decreased serum magnesium values in healthy subjects. Under certain circumstances, serum magnesium is a sensitive indicator of magnesium nutritional status.[120] Hypomagnesemia usually follows with severe magnesium depletion (serum magnesium under 1.7 mg/dL). In clinical conditions such as in diabetes mellitus,[121] in alcoholism,[122] or in malabsorption,[123] etc., magnesium concentration in blood cells, bone cells, and muscle cells are very low, but serum magnesium concentration is within normal range. Thus, for magnesium nutritional assessment of humans serum magnesium concentration is not suitable and intracellular magnesium may be a better indicator.[124,125] Ryzen and colleagues also said that magnesium deficiency might play a role in the pathogenesis of atherosclerosis, cardiac arrhythmias, and coronary spasm. Because less than 1% of magnesium (Mg) is extracellular, the serum magnesium (sMg) does not always accurately reflect intracellular magnesium stores.[126]

The 8251-subject, 10-year followup in the National Health and Nutrition Examination Survey (NHANES I) was to assess the important roles of modifiable dietary and behavioral characteristics in the causation and prevention of coronary heart disease. Serum magnesium levels in 492 coronary heart disease patients selected from this group were under 1.9 mg/dL with risk factor for cardiovascular disease.[127] Based these results, we must discuss whether it is suitable or not for the normal range of serum magnesium level to be 1.8 to 2.3 mg/dL.[128]

Serum Ionized Magnesium

Ion-specific electrodes were coming available for determining of ionized magnesium in serum or plasma that is physiologically active. Ionized magnesium value in serum may be a better index for magnesium status than the total magnesium concentration, but may not to be most suitable to assess magnesium status and more studies are needed.[129–131]

Intracellular Magnesium

The total magnesium content of tissues such as red blood cells, skeletal muscle, bone, blood cells, and lymphocytes have been pointed out for index of magnesium nutritional status.[132–134] Determination of intracellular magnesium level should be a physiologically adequate measurement of magnesium status, that is, it plays a critical role in enzyme activation within the cell. But tissue and cell samples are not routinely collected in clinical spots. Between serum magnesium concentration and intracellular magnesium level a good correlation was generally not found.[135–138]

Lymphocyte magnesium concentration does not good agreement with serum or red blood cell magnesium levels and also does not reflect muscle magnesium.[139,140] Lymphocyte and skeletal muscle magnesium correlated well in healthy subjects, but not in patients with congestive heart failure.[141,142] For magnesium assessment, lymphocyte or muscle magnesium content may be best, but more studies will be needed. The blood cells provide the most accessible source of cells for determining intracellular magnesium. Magnesium content of lymphocytes correlates with skeletal and cardiac magnesium content[143] and intracellular ionized magnesium level in patients with high risk for magnesium depletion are lower than in healthy subjects.[125,144]

Magnesium Tolerance Test

A more intensive intervention test is the magnesium load or tolerance test.[145] This assessment tool is proper for adults but not for infants and children.[146] In this test, magnesium content in 24-h urine after magnesium loading by intravenous infusion is measured. In normal subjects, 80% of loaded magnesium excretes in urine within 24 h, and retention under 30% of loaded magnesium suggests magnesium depletion. Its application has limits because this test needs a prolonged intravenous infusion.

Magnesium Balance Studies

In the past, the measure of adequate dietary magnesium has been the dietary balance study.[147–151] The best available method for estimating magnesium requirements is the dietary balance approach.[125] The balance study is based on individuals who consume a diet with known magnesium content. The quality of data depends upon both methodological aspects, such as selection of test diet, total collection urine, feces, sweat, etc., and physiological features of the test subjects, such as age, health status, prior nutrient intake, etc. To control the errors and variability associated with a balance study, the Food and Nutrition Board established minimum criteria for balance studies to be included for the Recommended Dietary Allowances in each country. The important items for a balance study are an adaptation period of at least 12 days before the balance study and preparing two dietary-magnesium-level diets in the studies. Magnesium intake is related to energy intake, so that magnesium needs a cofactor of energy metabolism, and in magnesium balance studies energy requirements must be considered.[152,153]

Dietary Requirements and Food Sources of Magnesium

The U.S. Recommended Dietary Allowances (RDA) for magnesium was revised in 1997.[125] The U.S. magnesium RDA for adult women is 320 mg/day and for adult men is 420 mg/day.[154] In 2005, the Japanese RDA was also revised; for adult women it is 240 mg/day and for adult men it is 310 mg/day.[155] The RDA for magnesium in various countries are: women, 270 mg/day, men, 300 mg/day in England; women, 270 mg/day, men, 320 mg/day in Australia; women and men, 400 mg/day in Russian; and women, 300 mg/day, men, 350 mg/day in Thailand.

Under increasing of death by cardiovascular diseases, magnesium status (high magnesium intake) is very important. For high magnesium intake, food containing high levels of magnesium and meals with low cooking loss must be prepared. Food with high magnesium content are whole grains, fish, algae, pulses (especially tofu), green leafy vegetables, etc. The poor magnesium foods are refined foods such as grains and wheat flour, etc., and may include meat.

Food with high levels of fiber, calcium, or phosphate may induce lower bioavailability of magnesium.

Toxicity of Magnesium

There may be no evidence generally in healthy people with natural meals to oversupplement magnesium. High magnesium intake with toxicity is only found while taking supplements. Upper limits for magnesium supplements are defined to include oversupplementation of magnesium into consideration by the Food and Nutrition Board's report.[124] Severe magnesium toxicity is rare. Upper limits for magnesium are not set up in the RDA for Japan.[155] There are some reports that much daily magnesium intake of 30 g/day results in symptoms of metabolic alkalosis and hypokalemia,[6] and by taking a single dose of up to 400 g, cardiorespiratory arrest and paralytic ileus can occur. When serum magnesium concentration are up to 4.8 to 8.4 mg/dL, cardiac and neurological symptoms will appear.[156]

References

1. McCollum EV, Orent ER. Effects on the rat of deprivation of magnesium. *J Biol Chem* 1931;92.
2. Shils ME. Magnesium. In: ME Shils, Olson JA, Shik M, eds. *Modern Nutrition in Health and Disease*, 8th edn. Lea & FEBIGER; 1994:164–184.
3. Hauary VG, Cantarow A. Variations of serum magnesium in 52 normal and 440 pathologic patients. *J Lab Clin Med* 1942;27:616–622.
4. Flink EB, Stutzman FL, Anderson AR, Konig T, Fraser R. Magnesium deficiency after prolonged parenteral fluid administration and after chronic alcoholism complicated by delirium tremens. *J Lab Clin Med* 1954;43:169–183.
5. Wallach S. Effects of magnesium on skeletal metabolism. *Magnes Trace Elem* 1990;9:1–14.
6. Gupta RK, Moore RD. 31P NMR studies of intracellular free Mg+2 in intact frog skeletal muscle. *J Biol Chem* 1980;255:3987.
7. Fleet JC, Cashman KD. Magnesium. In: Bowman BA, Russell RM, eds. *Present Knowledge in Nutrition*. 8th ed. International Life Sciences; 2001:292–301.
8. Frausto de Silva JJR, Williams RJP. *The Biological Chemistry of Magnesium: Phosphate Metabolism. The Biochemical Chemistry of the Elements*. Oxford: Oxford University Press; 1991:241–267.
9. Wacker WE, Parisi AF. Magnesium metabolism. *N Engl J Med* 1968;45:658–663, 712–717.
10. Garfinkel L, Garfinkel D. Magnesium regulation of the glycolytic pathway and the enzymes involved. *Magnesium* 1985;4:60–72.
11. Mildvan AS. Role of magnesium and other divalent cations in ATP-utilizing enzymes. *Magnesium* 1987;6:28–33.
12. Gunther T. Magnesium and regulation of Mg2+ efflux and Mg2+ influx. *Miner Electrolyte Metab* 1993;19:259–265.
13. White RE, Hartzell HC. Magnesium ions in cardiac function. Regulator of ion channels and second messengers. *Biochem Pharmacol* 1989;38:859–867.
14. Kayne LH, Lee DB. Intestinal magnesium absorption. *Miner Electrolyte Metab* 1993;19:210–217.

15. Meneely R, Leeper L, Chishan FK. Intestinal maturation: in vivo magnesium transport. *Pediatr Res* 1982;16:295–298.
16. Graham LA, Caesae JJ, Burgen ASV. Gastrointestinal absorption and excretion of Mg26 in man. *Metabolism* 1960;9:646–659.
17. Schwartz R, Spencer H, Welsh JJ. Magnesium absorption in human subjects from leafy vegetables, intrinsically labeled with stable 26Mg. *Am J Clin Nutr* 1984;39:571–576.
18. Fine KD, Santa Ana CA, Porter JI, Fordtran JS. Diagnosis of magnesium-induced diarrhea. *J Clin Invest* 1991;88:396–402.
19. Roth P, Werner E. Intestinal absorption of magnesium in man. *Int J Appl Radiat Isot* 1979;30:523–526.
20. Brannan OB, Vergne-Marini P, Pak CVC. Magnesium absorption in the human small intestine: results in normal subjects, patients with chronic renal disease, and patients with absorption hypercalciuria. *J Clin Invest* 1976;57:1412–1428.
21. Ross DB. In vitro studies on the transport of magnesium across the intestinal wall of the rat. *J Physiol* 1962;160:417–428.
22. Meneely R, Leeper L, Chishan FK. Intestinal maturation: in vivo magnesium transport. *Pediatr Res* 1982;16:295–298.
23. Krejs GJ, Nicar MJ, Zerwekh HE, Normal DA, Kane MG, Pak CY. Effects of 1,25-dihydroxyvitamin D3 on calcium and magnesium absorption in the healthy human jejunum and ileum. *Am J Med* 1985;75:975–976.
24. Hardwick LL, Jones MR, Brautbar N, Lee DB. Magnesium absorption: metabolisms and the influence of vitamin D, calcium and phosphate. *J Nutr* 1991;121:13–23.
25. Levine BS, Brautbar N, Walling MW, Lee DB, Cobum JW. Effects of vitamin D and diet magnesium on magnesium metabolism. *Am J Physiol* 1980;239:E515–E523.
26. Leichsenring JM, Norris LM, Lamison SA. Magnesium metabolism in college women; observations on the effect of calcium and phosphorus intake levels. *J Nutr* 1991;45:477–485.
27. Hendix JZ, Alcock NW, Archibald RM. Compertition between calcium, strontium, and magnesium for absorption in the isolated rat intestine. *Clin Chem* 1963;9:734–744.
28. O'Dell BL, Morris ER, Regan WO. Magnesium requirement of guinea pigs and rats: effect of calcium and phosphorus and symptoms of magnesium deficiency. *J Nutr* 1960;70:103–110.
29. Siener R, Hesse A. Influence of a mixed and vegetarian deit on urinary magnesium excretion and concentration. *Br J Nutr* 1995;73:783–790.
30. Quamme GA, Dirks JH. The physiology of renal magnesium handling. *Renal Physiol* 1986;9:257–269.
31. Quamme GA. Renal magnesium handling: new insights in understanding old problems. *Kidney Int* 1997;52:1180–1195.
32. Quamme GA. Control of magnesium transport in the thick limb. *Am J Physiol* 1989;256:F197–F210.
33. Brown EM, Herbert SC. A cloned Ca2+-sensing receptor; a mediator of direct effects of extracellular Ca2+ on renal function. *J Am Soc Nephrol* 1996;6:1530–1540.
34. Simon DB, Lu Y, Choate KA, et al. Paracellin-1, a renal tight junction protein required for paracellular Mg2+ resorption. *Science* 1999;285:103–106.
35. Quamme GA, Dirks JH. The physiology of renal magnesium handling. *Renal Physiol* 1986;9:257–269.

36. De Rouffignac C, Quamme GA. Renal magnesium handling and its hormonal control. *Physiol Rev* 1994;74:305–322.
37. Fitzgerald MG, Fourman P. An experimental study of magnesium deficiency in man. *Clin Sci* 1956;15:635.
38. Rude RK, Bethune JE, Singer FR. Renal tubular maximum for magnesium in normal, hyper-parathyroid and hypo-parathyroid man. *J Clin Endocinol Metab* 1980;51:1425–1451.
39. Abbott L, Nadler J, Rude RK. Magnesium deficiency in alcoholism: Possible contribution to osteoporosis and cardiovascular disease in alcoholics. *Alcohol Clin Exp Res* 1994;18:1976–1082.
40. Ryan MP. Diuretics and potassium/magnesium depletion. Diuretics for treatment. *Am J Med* 1987;82:38–47.
41. Kunkei HO, Pearson PB. Magnesium in the nutrition of the rabbit. *J Nutr* 1948;36:1948.
42. Orent E, Kruse HD, McCollum EV. Studies on magnesium deficiency in animals. II. Species variation in symptomatology of magnesium deprivation. *Am J Physiol* 1931;101:454.
43. Blaxter KL, Rook JAF, MacDonald AM. Experimental magnesium deficiency in calves. *J Comp Pathol Therap* 1954;64:157.
44. Alcock NW, Shils ME. Comparsion of magnesium deficiency in the rat and mouse. *Proc Soc Exp Biol Med* 1974;146:137–141.
45. Kimura M, Harada T, Itokawa Y. Changes in blood pressure, body temperature, serotonin metabolism and tissue mineral levels in magnesium deficient rats. *JJSMgR* 1983;2:7–13.
46. Kimura M, Yokoi T, Hayase M, Itokawa Y. Effects of iron intake on serum biochemical parameters in magnesium deficient rats. *JJSMgR* 1992;11:19–29.
47. Alcock NW, Shils ME, Lieberman PH, Erlandson RA. Thymic changes in the magnesium-depleted rat. *Cancer Res* 1973;33:2196–2204.
48. Rude RK. Magnesium deficiency: a cause of heterogeneous disease in humans. *J Bone Miner Res* 1998;13:749–758.
49. Al-Ghamdi SMG, Cameron EC, Sutton RAL. Magnesium deficiency: pathophysiologic and clinical overview. *Am J Kidney Dis* 1994;24:737–752.
50. Milla PJ, Aggett PJ, Wolff OH, Harries JT. Studies in primary hypomagnesaemia: evidence for defective carrier-mediated small intestinal transport of magnesium. *Gut* 1979;20:1028–1033.
51. National Institute for Biotechnology Information. Online Mendelian genetics in man (OMIM). Available at http://www.ncbi.nlm.nih.gov/omim. Accessed 19 January 2001.
52. Fleet JC, Cashman KD. Magnesium. In: Bowman BA, Russell RM, eds. *Present Knowledge in Nutrition*, 8th edn. International Life Sciences; 2001:292–301.
53. Flink EB. Magnesium deficiency in alcoholism. *Alcohol Clin Exp Res* 1986;10: 590–594.
54. Altura BM, Altura BT. Biochemistry and pathophysiology of congestive heart failure: is there a role for magnesium? *Magnesium* 1986;5:134–143.
55. Dyckner T. Relation of cardiovascular disease to potassium and magnesium deficiencies. *Am J Cardiol* 1990;65:44K–46K.
56. Seelig M. Cardiovascular consequences of magnesium deficiency and loss: pathogenesis, prevalence and manifestations—magnesium and chloride loss in refractory potassium repletion. *Am J Cardiol* 1989;63:4G–21G.

57. Kobayashi J. On geographical relationship between the chemical nature of river water and death-rate from apoplexy. *Ber Ohara Inst* 1957;11:13–21.
58. Marier JR, Neri LC. Quantifying the role of magnesium in the interrelationship between human mortality/morbidity and water hardness. *Magnesium* 1985; 4:53–59.
59. Schroeder HA. Relationship between mortality from cardiovascular diseases and treated water supplies. *J Am Med Assoc* 1960;172:1902–1908.
60. Karppanen H, Pennanen R, Passinen L. Minerals, coronary heart diseases and sudden coronary death. *Adv Cardiol* 1978;25:9–24.
61. Davis WH, et al. Monotherapy with magnesium increases abnormally low/high density lipoprotein cholesterol. A clinical assay *Curr Therap Res* 1984;36: 341–346.
62. Seelig MS. ISIS 4: clinical controversy regarding magnesium infusion, thrombolytic therapy, and acute myocardial infarction. *Nutr Rev* 1995;53:261–264.
63. Seelig MS, Elin RJ, Antman EM. Magnesium in acute myocardial infarction: still an open question. *Can J Cardiol* 1998;14:745–749.
64. Seelig MS. Interrelationship of magnesium and congestive heart failure. *Wien Med Wochenschr* 2000;150:335–341.
65. Cunnane SC, Soma M, McAdoo KR, Horrobin DF. Magnesium deficiency in the rat increases tissue levels of docosahexaenoic acid. *J Nutr* 1985;115:1498–1503.
66. Gueux E, Mazur A, Cardot P, Rayssiguier Y. Magnesium deficiency affects plasma lipoprotein composition in rats. *J Nutr* 1991;121:1222–1227.
67. Nigam S, Averdunk R, Gunther T. Alteration of prostanoid metabolism in rats with magnesium deficiency. *Prostaglandins Leukot Med* 1986;23:1–10.
68. Soma M, Cunnane SC, Horrobin DF, Manku MS, Honda M, Hatano M. Effects of low magnesium diet on the vascular prostaglandin and fatty acid metabolism in rats. *Prostaglandins* 1988;36:431–441.
69. Wallach S. Availability of body magnesium during magnesium deficiency. *Magnesium* 1988;7:262–270.
70. Altura BM, Altura BJ, Gebrewold J, Bloom S. Magnesium deficiency and hypertension: correlation between magnesium-deficient diets and microcirculatory changes in situ. *Science* 1984;223:1315–1317.
71. Mizushima S, Cappuccio FP, Nichols R, Elliot P. Dietary magnesium intake and blood pressure: a qualitative overview of calcium, potassium, and magnesium intake and risk of stroke in women. *Stroke* 1999;30:1772–1779.
72. Tillman DM, Semple PF. Calcium and magnesium in essential hypertension. *Clin Sci (Lond)* 1988;75:395–402.
73. Rude R, Manoogian C, Ehrlich L, DeRusso P, Ryzen E, Nadler J. Mechanisms of blood pressure regulation by magnesium in man. *Magnesium* 1989;8:266–273.
74. Nadler JL, Goodson S, Rude RK. Evidence that prostacyclin mediates the vascular action of magnesium in humans. *Hypertension* 1987;9:379–383.
75. Dechaux M, Kindermans C, Laborde K, Blazy I, Sachs C. Magnesium and plasma renin concentration. *Kidney Int Suppl* 1988;25:S12–S13.
76. Sanjuliani AF, de Abreu Fagundes VG, Francischetti EA. Effects of magnesium on blood pressure and intracellular ion levels of Brazilian hypertensive patients. *Int J Cardiol* 1996;56:177–183.
77. Touyz RM, Panz V, Milne FJ. Relations between magnesium, calcium, and plasma renin activity in black and white hypertensive patients. *Miner Electrolyte Metab* 1995;21:417–422.

78. Dyckner T, Wester PO. Effect of magnesium on blood pressure. *Br Med J* 1983;286:1847–1849.
79. Zemel PC, Zemel MB, Urberg M, Douglas FL, Geiser R, Sowers JR. Metabolic and hemodynamic effects of magnesium supplementation in patients with essential hypertension. *Am J Clin Nutr* 1990;51:665–669.
80. Overlack A, Zenzen JG, Ressel C, Muller HM, Stumpe KO. Influence of magnesium on blood pressure and the effect of nifedipine in rats. *Hypertension* 1987;9:139–143.
81. Rayssinguier Y, Mbega JD, Durlach V, et al. Magnesium and blood pressure. I. Animal studies *Magnes Res* 1992;5:139–146.
82. Mark SD, Wang W, Fraumeni JF Jr, et al. Lowered risks of hypertension and cerebrovascular disease after vitamin/mineral supplementation: the Linxian Nutrition Intervention Trial. *Am J Epidemiol* 1996;143:658–664.
83. Moore TJ, Vollmer WM, Appel LJ, et al. Effect of dietary patterns on ambulatory blood pressure: results from the Dietary Approaches to Stop Hypertension (DASH) Trial. DASH Collaborative Research Group. *Hypertension* 1999;34:472–477.
84. Butler AM. Diabetic coma. *N Engl J Med* 1950;243:648–659.
85. Kato WH, Folsom AR, Nieto FJ, et al. Serum and dietary magnesium and the risk for type 2 diabetes mellitus: the Atheroaclerosis Risk in Community Study. *Arch Inter Med* 1999;159:2151–2159.
86. Rosolova H, Mayer O Jr, Reaven G. Effect of variations in plasma magnesium concentration on resistance to insulin-mediated glucose disposal in nondiabetic subjects. *J Clin Endocrinol Metab* 1997;82:3783–3785.
87. Paolisso G, Scheen A, Cozzolino D, et al. Changes in glucose turnover parameters and improvement of glucose oxidation after 4-week magnesium administration in elderly noninsulin-dependent (type II) diabetic patients. *J Clin Endocrinol Metab* 1994;78:1510–1514.
88. Mimouni F, Miodovnik M, Tsang RC, Holroyde J, Dignan PS, Siddiqi TA. Decreased maternal serum magnesium concentration and adverse fetal outcome in insulin-dependent diabetic women. *Obstet Gynecol* 1987;70:85–88.
89. Balon TW, Gu JL, Tokuyama Y, Jasman AP, Nadler JL. Magnesium supplementation reduces development of diabetes in a rat model of spontaneous NIDDM. *Am J Physiol* 1995;269:E745–E752.
90. Wallach S. Effects of magnesium on skeletal metabolism. *Magnes Trace Elem* 1990;9:1–14.
91. Wallach S. Relation of magnesium to osteoporosis and calcium urolithiasis. *Magnes Trace Elem* 1991–1992;10:281–286.
92. Boncimino K, McMahon DJ, Addesso V, Bilezikian JP, Shane E. Magnesium deficiency and bone loss after cardiac transplantation. *J Bone Miner Res* 1999;14:295–303.
93. Stendig-Lindberg G, Tepper R, Leichter I. Trabecular bone density in a two year controlled trial of peroral magnesium in osteoporosis. *Magnes Res* 1993;6:155–163.
94. New SA, Bolton-Smith C, Grubb DA, Reid DM. Nutritional influences on bone mineral density: a cross-sectional study in premenopausal women. *Am J Clin Nutr* 1997;65:1831–1839.
95. Angus RM, Sambrook PN, Pocock NA, Eisman JA. Dietary intake and bone mineral density. *Bone Miner* 1988;4:265–277.

96. Tucker KL, Hannan MT, Chen H, Cupples LA, Wilson PW, Kiel DP. Potassium, magnesium, and fruit and vegetable intakes are associated with greater bone mineral density in elderly men and women. *Am J Clin Nutr* 1999;69:727–736.
97. Doyle L, Flynn A, Cashman K. The effect of magnesium supplementation on biochemical markers of bone metabolism or blood pressure in healthy young adult females. *Eur J Clin Nutr* 1999;53:255–261.
98. Anast CS, Winnacker JL, Forte LR, Burns TW. Impaired release of parathyroid hormone in magnesium deficiency. *J Clin Endocrinol Metab* 1976;42:707–717.
99. Shils ME. Magnesium, calcium, and parathyroid hormone interactions. *Ann N Y Acad Sci* 1980;355:165–180.
100. Rayssiguier Y, Thomasset M, Garel JM, Barlet JP. Plasma parathyroid hormone levels and intestinal calcium binding protein in magnesium deficient rats. *Horm Metab Res* 1982;14:379–382.
101. Anast CS, Forte LF. Parathyroid function and magnesium depletion in the rat. *Endocrinology* 1983;113:184–189.
102. Levi J, Massry SG, Coburn JW, Llach F, Kleeman CR. Hypocalcemia in magnesium-depleted dogs: evidence for reduced responsiveness to parathyroid hormone and relative failure of parathyroid gland function. *Metabolism* 1974;23:323–335.
103. MacManus J, Heaton FW, Lucas PW. A decreased response to parathyroid hormone in magnesium deficiency. *J Endocrinol* 1971;49:253–258.
104. Freitag JJ, Martin KJ, Conrades MB, et al. Evidence for skeletal resistance to parathyroid hormone in magnesium deficiency. Studies in isolated perfused bone. *J Clin Invest* 1979;64:1238–1244.
105. Johannesson AJ, Raisz LG. Effects of low medium magnesium concentration on bone resorption in response to parathyroid hormone and 1,25-dihydroxyvitamin D in organ culture. *Endocrinology* 1983;113:2294–2298.
106. Johannesson AJ, Raisz LG. Effects of low medium magnesium concentration on bone resorption in response to parathyroid hormone and 1,25-dihydroxyvitamin D in organ culture. *Endocrinology* 1983;113:2294–2298.
107. Anast CS, Winnacker JL, Forte LR, Burns TW. Impaired release of parathyroid hormone in magnesium deficiency. *J Clin Endocrinol Metab* 1976;42:707–717.
108. Anast CS, Forte LF. Parathyroid function and magnesium depletion in the rat. *Endocrinology* 1983;113:184–189.
109. Fleet JC, Cashman KD. Magnesium. In: Bowman BA, Russell RM, eds. *Present Knowledge in Nutrition*, 8th edn. International Life Sciences; 2001:292–301.
110. Fleet JC, Cashman KD. Magnesium. In: Bowman BA, Russell RM, eds. *Present Knowledge in Nutrition*, 8th edn. International Life Sciences; 2001:292–301.
111. Hirtz DG, Nelson K. Magnesium sulfate and cerebral palsy in premature infants. *Curr Opin Pediatr* 1998;10:131–137.
112. Paneth N, Jetton J, Pinto-Martin J, Susser M. Magnesium sulfate in labor and risk of neonatal brain lesions and cerebral palsy in low birth weight infants. The Neonatal Brain Hemorrhage Study Analysis Group. *Pediatrics* 1997;99:E1.
113. Mauskop A, Altura BM. Role of magnesium in the pathogenesis and treatment of migraines. *Clin Neurosci* 1998;5:24–27.
114. Pfaffenrath V, Wessely P, Meyer C, et al. Magnesium in the prophylaxis of migraine—a double-blind placebo-controlled study. *Cephalalgia* 1996;16:436–440.

115. Rowe BH, Bretzlaff JA, Bourdon C, Bota GW, Camargo CA Jr. Intravenous magnesium sulfate treatment for acute asthma in the emergency department: a systematic review of the literature. *Ann Emerg Med* 2000;36:181–190, 234–236. *Ann Emerg Med* 37:552–553.
116. de Onis M, Villar J, Gulmezoglu M. Nutritional interventions to prevent intrauterine growth retardation: evidence from randomized controlled trials. *Eur J Clin Nutr* 1998;52(Suppl 1):S83–93.
117. Rudman D, Dedonis JL, Fountain MT, Chandler JB, Gerron GG, Fleming GA, Kutner MH. Hypocitraturia in patients with gastrointestinal malabsorption. *N Engl J Med* 1980;303:657–661.
118. Quamme GA. Laboratory evaluation of magnesium status. Renal function and free intracellular magnesium concentration. *Clin Lab Med* 1993;13:209–223.
119. Elin RJ. Assessment of magnesium status. *Clin Chem* 1987;33:1965–1970.
120. Fatemi S, Ryzen E, Flores J, Endres DB, Rude RK. Effect of experimental human magnesium depletion on parathyroid hormone secretion and 1,25-dihydroxyvitamin D metabolism. *J Clin Endocrinol Metab* 1991;73:1067–1072.
121. Nadler JL, Malayan S, Luong H, Shaw S, Natarajan RD, Rude RK. Intracellular free magnesium deficiency plays a key role in increased platelet reactivity in type II diabetes mellitus. *Diabetes Care* 1992;15:835–841.
122. Abbott L, Nadler J, Rude RK. Magnesium deficiency in alcoholism: possible contribution to osteoporosis and cardiovascular disease in alcoholics. *Alcohol Clin Exp Res* 1994;18:1076–1082.
123. Rude RK, Olerich M. Magnesium deficiency: possible role in osteoporosis associated with gluten-sensitive enteropathy. *Osteoporos Int* 1996;6:453–461.
124. Reinhart RA. Magnesium metabolism. A review with special reference to the relationship between intracellular content and serum levels. *Arch Intern Med* 1988;148:2415–2420.
125. Food and Nutrition Board. Magnesium. Dietary Reference Intakes: Calcium, Magnesium, Phosphorus, Vitamin D, and Fluoride. Washington, DC: National Academy Press; 1997:190–249.
126. Ryzen E, Elkayam U, Rude RK. Low blood mononuclear cell magnesium in intensive cardiac care unit patients. *Am Heart J* 1986;111:475–480.
127. Gartside PS, Glueck CJ. The important role of modifiable dietary and behavioral characteristics in the causation and prevention of coronary heart disease hospitalization and mortality: the prospective NHANES I follow-up study. *J Am Coll Nutr* 1995;14:71–79.
128. Elin RJ. Assessment of magnesium status. *Clin Chem* 1987;33:1965–1970.
129. Mimouni FB. The ion-selective magnesium electrode: a new tool for clinicians and investigators. *J Am Coll Nutr* 1996;15:4–5.
130. Altura BT, Shirey TL, Young CC, et al. A new method for the rapid determination of ionized Mg^{2+} in whole blood, serum and plasma. *Methods Find Exp Clin* 1992;14:297–304.
131. Schuette SA, Ziegler EE, Nelson SE, Janghorbani M. Feasibility of using the stable isotope ^{25}Mg to study Mg metabolism in infants. *Pediatr Res* 1990;27:36–40.
132. Nadler JL, Malayan S, Luong H, Shaw S, Natarajan RD, Rude RK. Intracellular free magnesium deficiency plays a key role in increased platelet reactivity in type II diabetes mellitus. *Diabetes Care* 1992;15:835–841.

133. Abbott L, Nadler J, Rude RK. Magnesium deficiency in alcoholism: possible contribution to osteoporosis and cardiovascular disease in alcoholics. *Alcohol Clin Exp Res* 1994;18:1076–1082.
134. Rude RK, Olerich M. Magnesium deficiency: possible role in osteoporosis associated with glutensensitive enteropathy. *Osteoporos Int* 1996;6:453–461.
135. Ryzen E, Elkayam U, Rude RK. Low blood mononuclear cell magnesium in intensive cardiac care unit patients. *Am Heart J* 1986;11:475–480.
136. Elin RJ, Hosseini JM. Magnesium content of mononuclear blood cells. *Clin Chem* 1985;31:377–380.
137. Reinhart RA. Magnesium metabolism. A review with special reference to the relationship between intracellular content and serum levels. *Arch Intern Med* 1988;148:2415–2420.
138. Ryzen E. Magnesium homeostasis in critically ill patients. *Magnesium* 1989;8:201–212.
139. Alfrey AC, Miller NL, Butkus D. Evaluation of body magnesium stores. *J Lab Clin Med* 1974;84(2):153–162.
140. Wester PO, Dyckner T. Diuretictreatment and magnesium losses. *Acta Med SCAND* 1980;647:145–152.
141. Dyckner T, Wester PO. Skeletal muscle magnesium and potassium determinations: correlation with lymphocyte contents of magnesium and potassium. *J Am Coll Nutr* 1985;4:619–625.
142. Ralston MA, Murnane MR, Kelley RE, Altschuld RA, Unverferth DV, Leier VC. Magnesium content of serum, circulating mononuclear cells, skeletal muscle, and myocardium in congestive heart failure. *Circulation* 1989;80:573–580.
143. Fatemi S, Ryzen E, Flores J, Endres DB, Rude RK. Effect of experimental human magnesium depletion on parathyroid hormone secretion and 1,25-dihydroxyvitamin D metabolism. *J Clin Endocrinol Metab* 1991;73:1067–1072.
144. Hua H, Gonzales J, Rude RK. Magnesium transport induced ex vivo by a pharmacological dose of insulin is impaired in non-insulin-dependent diabetes mellitus. *Magnes Res* 1995;8:359–366.
145. Rude RK. Magnesium deficiency: a cause of heterogeneous disease in humans. *J Bone Miner Res* 1998;13:749–758.
146. Gullestad L, Dolva LO, Waage A, Falch D, Fagerthun H, Kjekshus J. Magnesium deficiency diagnosed by an intravenous loading test. *Scand J Clin Lab Invest* 1992;52:245–253.
147. Greger JL, Baier MJ. Effect of dietary aluminum on mineral metabolism of adult males. *Am J Clin Nutr* 1983;38:411–419.
148. Hunt MS, Schofield FA. Magnesium balance and protein intake level in adult human female. *Am J Clin Nutr* 1969;22:367–373.
149. Mahalko JR, Sandstead HH, Johnson LK, Milne DB. Effect of a moderate increase in dietary protein on the retention and excretion of Ca, Cu, Fe, Mg, P, and Zn by adult males. *Am J Clin Nutr* 1983;37:8–14.
150. Schwartz R, Spencer H, Welsh JJ. Magnesium absorption in human subjects from leafy vegetables, intrinsically labeled with stable 26Mg. *Am J Clin Nutr* 1984;39:571–576.
151. Schwartz R, Apgar BJ, Wien EM. Apparent absorption and retention of Ca, Cu, Mg, Mn, and Zn from a diet containing bran. *Am J Clin Nutr* 1986;43:444–455.
152. Clarkson PM, Haymes EM. Exercise and mineral status of athletes: calcium, magnesium, phosphorus, and iron. *Med Sci Sports Exerc* 1995;27:831–843.

153. Niekamp RA, Baer JT. In-season dietary adequacy of trained male cross-country runners. *Int J Sport Nutr* 1995;5:45–55.
154. Cleveland LE, Goldman JD, Borrud LG. Data tables: results from USDA's 1994 Continuing Survey of food intakes by Individuals and 1994 Diet and Health Knowleges Survey. Beltsville, MD: U.S. Department of Agriculture, 1996.
155. Ministry of Health, Labour, and Welfare, Japan. Dietary Reference Intakes for Japanese, 2005.
156. Rude RK, Bethune JE, Singer FR. Renal tubular maximum for magnesium in normal, hyperparathyroid, and hypoparathyroid man. *J Clin Endocrinol Metab* 1980;51:1425–1431.

8
Magnesium Requirement and Affecting Factors

Mamoru Nishimuta, Naoko Kodama, Eiko Morikuni,
Nobue Matsuzaki, Yayoi H. Yoshioka, Hideaki Yamada,
Hideaki Kitajima, and Hidemaro Takeyama

We performed 11 balance studies to learn the estimated average requirement (EAR) of magnesium (Mg). Magnesium intake was not correlated with Mg balance when all data was used ($n = 109$). However, Mg intake was correlated with calcium (Ca) and phosphorus (P) balances. During the analysis, we found a correlation between sodium intake and Ca and Mg balances. After excluding the data of the highest sodium (Na) intake study, Mg intake turned out to be correlated with Mg balance.

We also studied the effects of some lifestyles on urinary excretion of Ca and Mg. Ingestions of an energy source increased in urinary Ca and Mg. Risk factors for chronic degenerative diseases also increased in urinary Ca and Mg.

Usually, Ca/Mg molar ratio in urine is isomolar or less than 1. To keep plasma Ca and Mg constant, Ca is derived from the bone. If all Mg was released from the bone, where there is major pool of Mg, by the osteophagocytosis, more than 10 times of the amount of Ca must be released at the same time, because the contents of Ca in the bone is far higher than that of Mg. So, Mg in the cell is concluded to be the major source of the compensation.

It is generally believed that there are three levels of dietary intake of a nutrient: excess, adequate, and deficit. The border between excess and adequate is recognized as the upper limit, while the border between adequate and deficit is termed the *requirement*.[1,2]

To determine an upper limit and a requirement for a nutrient, it is necessary to understand the scientific evidence giving quantitative information about the dietary intake and the signs and symptoms of excess or deficit of the nutrient.[1,2]

For the minerals Na, Ca, Mg, and P, whose physiological pools include bone[3,4] and for which the signs and symptoms of deficit or excess are poor, determination of the requirement and the upper limit is difficult because of an absence of evidence.

In such cases, results of balance studies that show an EAR, that is, a balance between intake and output, are the sole sources of information available to determine the requirement.

Mineral Balance Studies

Subjects and Methods

From 1986 to 2000, 109 volunteers (23 males, 86 females), ranging from 18 to 28 years old, took part in mineral balance studies after written informed consent was obtained. The ethical committee, established by the National Institute of Health and Nutrition in 1990, approved the studies. All studies were carried out in the Humanities Ward of the National Institute of Health and Nutrition.

The duration of the balance study periods ranged from 5 to 12 days, with 2 to 4 days of adaptation period. Body temperature and weight in the morning were measured throughout the experiment in all subjects. However, menstrual cycle for female subjects was not taken into consideration in these studies.

In each study, the same quantity of the diet, which varied in each experiment, was supplied to each of the subjects during the balance period without consideration of body weight.

However, small changes to the diet were carried out during the adaptation period so as to ensure consumption of all food supplied.

The subjects ingested a coloring marker for their feces (Carmine, 0.3 g; Merk KgaA, Germany) just before breakfast in the morning at the beginning and the end of the balance period. In one subject (no. 8, Table 8.1), magnesium oxide was added to the diet.[5] In six subjects (nos. 4–7, 9, and 10; Table 8.1;

TABLE 8.1. Subjects and dietary intake of energy and minerals.

Exp. No.	sex	Subjects n	Duration day	Energy kcal/d	Protein g/d	Fat % of energy	Intake of minerals			
							Na (g/d)	K (g/d)	Ca (mg/d)	Mg (mg/d)
1	m	5	10	2150	71	24	3.20	1.96	676	154
1	m	5	10	2150	71	24	3.20	1.96	676	334*
2	f	11	8	1900	66	35	3.40	1.86	347	186$
3	f	12	8	1850	64	33	3.45	1.83	294	188$
4	f	12	10	1650	65	35	3.27	2.06	495	194
5	f	7	8	1800	76	26	3.06	2.20	653	216#
5	f	2	8	1800	76	26	3.06	2.20	653	216
6	f	8	8	1700	69	25	3.08	2.20	671	243#
7	f	8	12	1950	87	27	4.06	2.68	629	261#
8	f	12	8	1550	75	38	3.90	2.55	672	261#
9	f	8	12	1750	78	25	3.69	2.47	719	279#
10	f	6	10	1950	89	25	2.21	2.71	802	283#
11	m	13	5	3250	136	28	6.87	3.61	1131	379
Total		109								

Energy, protein and fat are calculated value, while minerals are measured ones.
*Mg (180 mg/d) was added to the diet as magnesium oxide (MgO).
$Low calcium study.
#Mineral lost during exercise was estimated (n = 49).

$n = 49$), sweat from the arm during exercise on a bicycle ergometer (intensity, 1–1.5 kp; velocity, 50–60 rpm; duration, 30–60 min/trial; frequency, once or twice a day; room temperature, 22–29º; relative humidity, 40%–65%) was collected to estimate element loss through sweat (Table 8.1).

The foodstuffs used in each study were selected from those commercially available. Some foodstuffs were avoided because of a heterogeneous content of nutrients revealed through chemical measurements taken before the studies. Both processed and nonperishable foodstuffs were purchased at the same time from the same lot before the experiments so as to ensure the same content of nutrients. Fresh foodstuffs were obtained from the same district by way of the same market.

Dietary menus were designed by a registered dietician so as to meet dietary allowances in Japan,[6] except for the low-calcium studies (subject nos. 1 and 2, Table 8.1) for which food composition tables were used.[7]

All foodstuffs were washed with ion-free water (passed through an ion-exchange resin) if necessary, weighed, cooked separately, and distributed uniformly to dishes for the subjects and diet sample(s).

The subjects were required to consume all of the diet. They were allowed no other food, but could drink as much ion-free water as they wanted. The weight of the water consumed was measured and recorded.

Statistics were obtained by StatView-J5.0.

Some indicators in this chapter are defined as follows:

Apparent Absorption = (Intake) − (Fecal Output) mg/kg body weight (BW)/day
Apparent Absorption (%) = ([Intake] − [Fecal Output])/(Intake) × 100%
Balance = (Intake) − ([Fecal Output] + [Urine Output] + [Sweat loss]*) mg/kg BW/day

Results

The dietary intake of Ca, Mg, and P ranged from 294 (1),131, 154 to 379, and 807(2), 198 mg/day, respectively (Table 8.1). In terms of body weight, the dietary intakes of Ca, Mg, and P were 4.83 to 23.58, 2.44 to 7.83, and 13.46 to 45.69 mg/kg BW/day, respectively. The relationships between dietary in-takes, apparent absorption, urine output, and balances for Mg are shown in Figure 8.1.

Dietary intake (Intake) of Mg was positively correlated with apparent absorption (AA; $r^2 = 0.451$), which was also correlated with both urine output (Urine; $r^2 = 0.486$) and balance (Balance; $r^2 = 0.349$). However, the Intake of Mg was not correlated with Balance. Using the above regression equations (Intake vs. AA and AA vs. Balance), the mean value of Mg Intake when balance is equal to 0 was indirectly calculated to be 4.395 mg/kg BW/day.[1]

However, intake of Mg was positively correlated, not only with balance of Ca, but also with balance of P. The mean value and upper limit of the 95% confidence interval (95% CI) for the regression equation between Mg intake and balances of Ca and P (when each balance was equal to zero) were 4.584 and 4.802 ($p = 0.0383$, against Ca balance), 4.554, and 4.785 ($p = 0.0024$, against P balance) mg/kg BW/day, respectively (Figure 8.2).[8]

8. Magnesium Requirement and Affecting Factors

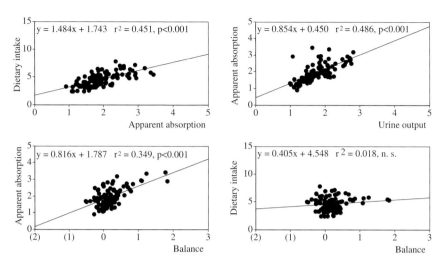

FIGURE 8.1. Dietary intake, apparent absorption, urine output, and balance of Mg (mg/kg BW/day; $n = 109$). Dietary intake of Mg was not correlated with Mg balance, although Mg absorption was positively correlated with both balance and dietary intake of Mg. Using relations both between the intake versus apparent absorption, and apparent absorption versus balance, the mean value of dietary intake of Mg at balance = 0 was calculated to be 4.395 mg/kg BW/day (= 1.484 × 1.787 + 1.743).

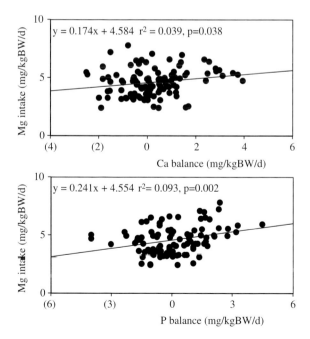

FIGURE 8.2. Correlations between Mg intake and Ca and P balances. Estimated average requirement (EAR) of Mg can be obtained the above equations.

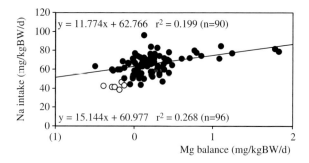

Figure 8.3. Relations between Na intake and Mg balance in Japanese young adults ($n = 96$ or 90). Data of the highest Na intake study (no. 11, Table 8.1) were omitted in this figure. Data of the lowest Na intake study (no. 10, Table 8.1) are shown as open circles. Regression lines were obtained with and without the lowest Na intake study ($n = 96$ and 90, both $p < 0.0001$).

On the other hand, Na intake influenced Mg balance. In the data of the combined studies ($n = 109$), the balance of Mg was not correlated positively with Na intake, whereas in the data that excluded the highest Na study (no. 11, Table 8.1; $n = 96$)[9] or the highest and lowest Na studies (no. 10, Table 8.1; $n = 90$),[10] the balance of Mg was correlated positively with Na intake ($p < 0.0001$)[11] as well as with Mg intake ($p = 0.003$; Figure 8.3).[12] The mean value and upper limit of the 95% CI for the regression equation between Mg intake and Mg balance when the balance is equal to 0 were 4.078 and 4.287 mg/kg BW/day,

Table 8.2. The mean value and upper limit of the 95% confidence interval for the regression equation between Mg intake and balance of minerals (Na, Ca, Mg and P).

($n = 109$)

Against	Intake range (mg/kgBW/d)	r^2	Mean	UL95%CI*	p
Na balance	38.56–142.23	0.129	4.143	4.458	0.0001
Ca balance	4.83–23.58	0.039	4.584	4.802	0.0383
Mg balance	2.44–7.83	0.018			NS
P balance	13.46–45.69	0.093	4.554	4.785	0.0024

*The upper limit of the 95% confidence interval for the regression equation.

($n = 90$)

Against	Intake range (mg/kgBW/d)	r^2	Mean	UL95%CI*	p
Na balance	43.71–96.40	0.001			NS
Ca balance	4.83–23.58	0.213	4.165	4.342	<.0001
Mg balance	2.44–6.42	0.141	4.078	4.287	0.003
P balance	13.46–45.69	0.027			NS

*The upper limit of the 95% confidence interval for the regression equation.

respectively ($n = 90$).[12] All EARs and the upper limit of the 95% CI for the regression equation between Mg intake and balances of Na, Ca, Mg, and P when the balance is equal to 0 are shown in Table 8.2.

Factors Affecting Urinary Excretion and/or Intestinal Absorption of Magnesium

It was reported that various nutrients and hormones influenced urinary divalent cation excretion.[13] Nishimuta and colleagues also demonstrated Mg uresis by risk factors for chronic degenerative diseases (Table 8.3).[14,15] In these experiments, Ca uresis is always associated with Mg uresis except for one study,[16] in which the subjects were poor in Mg.[17] These factors undoubtedly affect the balances of Ca and Mg.

Figure 8.4 shows circadian variations of urinary excretion for Ca and Mg under the normal diet (control) and at the butter (60 g) and egg (300 g) loading test in the morning. This suggested energy increases in urinary excretion of Ca and Mg. This concept is new and far different from that of Lindeman,[13] who thought that fat does not alter urinary calcium and magnesium excretions.

It was also suggested that urinary Ca/Mg ratio is stable and usually less than 1 (molar ratio).

TABLE 8.3. Effects of various nutrients, hormons, other agents and events on urinary divalent cation excretion.

A. Items that increase urinary calcium and magnesium
   1. Ethanol
   2. Glucose IV and oral
   3. Galactose
   4. Fluctose
   5. Protein (casein)
   6. Insulin
   7. Vasopressin
   8. Sodium lactate and pyruvate (vs. sodium bicarbonate)
   9. Tolbutamide
   10. Acid (NH4Cl) load
   11. Heavy exercise
   12. Mental stress
   13. Cold exposure
   14. Over eating
   15. Mild exercise (increase absorption)
   16. Sodium (increase absorption)

B. Items that decrease urinary calcium and magnesium
   1. Mild exercise
   2. Fasting
   3. Low energy diet

C. Items that do not alter urinary calcium and magnesium
   1. Xylose
   2. Fat (Lipomul)(should be deleted in this table)*
   3. Aldosteron
   4. Cortisone (with phosphate infusion)

*Because this is contradict to overeating (A-14).

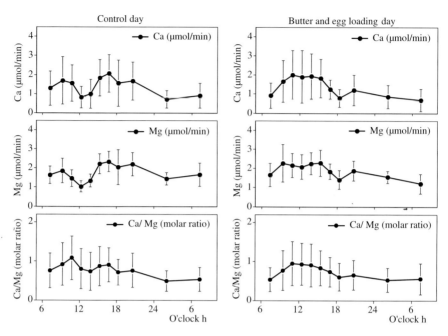

FIGURE 8.4. Circadian variation in urinary calcium (Ca) and magnesium (Mg) at control day (left) and butter-and-egg–loaded day. Eleven female students consumed an experimental diet (breakfast, 380 kcal; lunch, 730 kcal; supper, 390 kccal) at the control day (left). Next morning they consumed butter (60 g) and 5 boiled eggs (ca. 300 g) with a slice of bread (130 kcal), skip lunch, and eat same menu of supper as the control day (390 kcal) (right). Urinary Ca and Mg increased after diets, indicated by arrows.

Mechanism(s) to Maintain Plasma Levels of Calcium and Magnesium

To analyze the above-mentioned phenomena, characteristic of Mg and Ca as elements should be considered (Table 8.4).

TABLE 8.4. Speciation of the essential elements (20).

		Major elements (4) Minerals		H, C, N, O		
Speciation of essential minerals (16)				All elements except major elements		
			Intake/d	Intra-cellular minerals	Extra-cellular minerals	Others
Major minerals (7) Trace elements (9)			more than 100 mg less than 100 mg	K, Mg, P	Na, Cl, Ca	S
		Trace elements I (4) Trace elements II (5)	more than 1 mg less than 1 mg	Fe, Zn		Cu, Mn Co, Cr, I, Mo, Se
Bone minerals (5)				Mg, P, Zn	Na, Ca	

Concept (Intracellular, Extracellular, and Bone Minerals)

Minerals can be divided into two categories of intracellular and extracelluar minerals based on physiological sites of accumulation. Essential minerals rich in intracellular space compared with in extracellular one, such as potassium (K), Mg, P, iron (Fe), and zinc (Zn), are proposed as *intracellular minerals* (ICMs). Those rich in extracellular space, such as sodium chloride (NaCl) and Ca, are proposed *extracellular minerals* (ECMs).[3] Essential minerals, which compose the bone, and whose physiological pool is including the bone, are proposed as *bone minerals* (BMs). Bone minerals consist of both intracellular (Mg, P, and Zn) and extracellular (Na and Ca) minerals.[5] In these definitions, Mg is belongs to both intracellular and bone minerals, and Ca belongs to extracellular and bone minerals.

Homeostasis of Plasma Calcium and Magnesium after Isomolar Uresis of Calcium and Magnesium

Plasma contents in Mg and Ca are controlled within narrow ranges and have no circadian variations. Urine content in Mg and Ca are usually isomolar or high in Mg compared with that of Ca. Urinary excretion of both Mg and Ca have obvious circadian rhythm, higher in daytime and lower at night (Figure 8.4). Mechanism(s) to keep plasma Mg and Ca levels constant to compensate their urine loss may be important in considering both the metabolism of Mg and Ca and the etiology of chronic degenerative diseases. Under the condition of no food supply, Ca must be released from the bone, the sole Ca pool in the body. However, it is difficult to identify which organ, the bone or the cells, release Mg into the bloodstream to keep plasma Mg constant, because these two organs are both pools of Mg in the body. If all Mg was released from the bone, where there is major pool of Mg, by osteophagocytosis, more than 10 times the amount of Ca must be released at the same time from the bone, because the contents of Ca in the bone is far higher than that of Mg. So, Mg in the cell is concluded to be the major source of the compensation in case of enough Mg reservation. However, Mg in the bone may be released into bloodstream when Mg supply in the cell is short or when osteophagocytosis occurs by the unknown mechanism. In that case, excess Ca released from the bone enters into bloodstream, and then enters into the muscle, causing heart muscle infarction, skeletal muscle contracture (shoulder stiffness), and tonic contraction of blood vessels (hypertension). These are clearly illustrated the sign and symptoms of chronic degenerative diseases (Figure 8.5).

Excess Ca also forms stones outside of the bone as well as eminent Ca uresis without accompanied Mg uresis. In this case, molar ratio of Ca/Mg exceeds more than 1.

Molar ratio of Ca/Mg is a possible indicator for the Mg nutrition. If this value is more than 1, Mg in the bone is supposed to be resorbed with large amount of Ca.

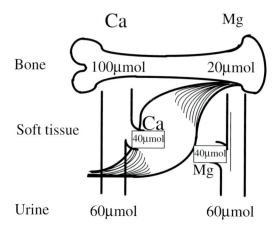

Figure 8.5. Trans-tissue transport of calcium (Ca) and magnesium (Mg), after isomolar uresis of Ca and Mg, induced by excess physical or mental load (stress). Here, isomolar (60 μmol) of Ca and Mg are supposed to be excreted. Those in the bone cannot explain all loss of the minerals, because Ca content is five times or much higher than that of Mg in the bone. Consequently, it concluded that some Mg is derived from the soft tissue. Actually, more Ca than that excreted are absorbed from the bone, and the rest of Ca enters into the soft tissue where Mg was deprived.

Lifestyle against Magnesium Deficit

Because most of foodstuffs come from the cells of animals and vegetables, dietary intake of Mg is also adequate unless there is refined food (refined cereals, sugar, alcohol) and high fat consumption, when dietary intake of energy is adequate. However, overeating (excess energy intake) is a risk factor for Mg deficit.

Moderate physical exercise as well as adequate rest and sleep under comfortable circumstances are important for the cells to accumulate intracellular minerals, including Mg.

Further, freedom from any kind of stress is beneficial for maintaining Mg status.

All studies reported in this article generally believe that a good lifestyle is important for good health. However, we reported them based on the evidence obtained from our experiments.

References

1. Nishimuta M, Kodama N, Morikuni E, et al. Balances of calcium, magnesium and phosphorus in Japanese young adults. *J Nutr Sci Vitaminol (Tokyo)* 2004;50:19–25.
2. Kodama N, Morikuni E, Matsuzaki N, et al. Sodium and potassium balances in Japanese adults. *J Nutr Sci Vitaminol (Tokyo)* 2005;51:161–168.

3. Nishimuta M. The concept (intra and extra cellular minerals). In: Collery P, Poirier, Manfait M, et al., eds. *Metal Ions in Biology and Medicine*. London: John Libbey; 1990:69–74.
4. Nishimuta M. The concept of intracellular-, extracellular- and bone-minerals. *BioFactors* 2000;12:35–38.
5. Takeyama H, Kodama N, Fuchi T, et al. Magnesium, calcium and phosphorus balances in young males at low dietary magnesium levels with or without magnesium supplementation. In: Smetana R, ed. *Advances in Magnesium Research*. London: John Libbey; 1997:355–363.
6. Ministry of Health and Welfare. *Dietary Allowances for the Japanese* [in Japanese]. 6th rev. ed. Tokyo: Daichi Shuppan; 1999.
7. Resources Council, Science and Technology Agency, Japan. *Standard Tables of Food Composition in Japan* [in Japanese]. 4th rev. ed. Tokyo: Ookurashou Insatukyoku; 1982.
8. Nishimuta M, Kodama N, Morikuni E, et al. Balance of magnesium positively correlates with that of calcium—Estimated Average Requirement of Mg-. *J Am Coll Nutr* 2004;23:768S–770S.
9. Nishimuta M, Kodama N, Hitachi Y, et al. A mineral balance study in male long distance runners [in Japanese]. *J Jap Soc Mg Res (JJSMgR)* 1991;10:243–253.
10. Kodama N, Nishimuta M, Suzuki K. Negative balance of calcium and magnesium under a relatively low sodium intake in human. *J Nutr Sci Vitaminol (Tokyo)* 2003;49:201–209.
11. Nishimuta M, Kodama N, Morikuni E, et al. Positive correlation between dietary intake of sodium and balances of calcium and magnesium in young Japanese adults—low sodium intake is a risk factor for loss of calcium and magnesium. J Nutr Sci Vitaminol (Tokyo) 2005;51:265–270.
12. Nishimuta M, Kodama N, Morikuni E, et al. Estimated Average Requirements of calcium and magnesium within a moderate and limited range of sodium intake in Humans. J Nutr Sci Vitaminol (Tokyo). In press.
13. Lindeman RD. Influence of various nutrients and hormones on urinary divalent cation excretion. Ann N Y Acad Sci 1969;162:802–809.
14. Nishimuta M, Kodama N, Ono K. Magnesium uresis by risk factors for chronic degenerative diseases. In: Itokawa Y, Durlach J, eds. Magnesium in Health and Disease. London: John Libbey; 1989:279–284.
15. Nishimuta M, Kodama N, Takeyama H, et al. Magnesium metabolism and physical exercise in humans. In: Theophanides T, Anastassopoulou J, eds. Magnesium: Current Status and New Developments. Dordrecht: Kluwer; 1997:109–113.
16. Nishimuta M, Tsuji E, Kodama N, et al. Magnesiuresis after butter and egg rich diet in young Japanese female [in Japanese]. J Jap Soc Mg Res (JJSMgR) 1986;5:53–60.
17. Nishimuta M, Kodama N, Morikuni E, et al. Effects of risk factors for chronic degenerative diseases on magnesium metabolism in human. In: Porr PJ, ed. New Data in Magnesium Research. London: John Libbey; 2006:1–3.

9 Experimental Data on Chronic Magnesium Deficiency

Pilar Aranda, Elena Planells, C. Sánchez, Bartolomé Quintero, and Juan Llopis

Many studies have shown that magnesium ions play an important role in the regulation of cellular and physiological metabolism in most systems within the organism. Moreover, epidemiological studies indicate that the amounts of magnesium consumed by considerable percentages of the population are below the recommended intakes, and that hypomagnesemia is present in 10% to 15% of the western population. However, the clinical manifestations of magnesium deficiency are difficult to define because depletion of this cation is associated with considerable abnormalities in the metabolism of many elements and enzymes. If prolonged, insufficient magnesium intake may be responsible for symptoms attributed to other causes, or whose causes are unknown. Many experimental models in laboratory animals have been used to investigate the alterations caused by magnesium deficiency, and these models can be divided into two main types: acute deficiency lasting a few days, and chronic deficiency in which a diet deficient in magnesium is given during a period of several weeks or for up to 1 year. Chronic studies aim to mimic the situation observed in western countries, that is, moderately low intakes of magnesium sustained over long periods. In this review we will summarize the results reported from long-term studies in experimental animals.

Magnesium (Mg) plays a structural and regulatory role in the organism. Among its functions, one of the most important is to act as the ion that activates a large number of enzyme systems, making Mg essential for the metabolism of many nutrients. Thus, Mg serves as an important link between transport systems and metabolism, and its concentration in the cytoplasm is probably regulated in a highly precise manner. Maguire[1] notes that "Ca^{2+} is more of an 'acute' signal, performing a type of all-or-none function. Perhaps Mg^{2+} fills the complementary role of a 'chronic' regulatory agent, setting the 'gain' or sensitivity of the overall response system."

From a physiological standpoint, the functions of Mg are well known and have been amply described in many publications.[2] In another area of research, most epidemiological studies have reported that in western countries, Mg intake is below the recommended allowances in considerable percentages of the population. According to data published to date, the amounts of Mg consumed are below 80% of the recommended intake in approximately 20% of the population.[2-4] In addition, hypomagnesemia (Mg < 0.75 mmol/L) has been found in substantial proportions of the adult population in Europe: 12.6% in Switzerland,[5] 21% in Austria,[6] and 14.5% in Germany.[7]

The clinical manifestations of Mg deficiency are difficult to define because depletion of this cation is associated with considerable abnormalities in the metabolism of many elements and enzymes.[8-11] However, if suboptimal intake is prolonged it can facilitate or cause the appearance of symptoms currently attributed to other causes, or whose causes are unknown. Some examples of this situation are the relationships between low intraerythrocyte levels of Mg and chronic fatigue,[12] migraine,[13] sleep alterations,[14,15] Alzheimer's disease,[16] and depression.[17,18]

Models Used in Chronic Deficiency Studies

A number of animal models have been used to study the alterations caused by nutritional Mg deficiency. Most such studies have centered on acute deficiency in rats or mice fed a diet deficient in Mg. However, studies of chronic deficiency, in which rats or mice (usually) are given an Mg-deficient diet for relatively prolonged periods, are less numerous. The duration of such studies has varied widely from several weeks[19] to 1 year.[20] These experimental models attempt to mimic, as far as possible, the situation observed in western countries: moderately deficient Mg intakes sustained over long periods.

In the experimental model we have used at the University of Granada, adult Wistar rats are given the diet recommended by the American Institute of Nutrition (AIN) without magnesium oxide in the mineral supplement for a period of 70 days. The results are compared with those in a group of pair-fed animals given the same diet adjusted to meet the requirements of this species for Mg intake.[21]

Our model has been used in a series of studies designed mainly to elucidate the digestive and metabolic alterations caused by nutritional Mg deficiency, and to study the effects of recovery from nutritional deficiency.

Effect of Chronic Magnesium Deficiency on Growth

Although an Mg-deficient diet is widely assumed to impair growth rate,[2] our research has led to different conclusions depending on the type of diet tested. When rats consumed the AIN-76 diet lacking in magnesium oxide, we found that rats given the deficient diet gained less weight, a finding attributable to the reduced Mg intake rather than to reduced food intake.[22] This finding was also reported by others.[23] However, when the dietary recommendations for this species were changed[21] and the new diet recommended by the AIN was tested (AIN-93G and AIN-93M), we found that the growth curves for control and Mg-deficient rats did not differ significantly during 10 weeks, regardless of whether the AIN-93G or AIN-93M diet was used (unpublished data). The likely explanation for this finding lies in the composition of the Mg-deficient diet, prepared without magnesium oxide in the mineral supplement. The resulting diet contains Mg from other ingredients, mainly macronutrients. The changes in the proportions of macronutrients and modifications in the mineral and vitamin supplements appear to be responsible for the reported changes in growth rate. However, it should be recalled that the degree of hypomagnesemia after feeding with an Mg-deficient diet for 70 days was similar in rats fed with the AIN-76 diet[22] and the AIN-93 diet.[24] Another finding of note was the clear decrease in mortality among animals fed with the Mg-deficient diet. In rats given the AIN-76 diet, mortality was 27%,[22] whereas in rats given the AIN-93 diet the mortality rate was similar to that for the control group (unpublished data).

Effect of Chronic Magnesium Deficiency on Mineral Absorption

Feeding with a low-Mg diet led to a clear increase in percentage absorption, although during the first few days of the experiment percentage Mg cation absorption decreased because of the decline in food intake. However, this situation was rapidly reversed, the changes in absorption reflecting the adaptation of the digestive system under conditions of Mg depletion.[22]

The Mg-deficient diet also affected the absorption of other nutrients, especially proteins and minerals. In this connection, earlier research has documented the interactions between calcium (Ca) and Mg in the intestine,[25] and decreased dietary Mg intake has been shown to lead to increased Ca absorption[26] despite alterations in 1,25-dihydroxyvitamin D synthesis.[27] Contradictory findings have been published regarding the intestinal interactions between phosphorus and Mg.[28,29] In our experiments with chronic Mg deficiency, we found that although absorption tended to decrease initially during the first few weeks, it later recovered and reached values approaching those in the control group.[30]

Although interactions in the intestine between Mg and zinc (Zn) were hypothesized,[31] the results obtained with our experimental model showed that short-term consumption of an Mg-deficient diet increased Zn absorption, and that the increase was sustained throughout the remainder of the study period.[32] Similar findings have been reported for iron (Fe)[33] and manganese (Mn) absorption.[34] However, no increase in copper (Cu) absorption was seen until the 10th week of Mg deficiency.[35]

We also investigated the effect on selenium (Se),[36] supplied in the diet in the form of tetravalent sodium selenite; we hypothesized that the digestive behavior of this element would differ from that of the divalent cations. Our findings showed that prolonged Mg deficiency decreased the absorption of Se.[36]

In general, deficient levels of Mg in the intestinal lumen, therefore, led to increases in the absorption of the divalent cations we investigated. The effects of Mg deficiency on digestive utilization may also be influenced by modifications in endogenous mineral excretion. It should be borne in mind that the method we used to measure absorption is indirect, and is based on determinations of mineral contents in the feed and in feces. The concentrations of these metals in enterocytes from the duodenum and proximal-most segment of the jejunum might provide a more accurate picture of the phenomena that occur in the digestive tract during Mg deficiency. These methodological considerations notwithstanding, in rats fed the deficient diet for 70 days the increases we found in enterocyte concentrations of all elements except Se[24] confirmed the findings obtained with indirect methods.

Digestive interactions have been reported between Mg–Ca, Fe–Ca, Fe–Mn, Zn–Cu, and other elements. Antagonisms may occur between elements whose electronic structure and states are similar.[37] However, our results show that the Mg-deficient diet changed the absorption of the divalent elements without causing interactions between them. This finding suggests that the enterocyte Mg carrier may be able to transport other cations with similar electronic structures, or that the increase in divalent cation absorption may be related to changes in the behavior of paracellular pathways.[38]

Effect of Chronic Magnesium Deficiency on Nutrient Metabolism

Intake of an Mg-deficient diet leads to mobilization of tissue Mg aimed at maintaining cation homeostasis. Under our experimental conditions, erythrocytes were the first cells to mobilize Mg cations in response to the dietary deficiency, and later studies have confirmed that both ionized and total intraerythrocyte Mg are good markers of Mg status,[39] although other authors have proposed that other markers able to more accurately reflect nutritional status for this cation should be sought.[40] After Mg from erythrocytes is used, Mg is mobilized from other stores in the body. After 35 days on an Mg-deficient diet, the rats showed evidence of mobilization from skeletal muscle, and after 49 days the cation was mobilized from bone and from other tissues.[22]

This situation not only affects the tissue distribution of Mg; in addition these changes in Mg concentrations in different tissues are accompanied by alterations in macronutrient and micronutrient metabolism, as summarized below.

Macronutrients

Proteins

An interesting observation documented in several countries was that the most common types of protein-energy malnutrition involve Mg deficiency.[41]

The addition of Mg to a perfusion solution was found to improve the absorption of amino acids.[42] In our experimental model we found that Mg deficiency reduces the nutritive utilization of protein (absorption and balance), apparently as a result of decreased protein absorption and synthesis.[43] The lower digestive utilization of protein may be related with a decrease in sodium (Na^+)—potassium (K^+) adenosine triphosphatase (ATPase) activity in the enterocyte, together with alterations in various membrane transport systems.[44] In addition, Mg deficiency reduces mesenteric circulation, which may also impede the absorption of peptides and aminoacids.[45]

Lipids

Substantial epidemiological and animal experimental evidence indicates that Mg deficiency is an important factor in the origin of cardiovascular diseases.[46–52] Studies with short-term experimental Mg deficiency in rodents have found hyperlipidemia, mainly due to hypertriglyceridemia, together with increased concentrations of free cholesterol and alterations in the content and distribution of plasma lipoproteins.[48,51,52] Experiments with chronic Mg deficiency confirmed the findings of acute deficiency studies with regard to the appearance of hypertriglyceridemia and increased free cholesterol and phospholipids, and also documented an increase in the levels of total cholesterol.[23,53]

Magnesium deficiency is thought to produce these results by reducing lecithin–cholesterol acyltransferase[48] and lipoproteinlipase[49] activity. This effect, together with a complex pattern of alterations in apolipoprotein gene expression, suggest a defect in catabolism as the major factor underlying alterations in the plasma lipoprotein profile.[52]

Studies of chronic Mg deficiency also confirmed that changes occurred in the fatty acid composition of tissues and erythrocyte membranes. The most notable change in tissue profiles was the increase in docosahexaenoic acid,[54] whereas in the erythrocyte membrane the main findings were a reduction in oleic acid and arachidonic acid content.[53] The decrease in arachidonic acid appears to be a result of impaired phospholipid incorporation into the membrane as a result of Mg deficiency,[55] a situation that allows the formation of greater amounts of eicosanoids derived from arachidonic acid, including thromboxane A_2, and thus in turn favors platelet aggregation and increases the susceptibility to vascular lesions.[47]

Glucose Homeostasis

Many studies have shown that Mg deficiency is linked to insulin resistance and type 2 diabetes mellitus.[56] Experiments with chronic Mg deficiency in rats have confirmed the relationship between Mg and alterations in glucose homeostasis.[57,58]

Micronutrients

Minerals

In rats fed an Mg-deficient diet for 70 days, the increase in mineral absorption described in the previous section is reflected as greater ion retention both in plasma and in various tissues. Plasma levels of all ions studied except Zn were significantly higher after feeding with an Mg-deficient diet.[32–36,59]

To investigate tissue changes in mineral concentrations we analyzed skeletal muscle (longissimus dorsi), heart, kidney, bone (femur), and liver. In rats fed for 10 weeks with an Mg-deficient diet, skeletal muscle tissue showed increases in the concentration of Ca,[59] Cu,[35] and Mn,[34] whereas Zn[32] and Fe[33] showed a tendency to increase and phosphorus (P) concentrations decreased.[59] Heart tissues showed in increase in weight[32] and increases in Ca,[59] Fe,[33] and Mn concentrations,[34] but decreases in Zn[32] and Cu.[35] In a separate set of chronic deficiency studies, Laurant and coworkers[23] also found increases in weight and Ca content in the heart. The larger Ca, Fe, and Mn stores may facilitate the appearance of alterations in contractility and favor the formation of reactive oxygen species, which would account for the cardiac symptoms that appear in connection with Mg deficiency, as described in many published studies.[60–63]

The kidney is one of the organs that is most seriously affected by Mg deficiency. Mineral concentrations begin to increase clearly soon after the animal begins to consume an Mg-deficient diet.[59] The changes are most notable for calcium phosphate, because when Mg concentrations in the filtrate decrease, Ca tends to bind to phosphate and precipitation ensues.[59,64] We also found that Mg deficiency caused a clear increase in phosphate concentration in urine as a result of the decrease in resorption,[30] which in turn facilitates the formation of calcium salts. The formation of these precipitates may favor Fe,[33] Zn,[32] and Cu[35] retention in the kidney.

Bone tissue is one of the largest reservoirs of Mg, and its mobilization from bone because of nutritional deficiency can alter the structure of this tissue. Our research showed that dietary Mg deficiency increased bone Ca concentrations[59] and decreased P concentrations,[59] a finding which in our experimental model was apparently independent of circulating levels of parathyroid hormone (PTH). Although PTH initially decreases, it tends to recover normal values after 5 weeks of feeding with an Mg-deficient diet.[59] The changes thus appear to be the result of modifications in the functioning of osteoblasts, osteocytes, and osteoclasts rather than a result of hypercalcemia.[65]

Chronic Mg deficiency for 16 weeks led in rats to increased calcemia and bone Ca content, and to decreases in PTH and 1,25-dihydroxyvitamin D. Histomorphometric studies revealed alterations in bone formation and loss of bone mass [secondary to the increased release of substance P and tumor necrosis factor α (TNF-α)], and the authors concluded that Mg deficiency constituted a risk for osteoporosis.[66,67] In another study, rats were made Mg deficient for 1 year, after which no changes were found in calcemia or bone Ca and P content, although plasma concentrations of PTH were markedly increased. In these Mg-deficient animals there was a decrease in bone density, an increase in fragility, and alterations in bone architecture, and as in the study cited previously, the authors concluded that prolonged consumption of an Mg-deficient diet increased the risk of osteoporosis.[20] Results obtained with our own experimental model showed that after 10 weeks of feeding with an Mg-deficient diet, femoral concentrations of Zn[32,68] and Mn[34] had increased.

In the liver, Mg deficiency leads to clear increases in Fe[33,68] and Cu[35] content, whereas in the spleen only Fe content was increased.[33] The high Fe and Cu content in the liver may be responsible for the activation of hepatic oxidative stress and for alterations in lipid metabolism.[69,70]

Our experimental model also made it possible to analyze the effect of Mg deficiency on mineral content in brain tissue. Initially we measured the concentration of Mg,[22] Zn,[32] Ca, and P[59] in brain homogenates, and found no differences between pair-fed control rats and animals fed the Mg-deficient diet. In later analyses we separated the brain into striatum, hippocampus, cerebellum, and cortex, and found that feeding with an Mg-deficient diet for 10 weeks led to increases in mitochondrial Zn,[71] Cu,[72] and Se[73] content in the striatum and hippocampus. In both the striatum and hippocampus, we also found a significant increase in glutathione peroxidase activity.[73] However, we found no changes as a result of Mg deficiency in the cortex and cerebellum. These changes in mineral content in certain brain structures may be related to alterations in the nervous system caused by Mg deficiency.[14,15,74–76]

The overall picture to emerge from these findings is one that suggests that consumption of an Mg-deficient diet tends to increase the content of divalent cations in tissues, but not all tissues show the same sensitivity to this dietary deficiency.

Vitamins

Magnesium is known to be necessary for the correct metabolism of vitamins such as thiamin,[77] pyridoxine,[2] and vitamins D[27,78] and E.[69] In long-term experiments with rats, our group has shown that Mg deficiency leads to impaired pyridoxine status, possibly as a consequence of the inhibition of alkaline phosphate activity.[79] Studies with this same experimental model showed that Mg deficiency can also decrease plasma levels of vitamin C (control = 0.41 ± 0.01 mg/dL, $n = 10$; Mg deficient = 0.3 ± 0.03 mg/dL, $n = 10$; $p < 0.05$; unpublished data). However, we found no alterations in riboflavin status in association with Mg deficiency.[79]

Magnesium Deficiency and Carcinogenesis

Several authors have found that Mg deficiency increased free radical production, inflammatory processes, and apoptosis.[62,68,69,80,81] Because of the important role Mg plays in regulating cell cycles, enzymatic activity, and the process of DNA repair, low concentrations of Mg might facilitate the accumulation of mutations. There is clinical and epidemiological evidence that Mg deficiency can favor the presence of cancer.[82,83] Although little information is available regarding the influence of chronic Mg deficiency on cancer, we recently found that the skin tumor-promoting activity of *p*-hydroxybenzenediazonium (PDQ) is accelerated in Mg-deficient rats. These results support the idea that PDQ genotoxicity is related to peroxidative damage resulting from the formation in vivo of primary aryl radicals.[84]

Acknowledgments. We thank K. Shashok for translating parts of the original manuscript into English.

References

1. Maguire ME. Magnesium: a regulated and regulatory cation. In: Sigel H, Sigel A, eds. *Metal Ions in Biological Systems*. New York: Marcel Dekker; 1990:137–153.
2. Durlach J. *Magnesium in Clinical Practice*. Paris: John Libbey; 1988.
3. Durlach J, Bac P, Durlach V, Rayssiguier Y, Bara M, Guiet-Bara A. Magnesium status and ageing: an update. *Magnes Res* 1998;11:25–42.
4. Durlach J. New data on the importance of gestational Mg deficiency. *J Am Coll Nutr* 2004;23:694S–700S.
5. Laserre B, Spoerri M, Theubet MP, Moullet V. Magnesium balance in ambulatory care: the Donmag Study. II. Prevalence studies and clinical correlates. *Magnes Res* 1995;8:45–46.
6. Zirm B, Dreissger G, Scheucher G, et al. Twenty-one per cent of 3700 probands are hypomagnesemic: relationship of their serum magnesium subfractions. *Magnes Res* 1995;8:79.
7. Schimatschek HF, Rempis R. Prevalence of hypomagnesemia in an unselected German population of 16,000 individuals. *Magnes Res* 2001;14:283–290.
8. Shils ME. Magnesium. In: Ziegler EE, Filer LJ, eds. *Present Knowledge in Nutrition*. Washington, DC: ILSI Press; 1996:307.
9. Aranda P, Planell E, LLopis J. Magnesio. *Ars Pharmaceutica* 2000;41:91–100.
10. Johnson S. The multifaceted and widespread pathology of magnesium deficiency. *Med Hypotheses* 2001;56:163–170.
11. Kimura M, Honda K, Takeda A, Imanishi M, Takeda T. Developed determination method of ultra trace elements and ultratrace elements levels in plasma of rat fed low magnesium diet. *J Am Coll Nutr* 2004;23:748S–750S.
12. Keenoy BM, Moorkens G, Vertomen J, Noe M, Nève J, De Leeuw I. Magnesium status and parameters of the oxidant-antioxidant balance in patients with chronic fatigue: effects of supplementation with magnesium. *J Am Coll Nutr* 2000;3:374–382.

13. Thomas J, Millot JM, Sebille S, et al. Free and total magnesium in lymphocytes of migraine patients—effect of magnesium-rich mineral water intake. *Clin Chim Acta* 2000;295:63–75.
14. Depoortere H, Francon D, Llopis J. Effects of a Mg-deficient diet on sleep organization in rats. *Neuropsychobiology* 1991;27:237–245.
15. Durlach J. Circadian rhythms, magnesium status and clinical disorders: possible pathophysiological and therapeutical importance of various types of light therapy and of treatment through light deprivation, melatonin and their mimicking agents. In: Escanero JF, Alda O, Guerra M, Durlach J, eds. *Magnesium Research: Physiology, Pathology and Pharmacology.* Zaragoza, Spain: Prensas Universitarias de Zaragoza; 2003:9–30.
16. Lemke MR. Plasma magnesium decrease and altered calcium/magnesium ratio in severe dementia of the Alzheimer type. *Biol Psychiatr* 1995;37:341–343.
17. Widmer J, Henrotte JG, Raffin Y, Bovier PH, Hilleret H, Gaillard JM. Relationship between erythrocyte magnesium, plasma electrolytes and cortisol, and intensity of symptoms in mayor depressed patients. *J Affect Dis* 1995;34:201–209.
18. Singewald N, Sinner C, Hetzenauer A, Sartori SB, Murck H. Magnesium-deficient diet alters depression and anxiety related behavior in mice influence of desipramine and *Hypericum perforatum* extract. *Neuropharmacology* 2004;47:1189–1197.
19. Pages N, Bac P, Maurois P, Durlach J, Agrapart G. Comparison of a short irradiation (50 sec) by different wavelengths on audiogenic seizures in magnesium-deficient mice: evidence for a preventive neuroprotective effect of yellow. *Magnes Res* 2003;16:29–34.
20. Stendig-Lindberg G, Koeller W, Bauer A, Rob PM. Experimentally induced prolonged Magnesium deficiency causes osteoporosis in the rat. *Eur J Intern Med* 2004;15:97–107.
21. Reeves PG, Nielsen FH, Fahey GC. AIN-93 purified diets for laboratory rodents: final report of the American Institute of Nutrition ad hoc writing Committee on the reformulation of the AIN-76 rodent diet. *J Nutr* 1993;123:1939–1951.
22. Lerma A, Planells E, Aranda P, Llopis J. Evolution of Mg-deficiency in rats. *Ann Nutr Metab* 1993;37:210–217.
23. Laurant P, Dalle M, Berthelot A, Rayssiguier Y. Time-course of the change in blood pressure level in magnesium-deficient Wistar rats. *Br J Nutr* 1999;82:243–251.
24. Planells E, Sánchez-Morito N, Montellano MA, Aranda P, LLopis J. Effect of magnesium deficiency on enterocyte Ca, Fe, Cu, Zn, Mn and Se content. *J Physiol Biochem* 2000;56:217–222.
25. Hardwick LL, Jones N, Brautbar N, Lee DBN. Magnesium absorption: mechanism and the influence of vitamin D, calcium and phosphate. *J Nutr* 1991;121:13–23.
26. Alcock NW, McIntyre I. Interrelation between calcium and magnesium absorption. *Clin Sci* 1962;22:185–193.
27. Traba ML, De la Piedra C, Rapado A. Role of PTH and cAMP in the impaired synthesis of 1,25-dihydroxyvitamin D in magnesium deficient rats. *Med Sci Res* 1989;17:327–328.
28. Clark I, Rivera-Cordero F. Effect of endogenous parathyroid hormone on calcium, magnesium and phosphorus metabolism in rat. II. Alterations in dietary phosphate. *Endocrinology* 1974;95:360–369.

29. Greger JL, Smith SA, Snedecker SM. Effect of dietary calcium and phosphorus levels on the utilization of calcium, phosphorus, magnesium, manganese and selenium by adult males. *Nutr Res* 1981;1:315–325.
30. Planells E, Aranda P, Peran F, Llopis J. Changes in calcium and phosphorus absorption and retention during long-term magnesium deficiency in rats. *Nutr Res* 1993;13:691–699.
31. Yasui M, Ota K, Garruto RM. Aluminium decreases the zinc concentration of soft tissues and bones of rats fed a low calcium-magnesium diet. *Biol Trace Elem Res* 1991;31:293–304.
32. Planells E, Aranda P, Lerma A, Llopis J. Changes in bioavailability and tissue distribution of zinc caused by magnesium deficiency in rats. *Br J Nutr* 1994;72: 315–323.
33. Sanchez-Morito N, Planells E, Aranda P, Llopis J. Influence of magnesium deficiency on the bioavailability and tissue distribution of iron in the rat. *J Nutr Biochem* 2000;11:103–108.
34. Sanchez-Morito N, Planells E, Aranda P, Llopis J. Magnesium-manganese interactions caused by magnesium deficiency in rats. *J Am Coll Nutr* 1999;18:475–480.
35. Jiménez A, Planells E, Aranda P, Sánchez-Viñas M, Llopis J. Changes in bioavailability and tissue distribution of copper caused by magnesium deficiency in rats. *J Agric Food Chem* 1997;45:4023–4027.
36. Jiménez A, Planells E, Aranda P, Sánchez-Viñas M, Llopis J. Changes in bioavailability and tissue distribution of selenium caused by magnesium deficiency in rats. *J Am Coll Nutr* 1997;16:175–180.
37. Rucker RB, Lönnerdal B, Keen CL. Intestinal absorption of nutritionally important trace elements. In: Johnson LR, ed. *Physiology the Gastrointestinal Tract.* 3rd ed. New York: Raven Press; 1994:2183.
38. Hayashi H, Hoshi T. Properties of active magnesium flux across the small intestine of the guinea pig. *Jpn J Physiol* 1992;42:561–575.
39. Malon A, Brockmann C, Fijalkowska-Morawska J, Rob P, Majzurawska M. Ionized magnesium in erythrocytes the best magnesium parameter to observe hypo or hypermagnesemia. *Clin Chim Acta* 2004;349:67–73.
40. Franz KB. A functional biological marker is needed for diagnosing magnesium deficiency. *J Am Coll Nutr* 2004;23:738S–741S.
41. Wapnir RA. Protein Nutrition and Mineral Absorption. Boca Raton, FL: CRC Press; 1990:88–90.
42. Wozniak J, Oledzka R, Groszyk D. Effect of magnesium on the intestinal transport of amino acids in the presence of lead. *Bromato Chem Toksykologiezna* 1989;22:367–373.
43. Rico MC, Lerma A, Planells E, Aranda P, Llopis J. Changes in the nutritive utilization of protein induced by Mg deficiency in rats. *Int J Vit Nutr Res* 1995;65: 122–126.
44. Bara M, Guiet-Bara A, Durlach J. Regulation of sodium and potassium pathways by magnesium in cell membranes. *Magnes Res* 1993;6:167–177.
45. Altura BM, Altura BT, Gebrewold A, Ising H, Günther T. Magnesium deficiency and hypertension: correlation between magnesium-deficient diets and microcirculatory changes in situ. *Science* 1984;223:1315–1317.
46. Rayssiguier Y. Magnesium and lipid interrelationship in the pathogenesis of vascular diseases. *Magnes Bull* 1981;3:165–177.

47. Rayssiguier Y, Gueux E, Cardot P, Thomas G, Robert A, Trugnan. Variations of fatty acid composition in plasma lipids and platelet aggregation in magnesium deficient rats. *Nutr Res* 1986;6:233–240.
48. Gueux E, Rayssiguier Y, Piot MC, Alcindor L. Reduction of plasma LCAT activity by acute magnesium deficiency in the rat. *J Nutr* 1984;114:1479–1483.
49. Rayssiguier Y. Lipoprotein metabolism: Importance of magnesium. *Magnes Bull* 1986;8:186–193.
50. Gueux E, Mazur A, Cardot P, Rayssiguier Y. Magnesium deficiency affects plasma lipoprotein composition in rats. *J Nutr* 1991;121:1222–1227.
51. Altura BT, Bloom S, Barbour RL, Stempak JG, Altura BM. Magnesium dietary intake modulates blood lipids levels and atherogenesis. *Proc Natl Acad Sci U S A* 1990;87:1840–1844.
52. Nassir F, Mazur A, Giannoni F, Gueux E, Davidson NO, Rayssiguier Y. Magnesium deficiency modulates hepatic lipogenesis and apolipoprotein gene expression in the rat. *Biochim Biophys Acta* 1995;1257:125–132.
53. Lerma A, Planells E, Aranda P, Llopis J. Effect of magnesium deficiency on fatty acid composition of the erythrocyte membrane and plasma lipid concentration in rats. *J Nutr Biochem* 1995;6:577–581.
54. Cunnane S, Soma M, Mcadoo KR, Horrobin DF. Magnesium deficiency in the rat increases tissue levels of docosahexaenoic acid. *J Nutr* 1985;115:1498–1503.
55. Weis MT, Saunders C. Magnesium and arachidonic acid metabolism. *Magnes Res* 1993;6:179–190.
56. Suarez A, Pulido N, Casla A, Casanova B, Arrieta FJ, Rovira A. Impaired tyrosine-kinase activity of muscle insulin receptors from hypomagnesemic rats. *Diabetología* 1995;38:1262–1270.
57. Kimura Y, Murase M, Nagata Y. Change in glucose homeostasis in rats by long-term magnesium-deficient diet. *J Nutr Sci Vitaminol* 1996;42:407–422.
58. Venu L, Kishore YD, Raghunath M. Maternal and perinatal magnesium restriction predisposes rat pups to insulin resistance and glucose intolerance. *J Nutr* 2005;135:1353–1358.
59. Planells E, Llopis J, Perán F, Aranda, P. Changes in tissue calcium and phosphorus content and plasma concentration of parathyroid hormone and calcitonin after long-term magnesium deficiency in rats. *J Am Coll Nutr* 1995;14:292–298.
60. Arsenian MA. Magnesium and cardiovascular disease. *Prog Cardiovasc Dis* 1993;35:271–310.
61. Nair RR, Nair P. Alterations of myocardial mechanics in marginal magnesium deficiency. *Magnes Res* 2002;15:287–306.
62. Kramer JH, Mak IT, Phillips TM, Wegliki. Dietary magnesium intake influences circulating pro-inflamatory neuropeptide levels and loss of myocardial tolerance to postischemic stress. *Exp Biol Med* 2003;228:665–673.
63. Touyz RM, Pu Q, He G, et al. Effects of low dietary magnesium intake on development of hypertension in stroke-prone spontaneously hypertensive rats: role of reactive oxygen species. *J Hypertens* 2002;20:2221–2232.
64. Straub B, Müller M, Schrader M, Goessl C, Heicappell R, Miller K. Intestinal and renal handling of oxalate in magnesium-deficent rats. Evaluation of intestinal in vivo ^{14}C-oxalate perfusion. *BJU Int* 2002;90:312–316.
65. Weaver VM, Welsh J. 1,25-dihydroxycholecalciferol supplementation prevents hypocalcemia in magnesium-deficient chicks. *J Nutr* 1993;123:764–771.

66. Rude RK, Kirchen ME, Gruber HE, Meyer MH, Luck JS, Crawford DL. Magnesium deficiency-induced osteoporosis in the rat: uncoupling of bone formation and bone resorption. *Magnes Res* 1999;12:257–267.
67. Rude RK, Gruber HE, Norton HJ, Wei LY, Frausto A, Mills BG. Bone loss induced by dietary magnesium reduction to 10% of the nutrient requirement in rats is associated with increased release of substance P and tumor necrosis factor-α. *J Nutr* 2004;134:79–85.
68. Kimura M, Itokawa Y. Inefficient utilization of iron and minerals in magnesium deficient rats. In: Itokawa Y, Durlach J, eds. *Magnesium in Health and Disease*. London: John Libbey; 1989:95–102.
69. Günther T, Höllriegl V, Massh A, Vormann J, Bubeck J, Classen HG. Effect of Desferrioxamine on lipid peroxidation in tissues of magnesium-deficient and vitamin E-depleted rats. *Magnes Bull* 1994;16:119–125.
70. Chakraborti S, Chakraborti T, Mandal M, Mandal A, das S, Ghosh S. Protective role of magnesium in cardiovascular diseases: a review. *Mol Cell Biochem* 2002;238:163–179.
71. Planells E, Sánchez-Morito N, Moreno MJ, Aranda P, LLopis J. Effect of chronic magnesium deficiency on mitochondrial Zn content in different rat brain structures. In: Centeno JA, Collery P, Vernet G, Finkelman RB, Gibb H, Etienne JC, eds. *Metal Ions in Biology and Medicine*. Vol 6. Paris: John Libbey; 2000:154–157.
72. Planells E, Sánchez-Morito N, Moreno MJ, Aranda P, LLopis J. Effect of chronic magnesium deficiency on mitochondrial Cu content in different rat brain structures. In: Rayssiguier Y, Mazur A, Durlach J, eds. *Advances in Magnesium Research: Nutrition and Health*. Paris: John Libbey; 2001:423–426.
73. Planells E, Aranda P, Llopis J. Influence of chronic magnesium deficiency on mitochondrial Se content and glutathione peroxidase activity in different rat brain structures. *1st International FESTEM Congress on Trace Elements and Minerals in Medicine and Biology. Book of Abstracts*. 2001:135.
74. Durlach J, Bac P, Durlach V, Bara M, Guiet-Bara A. Neurotic, neuromuscular and autonomic nervous form of magnesium imbalance. *Magnes Res* 1997;10:169–195.
75. Gong H, Amemiya T, Takaya K. Retinal changes in magnesium-deficient rats. *Exp Eye Res* 2001;72:23–32.
76. Durlach J, Pagès N, Bac P, Bara M, Guiet-Bara A. Importance of magnesium depletion with hypofunction on the biological clock in the pathophysiology of headaches with photophobia, sudden infant death and some clinical forms of multiple sclerosis. *Magnes Res* 2004;17:314–326.
77. Itokawa Y, Kimura M. Effect of Mg deficiency on thiamine metabolism. *Magnes Bull* 1982;4:5–8.
78. Risco F, Traba ML. Bone specific binding sites for 1,25 (OH)2D3 in magnesium deficiency. *J Physiol Biochem* 2004;60:199–203.
79. Planells E, Lerma A, Sánchez-Morito N, Aranda P, Llopis J. Effect of magnesium deficiency on vitamin B2 and B6 status in the rat. *J Am Coll Nutr* 1997;16:4352–4356.
80. Günther T, Schümann K, Vormann J. Tumor necrosis factor-α, prostanoids and immunoglobins in magnesium deficiency. *Magnes Bull* 1995;17:109–114.
81. Weglicki WB, Dickens BF, Wagner TL, Chmielinska JJ, Phillips TM. Immunoregulation by neuropeptides in magnesium deficiency: ex vivo effect of enhanced substance P production on circulating T lymphocytes from magnesium-deficient rats. *Magnes Res* 1996;9:3–11.

82. Durlach J, Bara M, Guit-Bara A. magnesium and its relationship to oncology. In: Sigel H, Sigel A, eds. *Metal Ions in Biological Systems*. Vol 26. New York: Marcel Dekker; 1990:549–578.
83. Anastassopoulou J, Theophanides T. Magnesium-DNA interactions and the possible relation of magnesium to carcinogenesis. Irradiation and free radicals. *Crit Rev Oncol Hematol* 2002;42:79–91.
84. Quintero B, Planells E, Cabeza MC, et al. Tumor-promoting activity of *p*-hydroxybenzenediazonium is accelerated in Mg-deficient rats. *Chem Biol Interac* 2006;159:186–195.

10
Clinical Forms of Magnesium Depletion by Photosensitization and Treatment with Scototherapy

Jean Durlach

The clinical forms of magnesium depletion due to dysregulation biorhythms may result from the association of a low magnesium intake with a dysregulated biorhythm: either hyperfunction or hypofunction of the biological clock. Their main biological marker is the production of melatonin.[1] The biological dysfunction of the biological clock may be primary (e.g., impaired maturation of photoneuroendocrine system in clinical forms of sudden infant death due to magnesium depletion by hypofunction of the biological clock)[2,3] or secondary to light hypersensitivity. The organism responds to the pathogenic effect of this hypersensitivity with a protective reactive photophobia whose mechanism is still unclear.[1,4-6]

The aim of this study is (1) to stress the importance of the main chronopathological form of magnesium depletion [with a decrease of melotonin (MT) product] secondary to photosensitization: that is, the clinical forms of photosensitive magnesium depletion with photophobia; and (2) to define its treatment, involving diverse forms of scototherapy, appropriate in the control of specifically photosensitive magnesium depletion.

Clinical Forms of Magnesium Depletion by Photosensitization

Biologically and clinically, this chronopathological form of magnesium depletion secondary to photic hypersensitivity associates stigma of magnesium depletion with stigma of hypofunction of the biological clock with decreased production of melatonin (↓MT) and with central and peripheral nervous hyperexcitability (NHE).

The manifestations may be psychic, algic, dyssomniac, or neuromuscular.

Magnesium Depletion[4]

Magnesium deficit involves two types of dysregulation, distinct from each other: magnesium depletion and magnesium deficiency.

Magnesium deficiency is linked to an insufficient intake that may be corrected, over a long period of time, through a physiological nutritional oral magnesium supplementation. Magnesium depletion is due to a dysregulation of the magnesium status that cannot be corrected through nutritional supplementation only, but requests the most specific correction of the dysregulating mechanisms.

There exist as many clinical forms of magnesium depletion as possibilities of dysregulation of the magnesium status. But in both clinical therapeutics and in animal experiments, the dysregulating mechanisms of magnesium depletion associate a reduced magnesium intake with various types of stress.[2] Among these, biological clock dysrhythmias are found.

Clinical Forms of Photosensitive Magnesium Depletion

Clinical forms of magnesium depletion by photosensitization may be algic, neuromuscular, dyssomniac, or psychic, but present the same stigma: biologically, markers of magnesium depletion and of decreased melatonin production, and clinically, photophobia.

Headache Due to Photosensitive Magnesium Depletion[4–11]

The association of the stigma of magnesium depletion with clinical and paraclinical data of photosensitivity (visual stress tests, electroencephalographic, and cerebrovascular photic driving) in primary headache patients demonstrates that the new concept of photosensitive magnesium depletion is fully justified and agrees with the so-called continuum severity theory: the lower limit of the continuum is tension type headache (TTH), which progressively evolves into migraine (M) without aura and finally into migraine with aura. The interictal hallmark of these cephalalgic patients is sensitization (or potentiation) instead of physiological habituation. Comorbidity with the other forms (neuromuscular, dyssomniac, and psychic) of photosensitive magnesium depletion may be observed.

Neuromuscular Forms of Photosensitive Magnesium Depletion[4,5,12–16]

The central and peripheral manifestations of neuromuscular photosensitive magnesium depletion are represented by photosensitive epilepsy which may be either generalized or focal, authentified through electroencephalogram (EEG) with intermittent light stimulation (ILS), with its corresponding form observed among television (TV) viewers and videogame players. Some migraine equivalents are often associated in this context.

The neuromuscular form of photosensitive magnesium depletion may also clinically appear as chronic fatigue syndrome or as fibromyalgia.

Dyssomnia Due to Photosensitive Magnesium Depletion[4,17–22]

Dyssomnia due to photosensitive magnesium depletion is mainly represented by the delayed sleep phase syndrome (DSPS). It is observed, for example, in jet lag, night-work disorders, or insomnia of elderly patients. Some chronopathological forms of sudden infant death syndrome, of multiple sclerosis, and of asthma may be also associated here.[2–4,22]

Psychic Forms Due to Photosensitive Magnesium Depletion

Psychic forms of photosensitive magnesium depletion essentially express themselves in the form of anxiety: either generalized anxiety disorder (GAD) or panic attacks (PA).[4,5]

These four clinical forms of photosensitive magnesium depletion may coexist (Figure 10.1) with the same association of clinical and biological characteristics: mainly, with aspecific clinical symptoms of NHE, with photophobia, and the tinted glass sign, the symptomatology being mainly diurnal and observed in spring and summer, particularly in sunny countries when light hyperstimulation is obviously maximum.[4,5,23,24]

This new concept of photosensitive magnesium depletion may induce therapeutic consequences through the association of an aspecific treatment of magnesium depletion with a specific treatment of photosensitivity through scototherapy.

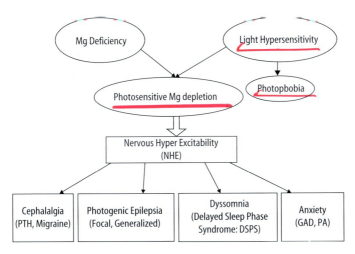

FIGURE 10.1. The four main clinical forms of photosensitive magnesium depletion. The sum of a nutritional magnesium deficiency (deficient magnesium intake) and of a high light hypersensitivity (with reactive photophobia) induces four main clinical forms of photosensitive magnesium depletion: nervous hyperexcitability with stigma of magnesium deficit and of hypofunction of the biological clock. A large number of primary headache patients [currently classified as tension type headache (TTH) or migraine (M)] exhibit hypersensitivity to light associated with magnesium deficit. TTH and M agree with the so-called continuum severity theory. The interictal hallmark is sensitization (or potentiation) instead of habituation. Photosensitive magnesium depletion may also induce photogenic epilepsia (focal or generalized), delayed sleep phase syndrome (DSPS), and anxiety (generalized or attacks: GAD to PA). Comorbidity of these main clinical forms, cephalalgic, convulsive, dysomniac, and psychic, of photosensitive magnesium depletion is frequent.

Treatment of Magnesium Depletion by Photosensitization

Symptomatic medications classically used for the treatment the various clinical forms of photosensitive magnesium depletion—antalgic drugs, β blockers, ergot derivatives, triptans, anxiolytic drugs, anti-convulsants, hypnotics, sedatives—although useful will not be considered in this study.

The following therapeutic approach only concerns the aspecific treatment of magnesium depletion and the specific treatment of photosensitivity through scototherapy (darkness therapy).

Aspecific Treatment of Magnesium Depletion[5,22–28]

Preventive Treatment

Preventive treatment is more efficient than curative treatment. Because magnesium depletion is usually due to both primary chronic magnesium deficiency and stress, a prophylactic treatment must rely on a balanced magnesium status and the most specific possible anti-stress treatment.

To insure a balanced magnesium status, in case of chronic primary magnesium deficiency, atoxic nutritional magnesium supplementation will be administered via the diet or with supplemental magnesium salts. The dietetic supplement should have a high magnesium density with the greatest availability. Magnesium in drinking water is of particular interest as it associates a high bioavailability with the lowest nutritional density. The magnesium salt used should be hydrosoluble and the properties of the anion should be considered.

A specific anti-photosensitive drug that could be used as a preventive treatment of photosensitive magnesium depletion cannot be currently found.

Curative Treatment

The aspecific treatments of magnesium depletion which are available, such as pharmacological doses of parenteral magnesium, may be used. Several studies have shown that 1 g of intravenous magnesium sulfate has been considered as efficient, safe, and well tolerated in migraine headache, but its efficiency as an antalgic drug and as an anti-migraine treatment remains controversial.

Pharmacological doses of magnesium salts may induce a toxicity that varies according to the nature of anions. For example, the effects of $MgCl_2$ and $MgSO_4$ on the ionic transfer components through isolated amniotic membrane have been studied and revealed major differences. $MgCl_2$ interacts with all the exchangers, whereas the effects of $MgSO_4$ are limited to paracellular components. $MgCl_2$ mainly increases the ionic flux ratio of this asymmetric human membrane while $MgSO_4$ decreases it, with many deleterious fetal consequences.

It seems, therefore, necessary to determine the therapeutic index (LD 50/ED 50) of the various available magnesium salts before pharmacological use. The selection of one magnesium salt among others should take into account reliable pharmacological and toxicological data and the comparative therapeutic index of the various salts: the larger its value, the greater the safety margin. This logical prerequisite is lacking in most protocols: MgSO₄ is just routinely used without justifcation.

Specific Treatment of Photosensitivity: Scototherapy[5,22,29-34]

The reactive response to photosensitivity induces hypofunction of the biological clock (hBC). But, in case of photosensitive headache, hBC is aggravated as the repetitive stimulating effects of light induce potentiation (sensitization)—sometimes with generalization—instead of habituation.[3]

Treatment of photosensitivity—so-called scototherapy (darkness therapy)—mirrors phototherapy, the treatment of hyperfunction of the biological clock.

Scototherapy aims either at stimulating the biological clock, or at palliating its hypofunction.

Stimulation of the biological clock may be obtained through physiological, psychotherapic, physiotherapic, or pharmacological strategies.

Palliative treatments of hBC are currently dependent on melatonin, its analogs, or its precursors.

Stimulating Scototherapies

Physiological Scototherapy (Darkness Therapy)

The best physiological stimulation of the biological clock is induced by light deprivation. It may be obtained by placing the patient in a closed room, in a totally dark environment, with an eye mask on. This genuine scototherapy may be used in acute indications, but should be of short duration. It is not compatible with any activity and is frequently associated with induction of bed rest, inactivity, and sleep. Prolonged exposure to darkness may induce darkness adaptation with a gain in photosensitivity, the reverse of the scototherapeutic effect.

Relative scototherapy may be obtained by wearing dark goggles or dark sunglasses but the number of lux passing through is not negligible. This relative darkness therapy may be used as an accessory treatment in the restoration of a light/dark schedule: a transition before a totally dark environment. A successful double-blind study demonstrated a significant difference between placebo and salicoside (salicin) in association with a photoprotective mask in treating the two main clinical forms of photosensitive headaches: M and TTH. The good results of this controlled clinical trial have not been confirmed.

Chromatotherapy uses a short exposure (4 min) to a precise yellow wavelength, once a week, for the treatment of hBC. This method, even though successfully used in practice, has not been validated yet.

Some studies have reported benefit from using colored filters in headache patients, in childhood migraine particularly. A double-masked randomized controlled study with crossover design compared the effectiveness of precision ophthalmic tints (optimal tint) or glasses that provided a slightly different color (control tint). Using individually prescribed colored filters selected by each migraineur seems more helpful than the conventional practice of using a neutral grey or sometimes brown tint, but the effect is statistically marginal; it is suggestive rather than conclusive.

Psychotherapic Scototherapies

Cognitive behavioral strategies have been efficient for the treatment of photosensitivity. The treatment was to gradually increase exposure to computer monitor and TV screen photostimulation. This desensitization procedure resulted in a complete removal of the patient's phobic anxiety of photostimulation and of avoidant behavior. This behavioral therapy has been used in photosensitive epilepsy. It is akin to deconditioning techniques used as a nonpharmacological approach to prevent photosensitive headache. For example, variable frequency photostimulation (VFP) googles, that is, a portable stroboscope using red light-emitting diodes (LED) to illuminate the right eye and the left eye alternately were used with limited efficacy. Various biofeedback treatments for migraine were disappointing because a reduction in the number of migraine headaches was observed, but with no change in the intensity, duration, or disability of the headaches.

The concept of headache due to photosensitive magnesium depletion places this clinical form of headache among the indications of psychotherapic scototherapies.

Physiotherapic Scototherapies

Magnetic fields may be used to stimulate the biological clock in a variety of treatment methods using very weak (picotesla), extremely low frequency (2–7 Hz) electromagnetic fields. Transcranial treatment with alternative current pulsed electromagnetic fields of picotesla flux density may stimulate various brain areas (the hypothalamus particularly) and the pineal gland (which functions as a magneto receptor). Several studies concern its use for treatment of headaches. A double-blind placebo controlled trial has shown that this physiotherapy can alleviate symptoms of M but not of TTH. Electromagnetic fields for at least 3 weeks may be considered as an effective, short-term intervention for migraine, although the clinical effects were small.

Pharmacological Scototherapies

Three pharmacological agents may stimulate the biological clock: magnesium, L-tryptophan, and taurine.

To stimulate the BC, it seems well advised to facilitate the neural function of suprachiasmatic nuclei (↑SCN) and of the hormonal pineal production (↑MT). The deleterious effects of light and those of magnesium deficiency are often found together and may be partly palliated by nutritional magnesium supply (↑Mg), providing the best possible link between photoperiod and magnesium status. Palliative nutritional magnesium supplementation is efficient and atoxic when magnesium deficiency is present, but when there is a balanced magnesium status, it is illogical and inefficient. Pharmacological use of magnesium (high oral doses or parenteral administration) is uncertain and susceptible of inducing toxicity. Many data remain imprecise, such as nature and doses of the magnesium salts, oral or parenteral routes, association with magnesium fixing agents.

L-tryptophan (or 50H-tryptophan) may stimulate the tryptophan pathway. But it is unspecific as it does not only concern melatonin production, but also serotonin synthesis. L-tryptophan supplementation may induce toxicity, eosinophilia–myalgia syndrome particularly,[35] and the more so as the occurrence may be facilitated by magnesium deficiency.[36]

Taurine is a sulphonated aminoacid that is present in the whole body in high concentrations, particularly in the brain. It has multiple functions in cell homeostasis, such as membrane stabilization, buffering, osmoregulation, and antioxidant activities together with effects on neurotransmitter release and on receptor modulation.

Taurine may act as a protective inhibitory neuromodulator that participates in the functional quality of the neural apparatus and in melatonin production and action. It plays a role in the maintenance of homeostasis in the central nervous system, particularly during central nervous hyperexcitability. Taurine, a volume-regulating amino acid, is released upon excitotoxicity-induced cell swelling. It has an established function as an osmolyte in the central nervous system.

In the course of magnesium deficit, the organism appears to stimulate taurine mobilization into playing the role of a magnesium vicarious agent. This compensatory action is usually rather limited. However, it allows us to observe the latent form of the least severe form of magnesium deficiency.[31–34] During migraine, the typical form of headache due to photosensitive magnesium depletion, taurine mobilization may be considered as a defensive reaction but it is less effective than in case of magnesium deficiency.[5,34]

Magnesium, tryptophan, and taurine may be used to stimulate the biological clock, but their efficiency seems limited. Palliative treatments of hypofunction of the biological clock may be necessary.

Substitutive Scototherapy (Darkness Mimicking Agents)

Mechanisms of the Action of Darkness

The mechanisms of action of darkness appear to be the reverse of those described with bright light, where direct cellular effects (membraneous and redox) and neural-mediated effects intervene.

Increased production of melatonin (↑MT) constitutes the best marker of darkness, but it is only an accessory mechanism in its action.

The main central neural mechanisms of darkness therapy associate decreased serotoninergy (↓ 5HT), which could account for the anti-migraine effect, with stimulation of inhibitory neuromodulators (gamma-aminobutyric acid, taurine, kappa opioids: ↑GABA, ↑TA, ↑kO) and of anti-inflammatory and anti-oxidative processes. These effects may induce neural hypoexcitability, that is, sedative and anti-convulsant effects.

Humoral transduction may reinforce these last effects by decreasing neuroactive gases (↓CO, ↓NO) through binding of CO with hemoglobine (Hb) and by increasing melatonin, bilirubin, and biliverdin, three antioxidants which are able to quench NO.

Apart from the exception of decreased serotonergy, these systemic effects of darkness are similar to those of magnesium.

Substitutive scototherapy should palliate all the mechanisms of action of darkness, but the only available scotherapic mimicking agents are at present melatonin (its analogs and its precursors, L-tryptophan, 5 hydroxytryptophan).

Melatonin: An Accessory Scototherapic Agent

Melatonin is the prototype of scototherapic agents. But, although its production is the best marker of photoperiod, melatonin appears to be only an accessory factor among the mechanisms of photoperiod actions. Most of the other mechanisms of the effects of darkness have been overlooked, which may account for the controversy around the therapeutic efficiency of MT. Its posology varies from physiological doses (around 3 mg per day) to pharmacological doses, usually 3 mg/per dose and even up to 300 mg as a contraceptive, which testifies to the weak toxicity of the hormone. In case of chronopathology with decreased MT production, MT constitutes a partial substitutive treatment of its deficiency.[5-22]

At the present time, substitutive scotherapy using melatonin or its analogs and its precursors as a partial substitutive treatment of hBC is possible, though melatonin is only an accessory mechanism of the action of darkness.

Conclusion

The treatment of photosensitive magnesium depletion must, at present, associate the classical symptomatic treatments; a balanced magnesium status (through nutritional or carefully checked pharmacological magnesium supplementation); control of photosensitivity through the various types of scototherapy: physiological, psychotherapic, physiotherapic, or pharmacological, stimulating scotherapy or palliative scotherapy, today through MT, the only available partial scototherapic-mimicking agents.

Further research should take into consideration other agents with more efficient darkness-mimicking properties. A new model of photosensitive magnesium depletion with potentiation is currently described.[37] This test should be a useful tool for discriminating the most efficient scototherapic agent in photosensitive magnesium-depleted mice.

References

1. Durlach J, Pagès N, Bac P, Bara M, Guiet-Bara A, Agrapart C. Chronopathological forms of magnesium depletion with hypofunction or with hyperfunction of the biological clock. *Magnes Res* 2002;15:263–268.
2. Durlach J, Pagès N, Bac P, Bara M, Guiet-Bara A. Magnesium deficit and sudden infant death syndrome (SIDS): SIDS due to magnesium deficiency and SIDS due to various forms of magnesium depletion: possible importance of the chronopathological forms. *Magnes Res* 2002;15:269–278.
3. Durlach J, Pagès N, Bac P, Bara M, Guiet-Bara A. New data on the importance of gestational magnesium deficiency. *Magnes Res* 2004;17:116–125.
4. Durlach J, Pagès N, Bac P, Bara M, Guiet-Bara A. Importance of magnesium depletion with hypofunction of the biological clock in the pathophysiology of headaches with photophobia, sudden infant death and some clinical forms of multiple sclerosis. *Magnes Res* 2004;17:314–326.
5. Durlach J, Pagès N, Bac P, Bara M, Guiet-Bara A. Headache due to photosensitive magnesium depletion. *Magnes Res* 2005;18:109–122.
6. Drummond PD. A quantitative assessment of photophobia in migraine and tension headache. *Headache* 1986;26:465–469.
7. Thompson RF, Spencer WA. Habituation: a model phenomenon for the study of neuronal substrates behaviour. *Psychol Rev* 1966;73:16–43.
8. Monnier M, Boehmer A, Scholer A. Early habituation, dishabituation and generalization induced in the visual centres by colour stimuli. *Vision Res* 1976;16:1497–1504.
9. Marcus DA. Migraine and tension type headaches: the questionable validity of the current classification system. *Clin J Pain* 1992;8:28–36.
10. Schoenen J, Wang W, Albert A, Delwaide PJ. Potentiation instead of habituation characterizes visual evoked potentials in migraine patients between attacks. *Eur J Neurol* 1995;2:115–122.
11. Vingen JV, Sand T, Stovner LJ. Sensitivity to various stimuli in primary headaches: a questionnaire study. *Headache* 1999;3:552–558.
12. Salas-Puig J, Parra J, Fernandez-Torre JL. Photogenic epilepsy. *Rev Neurol* 2000;30:S81–S84.
13. Harding GFA. TV can be bad for your health. *Nat Med* 1998;4:265–267.
14. Durlach J. Chronic fatigue syndrome and chronic primary magnesium deficiency. *Magnes Res* 1992;5:68.
15. Sandrini G, Proietti Cecchini A, Nappi G. Chronic fatigue syndrome: a borderline disorder [abstract]. *Funct Neurol* 2002;17:51–52.
16. Wikner J, Hirsh U, Nettenberg L, Röjdmark S. Fibromyalgia: a syndrome associated with decrease nocturnal MT secretion. *Clin Endocrinol (Oxf)* 1998;49:179–183.
17. Yasui M, Yase Y, Ando K, Adachi K, Mukoyama M, Ohsugi K. Magnesium concentration in brains from multiple sclerosis patients. *Acta Neurol Scand* 1990;81:187–200.
18. Yasui M, Ota K. Experimental and clinical studies on dysregulation of magnesium metabolism and the aetiopathogenesis of multiple sclerosis. *Magnes Res* 1992;5:295–302.
19. Stelmasiak Z, Solski J, Jakubowska B. Magnesium concentration in plasma and erythrocytes in multiple sclerosis. *Acta Neurol Scand* 1995;92:109–111.

20. Pugliatti M, Sotgiu S, Solinas G, Castiglia P, Rosati G. Multiple sclerosis prevalence among Sardinians: further evidence against the latitude gradient theory. *Neurol Sci* 2001;22:163–165.
21. Zorzon M, Zivadinos R, Nasuelli D, et al. Risk factors of multiple sclerosis: a case control study. *Neurol Sci* 2003;24:242–247.
22. Durlach J, Pagès N, Bac P, Bara M, Guiet-Bara A. Magnesium depletion with hypo- or hyper-function of the biological clock may be involved in chronopathological forms of asthma. *Magnes Res* 2005;18:19–34.
23. Zurak N. Role of the suprachiasmatic nucleus in the pathogenesis of migraine attacks. *Cephalalgia* 1997;17:723–728.
24. Salvesen R, Bakkelund SI. Migraine as compared to other headaches is worse during midnight-sun summer than during polar night. A questionnaire study in an Artic population. *Headache* 2000;40:824–829.
25. Durlach J. *Magnesium in Clinical Practice*. London: John Libbey; 1988.
26. Durlach J, Durlach V, Bac P, Bara M, Guiet-Bara A. Magnesium and therapeutics. *Magnes Res* 1994;7:313–328.
27. Durlach J, Bara M. *Le Magnésium en Biologie et en Médecine*. Paris: EMinter Tec et Doc; 2000.
28. Durlach J, Guiet-Bara A, Pagès N, Bac P, Bara M. MgCl2 or MgSO4, a genuine question. *Magnes Res* 2005;18:187–192.
29. Refinetti R. Dark adaptation in the circadian system of the mouse. *Physiol Behav* 2001;74:101–107.
30. Refinetti R. Effects of prolonged exposure to darkness on circadian photic responsiveness in the mouse. *Chronobiol Int* 2003;20:417–440.
31. Durlach J, Rapin JR, Le Poncin-Lafitte M, Rayssiguier Y, Bara M, Guiet-Bara A. 3H-Taurine distribution in various organs of magnesium deficient adult rats. In: Halpern MJ, Durlach J, eds. *Magnesium Deficiency: Physiopathology and Treatment Implications*. Basel: Karger; 1985:46–53.
32. Durlach J, Bara M, Guiet-Bara A, Rinjard P. Taurine and magnesium homeostasis: new data and recent advances. In: Altura BM, Durlach J, Seelig MS, eds. *Magnesium in Cellular Processes and Medicine*. Basel: S Karger; 1987:219–238.
33. Durlach J, Poenaru S, Rouhani S, Bara M, Guiet-Bara A. The control of central neural hyperexcitability in magnesium deficiency. In: Essman WB, ed. *Nutrients in Brain Function*. Basel: Karger; 1987:48–71.
34. Baskin SI, Finney CM. Factors that modify the tissue concentration or metabolism of taurine. In: Schaffer SW, Baskin SI, Kocsis JJ, eds. *The Effects of Taurine on Excitable Tissues*. Jamaica, NY: Spectrum; 1981:405–418.
35. Sternberg EM. Pathogenesis of L-tryptophan eosinophilia-myalgia syndrome. *Adv Exp Med Biol* 1996;398:325–330.
36. Clauw DJ, Ward K, Wilson B, Katz P, Rajan S. Magnesium deficiency in the eosinophilia-myalgia syndrome. Report of clinical and biochemichal improvement with repletion. *Arthritism Rheum* 1994;37:1331–1334.
37. Bac P, Pagès N, Maurois P, German-Fattal M, Durlach J. A new actimetry-based test of photic sensitization on a murine photosensitive magnesium depletion model. *Methods Find Exp Clin Pharmacol* 2005;27:1–4.

11
New Data on Pharmacological Properties and Indications of Magnesium

Pedro Serrano, Maríasol Soria, and Jesús F. Escanero

Metabolic therapy involves the administration of a substance normally found in the organism in order to enhance a metabolic reaction within the cell. This may be achieved in two ways: First, for some systems, a substance can be given to achieve greater-than-normal levels in the body so as to drive an enzymatic reaction in a preferred direction. Second, metabolic therapy may be used to correct an absolute or relative deficiency of a cellular component. Thus, metabolic therapy differs greatly from most standard pharmacological therapy.[1]

Due to the fact that magnesium is an essential mineral in human nutrition, involved in over 300 metabolic reactions, the difficulty in ascertaining the underlying mechanisms of the metabolic therapy of this element may be evident. Magnesium is necessary for every major biological process, including the production of cellular energy and the synthesis of nucleic acids and proteins. Moreover, it is important for the electrical stability of cells, the maintenance of membrane integrity, muscle contraction, nerve conduction, and the regulation of vascular tone, among others.[2]

However, it must be borne in mind that magnesium is biologically tightly interlocked with calcium. In some reactions, such as the synthesis of nucleic acids and proteins, calcium and magnesium are antagonistic. Magnesium is necessary for these processes, while calcium may inhibit them. In others, the production of adenosine triphosphate or ATP, magnesium and calcium cooperate. Magnesium has been called nature's physiological calcium channel blocker because it appears to regulate the intracellular flow of calcium ions.

Alterations at Subcellular Level

In order to establish the bases that may explain the pharmacological action and properties of magnesium in the different cases in which it is used, it is necessary to ascertain the alterations generated or the physiological processes altered in the cases in which it is administered or the processes or mechanisms restored with its administration.

Intracellular Concentrations of Calcium and Magnesium

In the 1980s, a considerable number of clinical, epidemiological, and experimental studies were published in which the important role of magnesium in cardiovascular pathophysiology was underlined.[3] The period in which the role of intracellular ions in hypertension was revealed began in 1984 with the study of Erne and coworkers,[4] which showed that the free cytosolic calcium of the platelets correlated tightly and positively with the levels of blood pressure in normotense subjects and in patients with hypertension. In the same year, Resnick[5] reported that the levels of free intracellular magnesium in the hypertense patients were lower than those of normotense subjects and, in 1993,[6] he indicated that the levels of free intracellular magnesium in hypertension were inversely related to those of free calcium. In 1991, our group[7] reported that different anti-hypertensive treatments provoked increases of intralymphocytary magnesium as their most consistent manifestation.

The cellular bases of a large amount of syndromes, such as hypertension, seem to be found in modifications of the levels of free intracellular ionic calcium and magnesium, as reported by Lawrence and Resnick.[8] Since then the majority of cases in which magnesium administration is recommended have presented these alterations.

Different mechanisms may be altered by this imbalance in the concentration of these intracellular ions. Thus, the amount of acetylcholine released has been shown to be dependent upon Ca^{2+} and Mg^{2+} concentrations. While Ca^{2+} increases acetylcholine release during neuronal depolarization, Mg^{2+} decreases acetylcholine release by stabilizing the membranes of presynaptic vesicles.[9]

In relation to the blockage of the calcium channel, magnesium blocks the voltage-gated calcium channels[10,11] and, consequently, may modulate posttraumatic neurochemical changes mediated by these elevated intracellular calcium levels.

Elevations in intracellular calcium concentration not only occur through ionotropic receptors, but also through second-messenger–linked receptors. Neurotransmitters, hormones, and even mechanical damage also increase intracellular calcium concentration through the activation of phospholipase C and the subsequent hydrolysis of phosphatidyl inositol 4,5-bisphosphate into inositol 1,4,5-triphosphate (IP_3).[12]

Receptors (Mediators/Modulators)

Not only do the membrane-stabilizing properties of magnesium affect lipid peroxidation and the generation of reactive oxygen species, but they also impact upon the release of neurotransmitters and other mediators/modulators.[13] In addition to the actions controlled by the IP_3 as intracellular messenger, magnesium participates in the regulation of other mediators and/or second messengers. Glutamate release, like acetylcholine release, is also reduced by the administration of magnesium. Both of these transmitters are increased after trauma.[14] It is possible that the effect of magnesium on glutamate release indirectly arises largely due to the cation's membrane-stabilizing properties. Indeed, the NMDA channel has been implicated as a critical factor in the development of cellular injury following neurotrauma.[15] Some studies have shown that the magnesium cation is the endogenous regulator of NMDA-channel activity.[16,17] The specific effects of the ion on the activity of the NMDA channel have recently been reported to be critical to outcome following central nervous system trauma. This study demonstrates that the magnesium block of the NMDA channel is reduced after neural injury and that this reduction may be linked to either a decline in magesium levels, or a change in the structure of the NMDA channel.[18]

Membrane Enzymes

There is now strong evidence that magnesium has an important role in the regulation of the metabolism of arachidonic acid.[19] Depletion of intracellular Mg^{2+} has been shown to reduce exogenous arachidonic acid incorporation into tissue phospholipids, perhaps by reducing arachidonyl CoA synthesis and thus modifying the first phase in the incorporation of exogenous fatty acids into membrane phospholipids.[20]

Moreover, Mg^{2+} binding to its specific binding site of protein kinase C (PKC) causes reduced activity of this enzyme, suggesting that some of the effects of intracellular Mg^{2+} depletion may be mediated by PKC activation. Increased PKC-mediated phosphorylation of enzymes that are involved in arachidonic acid incorporation (CoA synthetase, lysophosphatidylcholine acyl transferase) reduces their activities and subsequently increases the concentration of the free arachidonic acid and its metabolites.[19] The fact that phospholipase-induced products may also act to chelate magnesium[21] suggests a possible role of phospholipase activity in the post-traumatic depletion of free magnesium and impairments of the cellular bioenergetic state.

In the sphingomyelin pathway, a extracellular stimuli [vitamin D_3, tumor necrosis factor α (TNF-α), endotoxin, interferon, IL-1] causes the activation of sphingomyelinase, which hydrolizes plasma membrane sphingomyelin to produce ceramide and phosphorylcholine. Ceramide appears to act as a lipid second messenger, playing important roles in a variety of fundamental biological processes, such as cell proliferation, cell differentiation, apoptosis, receptor functions, oncogenesis, immune functions, and inflammatory responses.[22] Sphingomyelinases are a group of phospholipases. Five types of sphingomyelinases have been identified. These sphingomyelinases differ in tissue distribution, cofactor dependence, mechanism of regulation, and involvement in diverse cellular responses.[23] The neutral Mg^+-dependent sphingomyelinase is a plasma membrane-bound enzyme that requires magnesium ions and a neutral pH environment[23] and plays an important regulatory role with a vasorelaxant response in vascular smooth muscle.

Membrane Stabilizing Agent

One of the consequences of calcium influx is the initiation of pathways involved in the breakdown of lipid membrane constituents and the subsequent accumulation of free fatty acids, in particular arachidonic acid.[19] Magnesium depletion has been reported to increase membrane turnover and fluidity and to increase lipid peroxidation.[24] Increased lipid peroxidation is indicative of oxidative stress, suggesting that magnesium may also play a central role in this process. The reactive oxygen species initiate further cell damage, in part through the peroxidation of membrane components. These free radicals, along with other damaging neurochemical events, have been shown to play an important role in the pathophysiology of CNS injury.[25,26]

Thus, the magnesium ion is able to bind electrostatically to the negatively charged groups in the membranes, proteins, and nucleic acids. Binding to the polar heads of the membrane phospholipids may change the local configurations and produce a general effect of electric "screening." Consequently, magnesium may influence the binding of other cations such as calcium an the polyamines which, depending on their concentration, may have antagonistic or collaborative effects. Generally, magnesium has a protective and stabilizing effect on the membrane that may be attributed to the electric effects and to the inhibition of phospholipase A_2.

Conclusion

It is probable that the therapeutic effects of magnesium have a common base through the phenomena of the membrane which, according to the specificities of the cells of each tissue, will be manifested in one way or another. We think that the final tendency of the metabolic therapy of magnesium should be oriented to the restoration of the intracellular ionic balance.

Therapeutic Indications of Magnesium

Magnesium Deficits: Deficiency and Depletion

The therapeutics of magnesium supplementation do take into account the basic distinction between two types of magnesium deficit: magnesium deficiency and magnesium depletion.[2] Magnesium deficiency is a type of magnesium deficit due to insufficient magnesium intake. It responds to simple magnesium supplementation. Magnesium depletion is due to a dysregulation in the mechanisms that control or disturb magnesium metabolism. This type of magnesium deficit responds to the correction of the pathogenic dysregulation. Thus, magnesium can be used preventively in order to avoid deficiencies in high-risk populations (competition sports, diuretic use, etc.) and therapeutically to treat deficiencies (nutritional magnesium therapy).

In isolated deficiencies of magnesium, up to 50 mmol of magnesium can be administered, according to individual requirements. A variety of magnesium salts can be used in clinical practice, although the most widely used is magnesium sulphate. In acute or severe hypomagnesemia, magnesium can be administered parenterally, usually as chloride or sulphate.[27] As an alternative, magnesium sulphate can be administered in an intramuscular or slow intravenous injection.

Magnesium Use in Cardiology

Myocardial Infarction

Magnesium may have an anti-arrhythmic effect on myocardium and may protect myocardium from reperfusion arrhythmia and myocardial spasms, given the physiological role of magnesium in the homeostasis of ions in the myocardial muscle cells.

Although intravenous magnesium salts have been used for cardiac arrhythmia during acute coronary syndromes, the clinical trials assessing this use present inconclusive results. Based on the promising previous experience in small trials,[28] the second Leicester Intravenous Magnesium Intervention Trial LIMIT-2 study[29,30] was designed to assess the usefulness of magnesium in suspected myocardial infarction. In this trial it was observed an early and long-term (mean follow up of 2.7 years) beneficial effect on mortality, administrating 8 mmol of intravenous magnesium prior to thrombolysis, followed by a perfusion of 65 mmol during the following 24 h. On the other hand, the Fourth International Study of Infarct Survival Collaborative Group (ISIS-4) trial, with a larger sample of patients (58,050 patients) did not find a significant benefit of magnesium, although the way of administering magnesium was slightly different.[31] Finally, the Magnesium in Coronaries (MAGIC) trial[32] was designed in an attempt to solve the controversy, administering magnesium sulphate in similar doses to the LIMIT-2 trial, but the treatment did not show any benefit. In addition, the study by Galloe[33] did not show any benefit of the use of oral magnesium among survivors of an acute myocardial infarction. Currently, the guideline of the European Society of Cardiology does not recommend the use of magnesium in ST-elevation myocardial infarction.[34] However, the American College of Cardiology/American Heart Association (ACC/AHA) guideline[35] recommends analyzing electrolytes and magnesium in the early laboratory evaluation of these patients. In this guideline, ACC/AHA recommends magnesium administration exclusively in two situations (class IIa recommendations): (1) In magnesium deficits, especially in patients receiving diuretics before the onset of STEMI (level of evidence C); (2) In patients with episodes of torsade de pointes–type ventricular tachycardia associated with a prolonged QT interval (level of evidence C), the AHA recommends administering 1 to 2 g of magnesium as an intravenous bolus over 5 min.

Arrhythmia

Low serum levels of magnesium have not been clearly shown to be associated with an increased risk of ventricular fibrillation, although tissue depletion of magnesium remains a potential risk factor. The ACC/AHA guideline[35] also recommends maintaining serum potassium levels at greater than 4.0 mEq/L and magnesium levels at greater than 2.0 mEq/L in patients with myocardial infarction (class IIa, level of evidence C), especially in those with refractory polymorphic ventricular tachycardia. Magnesium sulphate is indicated in some arrhythmia, such as torsade de pointes and arrhythmia associated to hypokaliemia. The current dose is 2 g of magnesium sulphate (8 mmol of magnesium), administered intravenously over 10 to 15 min, and repeated once if necessary. Magnesium (8 mmol) is recommended for refractory ventricular fibrillation if there is a suspicion of hypomagnesaemia, for example, patients on potassium-losing diuretics.[36] The European Society of Cardiology indicates normalizing magnesium levels when treating supraventricular tachycardia in patients with acute heart failure, especially in those patients with ventricular arrhythmias.[37]

Heart Failure

American College of Cardiology/American Heart Association[38] recommends for the initial assessment of patients with chronic heart failure measuring electrolytes (including calcium and magnesium), searching hypomagnesemia as a potential cause of heart failure.

Hypertension

The Seventh Report of the Joint National Committee on Prevention, Detection, Evaluation, and Treatment of High Blood Pressure[39] in the treatment of acute severe hypertension in pre-eclampsia recommends caution when using nifedipine with magnesium sulphate, as it can induce precipitous blood pressure drop. Neither this guide nor the 2003 World Health Organization (WHO)/International Society of Hypertension statement on management of hypertension[40] support the use of magnesium sulphate to lower blood pressure.

Magnesium Use in Gynecology

Pre-Eclampsia

Magnesium sulphate is indicated to prevent eclampsia in patients with pre-eclampsia. It has been showed in several studies that it is more effective than placebo,[41] phenytoin,[42] and nimodipine.[43] Magnesium sulphate decreases the risk of developing eclampsia around 50%, and also decreases maternal mortality. The WHO considers that magnesium sulphate is the elective drug for the prevention of eclampsia in patients suffering from pre-eclampsia.[44]

Eclampsia

Magnesium sulphate induces a fast effect preventing seizures, but also a lack of sedation in mother and child treated for eclampsia. It has been demonstrated that it is more efficacious than phenytoin,[45,46] diazepam,[45,47] and lytic cocktail.[48] In addition to a wide security therapeutic margin, calcium gluconate is an effective antidote in case of overdose. The WHO considers that magnesium sulphate is the elective drug for the treatment of eclampsia.[44]

Preterm Labor

Magnesium sulphate has been administered either intravenous or orally as acute tocolytic therapy for preventing premature birth in threatened preterm labor. It has a similar efficacy to the adrenergic B2 agonists as tocolytic agent,[49,50] but a systematic study concluded that it is not effective to delay or prevent premature birth.[51] On the other hand, it has been found that the use of magnesium sulphate is associated to cerebral palsy, mental retardation, and a higher neonatal and pediatric mortality,[51] especially in very low birthweight infants whose mothers received doses over 48 g of magnesium sulphate.[52-54]

Magnesium Use in Neumology

Asthma

The Global Initiative for Asthma (GINA)[55] of the National Heart, Lung, and Blood Institute (NHLBI) says that regular dietary supplementation with magnesium adds no clinical benefit to current standard asthma therapy. However, they recommend to consider magnesium as adjuvant therapy for severe episodes of asthma exacerbations. Moreover, they affirm that intravenous magnesium can help reduce hospital admission rates in selected groups of patients: adults with FEV1 25% to 30% predicted at presentation; adults and children who fail to respond to initial treatment; and children whose FEV1 fails to improve above 60% predicted after 1 h of care[56,57] (evidence A). The usual dose is 2 g given intravenously over 20 min. In addition, magnesium sulphate should be infused early in the resuscitation phase of severe asthma episodes, and these patients should have at least daily monitoring of metabolic parameters, especially serum potassium (evidence D).[55] On the other hand, one study[58] indicated that use of isotonic magnesium as an adjuvant to nebulized salbutamol results in an enhanced bronchodilator response in treatment of severe asthma. The National Asthma Education and Prevention Program[59] recommends magnesium sulphate to treat preterm labor in patients with asthma during labor and delivery.

Chronic Obstructive Pulmonary Disease

Intravenous magnesium sulphate has also been used with promising results in patients with severe worsening of chronic obstructive pulmonary disease receiving inhaled salbutamol.[60]

Pulmonary Hypertension

Promising results have been published recently for newborn patients with severe persistent pulmonary hypertension. In a recent clinical trial,[61] it has been treated successfully with a loading dose of 200 mg/kg magnesium sulphate was given over a period of 20 min, followed by a continuous infusion at the rate of 20 to 150 mg/kg/h to obtain a serum magnesium level between 3.5 and 5.5 mmol/L. Oxygen index (OI) and alveolar–arterial oxygen gradient (A-aDO2) showed significant improvement within 24 h of treatment, proving that magnesium sulphate was found may be a safe and effective pulmonary vasodilator.

Magnesium Use in Gastroenterology

Antacids

Some magnesium salts, such as carbonate, hydroxide, oxide, and trisilicate are widely used due to their antacid properties.

Constipation

Magnesium salts present with a laxative effect. The most widely used for this purpose is magnesium sulphate at oral doses of 5 to 10 g in 250 mL of water. This provides a rapid evacuation of intestine. Magnesium citrate, magnesium phosphate, and magnesium hydroxide are also used. The American Gastroenterology Association recommends milk of magnesia for the management of constipation[62] as one of the initial therapeutic options. However, the Rehabilitation Nursing Foundation[63] discourages the routine use of saline magnesium laxatives due to possible side effects such as abdominal cramping, water stools, and potential for dehydration and hypermagnesium. They only indicate the use of these laxatives in end-stage patients when other options have failed, and with an adequate prospective evaluation of magnesium levels.

Acute Colonic Pseudo-Obstruction

The American Society for Gastrointestinal Endoscopy[64] indicates that if the conservative treatment option is preferred, magnesium levels should be evaluated and corrected parenterally.

Acute Liver Failure

The American Association for the Study of Liver Diseases recommends analyzing magnesium in the initial laboratory analysis of the acute liver failure and to monitor it closely, because it is frequently low and may require repeated supplementation throughout the hospital course.[65]

Other Potential Indications of Magnesium

Magnesium's past and present indications are very diverse. Dehydrated magnesium sulphate has been used as a cream for the topic treatment of inflammatory diseases of the skin, such as anthrax and infection of the hair follicles, but the prolonged or repeated use may harm the surrounding skin.

Magnesium has been used or is in study for the treatment of a variety of conditions, including analgesia, headache, alcohol withdrawal delirium, chronic fatigue syndrome, fibromyalgia, osteoporosis, and many others.

Conclusion

Metabolic therapy of magnesium is vastly used, although the basis of this therapy remain partly unknown. Future research will make them more easily understood, and it is likely that the preventive use of magnesium will increase in hospital environments as well as in specific risk groups (athletes, patients on diuretics, etc.).

References

1. Hadj A, Pepe S, Marasco S, Rosenfeldt F. The principles of metabolic therapy for heart disease. *Heart Lung Circ* 2003;12:S55–S62.
2. Durlach J. *Magnesium in Clinical Practice*. London: John Libbey; 1998.
3. Altura BM, Zhang A, Altura BT. Magnesium, hypertensive vascular diseases, atherogenesis, subcellular compartmentation of Ca^{2+} and Mg^{2+} and vascular contractility. *Miner Electrolyte Metab* 1993;19:323–336.
4. Erne P, Bolli P, Burgissen E, Buhler F. Correlation of platelet calcium with blood pressure: effect of antihypertensive therapy. *N Engl J Med* 1984;319:1084–1088.
5. Resnick LM. Intracellular free magnesium in erythrocytes of essential hypertension: relation to blood pressure and serum divalent cations. *Proc Natl Acad Sci U S A* 1984;81:6511–6515.
6. Resnick LM. Ionic basis of hypertension, insulin resistance, vascular disease, and related disorders. *Am J Hypertens* 1993;6:123S–134S.
7. Rodríguez LM. *Estudio de Magnesio Intracelular en la Hipertensión Arterial Esencial Antes y Después del Tratamiento* [dissertation]. Zaragoza, Spain: Facultad de Medicina de Zaragoza; 1991.
8. Lawrence M, Resnick MD. Cellular calcium and magnesium metabolism in the pathophysiology and treatment of hypertension and related metabolic disorders. *Am J Med* 1992;93(suppl 12A):2A11S–2A20S.
9. James MFM. Clinical use of magnesium infusions in anesthesia. *Anesth Analg* 1992;74:129–136.
10. Iseri LT, French JH. Magnesium: natures physiologic calcium blocker. *Am Heart J* 1984;108:188–193.
11. Bara M, Guiet-Bara A. Magnesium regulation of Ca2+ channels in smooth muscle and endotelial cells of human allantochorial placental vessels. *Magnes Res* 2001;14:11–18.

12. Farese RV. The phosphatidate-phosphoinositide cycle: an intracellular messenger system in the action of hormones and neurotransmitters. *Metabolism* 1983;32: 628–641.
13. Vink R, Cernak I. Regulation of intracellular free magnesium in central nervous system injury. *Frontiers Biosci* 2000;5:656–665.
14. Hayes RL, Jenkins LW, Lyeth BG. Neurotransmitter mediated mechanisms of traumatic brain injury: acetylcholine and excitatory amino acids. In: Jane JA, Anderson DK, Tomer JC, Young W, eds. *Central Nervous System Trauma Status Report—1991*. New York: Mary Ann Liebert; 1992:173–188.
15. Faden AI, Demediuk P, Panter SS, Vink R. Excitatory aminoacids, N-methyl-D-aspartatereceptors and traumatic brain injury. *Science* 1989;244:798–800.
16. Nowak L, Bregstovski P, Ascher P, Herbert A, Prochiantz A. Magnesium gates glutamate-activated channels in mouse central neurons. *Nature* 1984;307:462–465.
17. Choi DW. Excitotoxic cell death. *J Neurobiol* 1992;23:1261–1276.
18. Zhang L, Rzigalinski BA, Ellis EF, Satin LS. Reduction of voltage-dependent Mg^{2+} blockade of NMDA current in mechanically injured neurons. *Science* 1996; 274:1921–1923.
19. Weis MT, Saunders C. Magnesium and arachidonic acid metabolism. *Magnes Res* 1993;6:179–190.
20. Reddy TS, Bazan NG. Kinetic properties of araquidonil-coenzyme A syntetase in rat brain micrososmes. *Arch Biochem Biophys* 1983;226:125–133.
21. Vink R. Phospholipase C activiti reduces free magnesium concentration. *Biochem Biophys Res Commun* 1989;165:913–918.
22. Hannum YA. The sphimgomyelin cycle: a prototypic sphingolipid signaling pathway. *Adv Lipid Res* 1993;25:27–40.
23. Liu B, Hannun YA. Inhibition of the neutral magnesium-dependent sphingomyelinase by glutathione. *J Biol Chem* 1997;272:16281–16287.
24. Gunther T, Hollriegl V, Vommann J, Bubeck J, Classen HG. Increased lipid peroxidation in rat tissues by magnesium deficiency and vitamin E depletion. *Magnes Bull* 1994;16:38–43.
25. Hall ED. The role of oxygen radicals in traumatic injury. Clinical implications. *J Emerg Med* 1993;11:31–36.
26. Awasthy D, Church DF, Torbati D, Carey ME, Prior W. Oxidative stress following traumatic brain injury in rats. *Surg Neurol* 1997;47:575–581.
27. Durlach J, Guiet-Bara A, Pages N, Bac P, Bara M. Magnesium chloride or magnesium sulfate: a genuine question. *Magnes Res* 2005;18:187–192.
28. Teo KK, Yusuf S, Collins R, et al. Effects of intravenous magnesium in suspected acute myocardial infarction: overview of randomised trials. *BMJ* 1991;303:1499–1503.
29. Woods KL, Fletcher S, Roffe C, Haider Y. Intravenous magnesium sulphate in suspected acute myocardial infarction: results of the second Leicester Intravenous Magnesium Intervention Trial (LIMIT-2). *Lancet* 1992;339:1553–1558.
30. Woods KL, Fletcher S. Long-term outcome after intravenous magnesium sulphate in suspected acute myocardial infarction: the second Leicester Intravenous Magnesium Intervention Trial (LIMIT-2). *Lancet* 1994;343:816–819.
31. Fourth International Study of Infarct Survival Collaborative Group. ISIS-4: a randomised factorial trial assessing early oral captopril, oral mononitrate, and intra-

11. New Data on Pharmacological Properties and Indications of Magnesium 137

venous magnesium sulphate in 58,050 patients with suspected acute myocardial infarction. *Lancet* 1995;345:669–685.

32. The Magnesium in Coronaries (MAGIC) Trial Investigators. Early administration of intravenous magnesium to high-risk patients with acute myocardial infarction in the Magnesium in Coronaries (MAGIC) trial: a randomised controlled trial. *Lancet* 2002;360:1189–1196.

33. Galloe AM, Rasmussen HS, Jorgensen LN. Influence of oral magnesium supplementation on cardiac events among survivors of an acute myocardial infarction. *BMJ* 1993;307:585–587.

34. Van de Werf F, Ardissino D, Betru A, et al. Management of acute myocardial infarction in patients presenting with ST-segment elevation. *Eur Heart J* 2003;24:28–66.

35. Antman EM, Anbe DT, Armstrong PW, et al. American College of Cardiology; American Heart Association; Canadian Cardiovascular Society. ACC/AHA guidelines for the management of patients with ST-elevation myocardial infarction—executive summary. A report of the American College of Cardiology/American Heart Association Task Force on Practice Guidelines (Writing Committee to revise the 1999 guidelines for the management of patients with acute myocardial infarction) [published correction appears in *J Am Coll Cardiol* 2005;45:1376]. *J Am Coll Cardiol* 2004;44:671–719.

36. Priori SG, Aliot E, Blomstrom-Lundqvist C, et al. Task Force on Sudden Cardiac Death of the European Society of Cardiology. *Eur Heart J* 2001;22:1374–1450.

37. Nieminen MS, Böhm M, Cowie MR, et al. Executive summery of the guidelines on the diagnosis and treatment of acute heart failure: The Task Force on Acute Heart Failure of the European Society of Cardiology. *Eur Heart J* 2005;26:384–416.

38. Hunt SA; American College of Cardiology; American Heart Association Task Force on Practice Guidelines (Writing Committee to Update the 2001 Guidelines for the Evaluation and Management of Heart Failure). ACC/AHA 2005 guideline update for the diagnosis and management of chronic heart failure in the adult: a report of the American College of Cardiology/American Heart Association Task Force on Practice Guidelines (Writing Committee to Update the 2001 Guidelines for the Evaluation and Management of Heart Failure). *J Am Coll Cardiol* 2005;46:e1–82. Euratum in J Am Coll Cardiol 2006;47:1503–1505.

39. Chobanian AV, Bakris Gl, Black HR, et al. Seventh report of the Joint National Committee on Prevention, Detection, Evaluation, and Treatment of High Blood Pressure. *JAMA* 2003;289:2560–2571.

40. 2003 World Health Organization (WHO)/International Society of Hypertension (ISH) statement on management of hypertension. *J Hypertens* 2003;21:1983–1992.

41. The Magpie Trial Collaborative Group. Do women with pre-eclampsia, and their babies, benefit from magnesium sulphate? The Magpie Trial: a randomised placebo-controlled trial. *Lancet* 2002;359:1877–1890.

42. Lucas MJ, Leveno KJ, Cunningham, FG. A comparison of magnesium sulfate with phenytoin for the prevention of eclampsia. *N Engl J Med* 1995;333:201–205.

43. Belfort MA, Anthony J, Saade GR, Allen JC Jr, et al. A comparison of magnesium sulfate with nimodipine for the prevention of eclampsia. *N Engl J Med* 2004;348:304–311.

44. World Health Organization (WHO). *Managing Complications in Pregnancy and Childbirth: A Guide for Midwives and Doctors: Headache, Blurred Vision, Convul-*

sions or Loss of Consciousness, Elevated Blood Pressure. Available from: http://www.who.int/reproductive-health/impact/Symptoms/Headache_blood_pressure_S35_S56.html (accessed June 2006).
45. The Eclampsia Trial Collaborative Group. Which anticolvulsant for women with eclampsia: evidence from the Collaborative Eclampsia Trial [published correction in Lancet 1995;346:358]. *Lancet* 1995;345:1455–1463.
46. Duley L, Henderson-Smart D. *Magnesium Sulphate Versus Phenytoin for Eclampsia.* Chichester: Wiley; 2004.
47. Duley L, Henderson-Smart D. *Magnesium Sulphate Versus Diazepam for Eclampsia.* Chichester: Wiley; 2004.
48. Duley L, Henderson-Smart D. *Magnesium Sulphate Versus Lytic Cocktail for Eclampsia.* Chichester: Wiley; 2004.
49. Wilkins IA, Lynch L, Mehalek KE, Berkowitz GS, Berkowitz RL. Efficacy and side effects of magnesium sulfate and ritodrine as tocolytic agents. *Am J Obstet Gynecol* 1988;159:685–689.
50. Chau AC, Gabert HA, Miller JM Jr. A prospective comparison of terbutaline and magnesium for tocolysis. *Obstet Gynecol* 1992;80:847–851.
51. Crowther CA, et al. Magnesium Sulphate for Preventing Preterm Birth in Threatened Preterm Labour. Chichester: Wiley; 2004.
52. Schendel DE, Beng CJ, Yeargm-Allsopp M, Boyle CA, Decoufle P. Prenatal magnesium sulfate exposure and the risk of cerebral palsy or mental retardation among very low-birth-weight children aged 3 to 5 years. *JAMA* 1996;276:1805–1810.
53. Mittendorf R, Dambrosia J, Dammann O. Association between maternal serum ionised magnesium levels at delivery and neonatal intraventricular hemorrhage. *J Pediatr* 2002;140:540–546.
54. Mittendorf R, Pryde P, Phoshnood B, Lee KS. If tocolytic magnesium sulfate is associated with excess total pediatric mortality, what is its impact? *Obstet Gynecol* 1998;92:308–311.
55. Global Initiative for Asthma (GINA), National Heart, Lung and Blood Institute (NHLBI). *Global Strategy for Asthma Management and Prevention.* Bethesda, MD: Global Initiative for Asthma (GINA), National Heart, Lung and Blood Institute (NHLBI); 2005.
56. Scarfone RJ, Loiselle JM, Joffe MD, et al. A randomized trial of magnesium in the emergency department treatment of children with asthma. *Ann Emerg Med* 2000;36:572–578.
57. Silverman RA, Osborn H, Runge J, et al. Acute Asthma/Magnesium Study Group. IV magnesium sulfate in the treatment of acute severe asthma: a multicenter randomized controlled trial. *Chest* 2002;122:489–497.
58. Hughes R, Goldkorn A, Masoli M, et al. Use of isotonic nebulised magnesium sulphate as an adjuvant to salbutamol in treatment of severe asthma in adults: randomised placebo-controlled trial. *Lancet* 2003;21;361:2114–2117.
59. National Asthma Education and Prevention Program. *Managing Asthma during Pregnancy: Recommendations for Pharmacologic Treatment.* Bethesda, MD: National Heart, Lung, and Blood Institute; 2005.
60. Skorodin MS, Tenholder MF, Yetter B. Magnesium sulfate in exacerbations of chronic obstructive pulmonary disease. *Arch Intern Med* 1995;155:496–500.
61. Chandran S, Haqueb ME, Wickramasinghe HT, et al. Use of magnesium sulphate in severe persistent pulmonary hypertension of the newborn. *J Trop Pediatr* 2004;50:219–223.

62. American Gastroenterology Association medical position statement: guidelines on constipation. *Gastroenterology* 2000;119:1761–1778.
63. Folden SL, Backer JH, Maynard F, et al. Practice Guidelines for the Management of Constipation in Adults. Glenview, IL: Rehabilitation Nursing Foundation; 2002.
64. Eisen GM, Baron TH, Dominitz JA, et al. Acute colonic pseudo-obstruction. *Gastrointest Endosc* 2002;56:789–792.
65. Polson J, Lee WM. American Associaton for the Study of Liver Diseases (AASLD) position paper: the management of acute liver failure. *Hepatology* 2005; 41:1179–1197.

Epidemiology

12
Magnesium Intake and the Incidence of Type 2 Diabetes

Fernando Guerrero-Romero and Martha Rodríguez-Morán

Low serum magnesium levels have been related to abnormalities in insulin action and decrease in insulin secretion, both involved in the pathophysiology of type 2 diabetes. In addition, hypomagnesemia is also related to the main risk factors for development of diabetes, such as obesity, low-grade chronic inflammatory syndrome, and aging.

Because customary diet is the main source of magnesium and because several prospective, interventional, and cross-sectional studies have demonstrated a strong association between low serum magnesium levels and metabolic disorders of glucose and insulin, has been hypothesized that low magnesium intake could be a risk factor for development of type 2 diabetes.

In this chapter, we review evidence from large prospective studies and cross-sectional and interventional clinical research to discuss the role of magnesium intake and the incidence of type 2 diabetes.

Prevalence of diabetes in adults worldwide was estimated to be 4.0% in 1995 and is expected rise to 5.4% by the year 2025. The number of adults with diabetes in the world will rise from 135 million in 1995 to 300 million in the year 2025. Estimates indicate that diabetes will increase 0.42% in the developed countries and 1.7% in the developing countries,[1] imposing a substantial public health burden.

The main associated risk factors for development of diabetes, such as obesity, low-grade chronic inflammatory syndrome, aging, diet, and family phenotype[2–10] are also associated with hypomagnesemia.[11–16] Additionally, low serum magnesium levels has been related with abnormalities in insulin action and insulin secretion, both involved in the pathophysiology of type 2 diabetes.[12] Furthermore, evidence from prospective, interventional, and cross-sectional studies[17–24] have demonstrated an association between low serum magnesium levels and disorders of glucose and insulin metabolism, suggesting that low serum magnesium intake could be a risk factor for development of type 2 diabetes.

In addition to clinical evidence emerged from clinical research, the background for supporting the hypothesis that hypomagnesemia is associated with incidence of type 2 diabetes derives from the role that magnesium exerts on energetic metabolism. In this regard, magnesium, the fourth most abundant cation in the body and the second most abundant intracellular cation,[25] is an essential cofactor crucial in multiple physiological processes in humans.[26,27] It is involved in more than 300 enzymatic reactions, including glycogen breakdown, fat oxidation, and adenosine triphosphate (ATP) synthesis,[28] and plays an essential role for maintaining insulin sensitivity in target tissues.[29] In addition, magnesium is also involved in the immune and hormonal function.[30] Intracellular abnormalities of magnesium homeostasis may decrease tyrosine kinase activity at insulin receptors, increasing insulin resistance,[31] and influence glucose-stimulated insulin secretion,[32] decreasing β cell insulin secretion, the main process involved in the genesis of type 2 diabetes. Furthermore, we have found that low serum magnesium levels are strongly related with elevated serum concentrations of both tumor necrosis factor α (TNFα)[33] and C-reactive protein,[14] suggesting that magnesium deficiency may also be involved in the triggering of low-grade inflammatory response and, through this pathway, with the development of diabetes.[14,33]

Magnesium is an important component of whole grains, nuts, green leafy vegetables, dried fruit, and shellfish, and is largely lost during the processing of foods.[25,34] In 1967 it was stated that magnesium intake of 0.30 to 0.35 mEq/kg per day in healthy people regularly produced positive magnesium balance.[35] The Recommended Dietary Allowance (RDA) is 400 mg for men aged 19 to 30 years and 420 mg for men more than 31 years old. The RDA is 310 mg for women aged 19 to 30 years and 320 mg for women more than 31 years old.

Although small amounts of magnesium are required in the diet in order to ensure the capacity for energy expenditure and to maintain optimal physiological function, usually dietary intake of magnesium is inadequate.[36-39] Recently, analyzing data from the National Health and Nutrition Examination Survey (NHANES III; 1999–2000), Ford[39] reported that mean magnesium intake was 290 ± 4 mg per day, and that the intake of magnesium decreased with increasing age. Ford[39] concluded that during 1999 to 2000, the diet of a large proportion of the U.S. population was inadequate in magnesium intake. The mean intake of magnesium estimated from the first phase of NHANES III, conducted from 1988 to 1991, was 361 mg and 291 for Caucasian and African American men, and 256 mg and 215 mg for Caucasian and African American women.[39] In addition, extensive analysis by Kawano and colleagues[40] showed that dietary magnesium intake has declined in the U.S population, from 475 to 500 mg/day in 1900 to 215 to 283 mg/day in 1990.[40]

Because dietary sources accounted for approximately 96% of total intake of magnesium[16] and individuals in westernized societies largely consume processed foods, it is difficult to reach the RDA through customary diet, and as consequence deficiency of magnesium is becoming more common.[41]

Magnesium deficiency has been associated with cardiovascular abnormalities, including hypertension, acute myocardial infarction, arrhythmia, dyslipidemia, and coronary artery disease[25,42,43] and macrovascular diabetic complications.[21,44-46] In addition, several large prospective cohorts[16-19,47] provide evidence to support an inverse association between dietary magnesium intake and incident type 2 diabetes (Table 12.1). However, results from these studies are controversial.

Salmeron and colleagues[48,49] analyzed, during 6 years of follow up, the relationship between glycemic diets, low fiber intake, and risk of type 2 diabetes in two large cohorts of 65,173 women 40 to 65 years of age and 42,759 men 40 to 75 years of age, free from cardiovascular disease, cancer, and diabetes at baseline. They reported an inverse relationship between dietary intake of magnesium and incident type 2 diabetes and concluded that their finding support the hypothesis that diets with high glycemic load and lower fiber content increase the risk of diabetes. These data are partially in accordance with previous reports by Colditz,[18] who analyzed, also during 6 years of follow up, data of 84,360 women from the Nurse's Health Study Cohort and found that intakes of fiber and carbohydrate were not related to risk of diabetes. Both Salmeron and colleagues[48,49] and Colditz and colleagues[18] reported that magnesium intake was inversely related to risk of diabetes. However, Kao and coworkers,[47] who conducted a prospective study of 15,792 individuals aged 45 to 64 years to examine the association between serum magnesium level and dietary magnesium intake and the subsequent risk for type 2 diabetes, found that individuals in the highest quartile of dietary magnesium intake appeared to be at lower risk than those in the lower quartiles; however, this relationship was not statistically significant and showed no evidences of gradedness.

Recently, the association between magnesium intake and the risk for diabetes has been examined in only sparse prospective studies.[16,17,19,50] Lopez-Ridaura and coworkers[17] followed, over 18 years, 85,060 women and 42,872 men free from diabetes, cardiovascular disease, and cancer at baseline. After adjusting for age Body Mass Index (BMI), physical activity, family history of diabetes, smoking, alcohol consumption, history of hypertension, and hypercholesterolemia, they found a relative risk (RR) for type diabetes of 0.66 (95% confidence interval (CI), 0.60–0.73; p for trend <0.001) in women, and 0.67 (95% CI, 0.56–0.80; p for trend <0.001) in men, comparing the highest with the lowest quintile of magnesium intake. Additionally, because the high proportion of subjects taking vitamins containing magnesium, Lopez-Ridaura and colleagues[17] also assessed the association between magnesium supplements and risk of diabetes, which was not statistical significant.

Meyer and colleagues[19] followed, over 6 years, 35,988 older Iowa women initially free of diabetes. Adjusted for age, total energy intake, BMI, waist-to-hip ratio (WHR), education, smoking, alcohol intake, and physical activity, the multivariate analyses RR was 1.0, 0.81, 0.82, 0.81, and 0.67 (p for trend = 0.0003) across quintiles of dietary magnesium intake.

Song and colleagues[16] followed over 6 years a cohort of 39,345 American women over 45 years old without previous diagnosis of diabetes, cardiovascular disease, and cancer in the Women's Health Study, found a significant inverse association between magnesium intake and risk of type 2 diabetes that was independent for age and BMI (p for trend = 0.007), but significantly

TABLE 12.1. Relative risk (95% CI) of incident type 2 diabetes across quintiles of dietary magnesium intake.

Reference	Quintile of Intake					p for Trend
	1	2	3	4	5	
Lopez-Ridaura[17,a]	1.0	0.80 (0.69–0.94)	0.84 (0.71–0.98)	0.56 (0.54–0.76)	0.56 (0.47–0.67)	<0.001
Song[16,b]	1.0	1.06 (0.88–1.28)	0.81 (0.66–1.0)	0.86 (0.70–1.06)	0.89 (0.71–1.10)	0.05
Colditz[18,c,d]	1.0	1.25	1.03	0.81	0.73	0.008
Meyer[19,e,f]	1.0	0.81 (0.68–0.96)	0.82 (0.68–0.98)	0.81 (0.67–0.97)	0.67 (0.55–0.82)	0.0003
Kao[47,g,h]	0.95 (0.52–1.74)	1.19 (0.72–1.96)	1.11 (0.74–1.67)	1.0	—	ns
Kao[47,h,i]	0.80 (0.56–1.14)	0.96 (0.70–1.31)	0.98 (0.75–1.29)	1.0	—	ns

Abbreviation: ns, not significant.
[a]Adjusted by age and energy intake.
[b]Adjusted by age, smoking, BMI, exercise, alcohol use, family history of diabetes, and total calories intake.
[c]Distribution by quintiles intake of energy.
[d]Adjusted for age, BMI, alcohol intake, family history of diabetes, prior weight change, and time period.
[e]Distribution by quintiles intake of grain, dietary fiber, and dietary magnesium.
[f]Adjusted for age, total energy intake, BMI, WHR, education, pack-years of smoking, alcohol intake, and physical activity.
[g]Distribution for quartiles of dietary magnesium intake, comparing the lowest with the highest quartile of dietary magnesium intake in African American individuals.
[h]Adjusted for age, sex, education, family history of diabetes, BMI, WHR, exercise, alcohol consumption, diuretic use, dietary calcium intake, potassium dietary intake, fasting insulin, and fasting glucose levels.
[i]Distribution for quartiles of dietary magnesium intake, comparing the lowest with the highest quartile of dietary magnesium intake in Caucasian individuals.

attenuated after adjustment for physical activity, alcohol intake, smoking, family history of diabetes, and total calorie intake at 1.0, 0.96, 0.76, 0.84, and 0.78 (p for trend = 0.05). Among overweight and obese women, the inverse trend in the multivariate-adjusted model was statistically significant (p for trend = 0.02).

Hodge and coworkers,[50] analyzing data from The Melbourne Collaborative Cohort Study, reported that the most consistent patterns among nutrients were observed for sugars, fiber, and magnesium, which decreased with increasing glycemic index. The highest quartile of magnesium intake compared with the lowest was inversely associated with incidence of diabetes [odds ratio (OR) 0.62; 95% CI, 0.43–0.90], a relationship that was attenuated after adjustment for BMI and WHR.

Although dietary information was collected using different questionnaires (a 131-item semiquantitative food frequency questionnaire by Song et al.; a 61-item semiquantitative food frequency questionnaire by Lopez-Ridaura et al.; and 127-item food frequency questionnaire by Meyer et al.), we analyzed data on magnesium intake and the risk for incidence of type 2 diabetes in women included in the Women's Health Study, the Iowa Women's Health Study, the Nurses' Health Study, and the Health Professionals' Follow-Up Study.[16,17,19] Comparing the highest with the lowest quintile of total magnesium intake the RR that estimates the risk for incident diabetes remained significant (0.69; 95% CI, 0.56–0.98; Figure 12.1), suggesting a protective role of dietary magnesium in the development of type 2 diabetes.

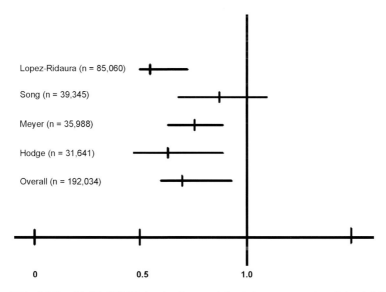

Figure 12.1. Relative risk (RR; 95% CI) showing the association between magnesium dietary intake and risk for type 2 diabetes among women.

Despite the high number of subjects included in these studies,[16,17,19] the estimation on the underlying association between magnesium dietary intake and risk of incident diabetes in women should be conservative. Errors in the measurement of exposition to risk (dietary magnesium intake) and outcome (incident diabetes) limit the reliability of estimated RRs. On this concern, although the prospective design reduce recall bias, the well-known limitation of food-frequency questionnaires, mainly their limitation to evaluate dietary changes during large follow-up periods, could be an important source of bias and misinterpretation. Furthermore, although the vast majority of dietary factors related to diabetes risk were considered, several additional nutrients that may influence diabetes risk were not included. On the other hand, because serum glucose levels were not measured, incident cases of diabetes were ascertained by self-reported questionnaire. Although diagnostic algorithm for type 2 diabetes was validated by review of medical records reporting a high rate of true-positive response, the true-negative and false-negative rates are not reported. Thus, the RR could be overestimated.

In addition to follow-up studies, Balon nd colleagues[51] analyzed the effects of a magnesium-supplemented diet in the male obese Zucker fatty rat. The rats maintained on the magnesium diet had markedly lower fasting and fed-state blood glucose concentrations and an improved glucose disposal. At the end of 6-weeks follow-up, all of the eight animals on the control diet group became diabetic, whereas diabetes developed in only one of eight animals on the magnesium diet. These data indicate that an increased dietary magnesium intake in male obese rats prevents deterioration of glucose tolerance, delaying the development of type 2 diabetes.[51]

Although experimental studies in humans have shown benefits of magnesium supplements on glucose metabolism and/or insulin sensitivity, the vast majority of these studies have been conducted in subjects with diabetes and their results are controversial.[24,52–59]

Whether magnesium supplementation decreases the risk for type 2 diabetes in nondiabetic subjects is unclear. Studies for testing the hypothesis that magnesium supplementation improves insulin secretion or insulin actions in healthy subjects are scarce (Table 12.2)[23,60–64]

Fung and colleagues,[61] after adjustment for age, obesity, total energy intake, physical activity, hours per week spent sitting outside work, alcohol intake, smoking, and family history of diabetes, showed that magnesium intake was inversely associated with fasting insulin concentration among nondiabetic women. Nadler and colleagues[60] found that dietary-induced isolated magnesium deficiency in lean subjects leads to a significant reduction in insulin sensitivity, suggesting that magnesium deficiency may be a common factor associated with insulin resistance. Humphries and coworkers,[64] in a sample of nondiabetic, young adult black Americans (aged 30 ± 3.4 years), showed a significant negative correlation of total dietary magnesium intake with the sum of insulin levels measured during an oral glucose tolerance test. Rosolova and colleagues,[62,63] assessing the relationship between the plasma magnesium

12. Magnesium Intake and the Incidence of Type 2 Diabetes

TABLE 12.2. Low serum magnesium levels and glucose-insulin changes in nondiabetic subjects.

Reference	n	Population	Design	Conclusion
Fung[61]	219	Women subsample of the Nurses' Health Study	Nested case-control study	Higher magnesium intake is associated with lower fasting insulin concentrations.
Nadler[60]	16	Normal subjects	Cross-sectional	Magnesium deficiency is associated with insulin resistance.
Humphries[64]	179	Young adult black Americans	Longitudinal	Total dietary magnesium intake is inversely related with insulin levels.
Rosolova[62]	18	7 mild hypertension, 7 no apparent disease, 2 chronic bronchitis, 1 vascular disease, 1 stable angina	Cross-sectional	Magnesium concentration is associated with insulin resistance, glucose intolerance, and hyperinsulinemia.
Rosolova[63]	98	Healthy subjects	Cross-sectional	Serum magnesium have a significant effect on insulin-mediated glucose disposal in insulin resistant subjects.
Guerrero-Romero[23]	63	Healthy Mexican subjects	Randomized double-blind placebo-controlled trial	Oral magnesium supplementation improves insulin sensitivity.

concentration and steady-state plasma insulin and glucose concentrations at the end of a 180-min infusion of octreotide, showed that a low magnesium concentration in nondiabetic subjects is associated with insulin resistance, glucose intolerance, and hyperinsulinemia, suggesting that variations in the serum magnesium concentration have a significant effect on insulin-mediated glucose disposal in healthy subjects, and that lower serum magnesium levels are associated with increased insulin resistance. These findings support the hypothesis that higher dietary magnesium intake may reduce the risk of developing type 2 diabetes mellitus.

Recently, we conducted a randomized double-blind placebo controlled trial[23] to test the hypothesis that oral magnesium supplementation modify insulin sensitivity in nondiabetic subjects. The causes of magnesium depletion linked to metabolic dysregulation, such as chronic diarrhea, alcohol intake, smoking, diabetes, high blood pressure, surgical stress, chronic diseases, diuretic therapy, and reduced renal function, were exclusion criteria. In addition, subjects who received magnesium supplementation previous to randomization were not included. At end of 3-month follow up, magnesium-supplemented subjects significantly reduced their fasting serum glucose, insulin levels, HOMA-IR index, serum triglycerides, total and low-density lipoprotein (LDL) cholesterol, and increased high-density lipoprotein (HDL) cholesterol levels, whereas in the control subjects there were no significant changes. These data supports the hypothesis that oral magnesium supplementation increases insulin sensitivity in nondiabetic subjects with decreased serum magnesium levels, but it clinical significance in the reduction of incidence of type 2 diabetes remain to be established.[23]

Taking into account the bias involved in measuring magnesium intake, dietary change during follow up, and diagnostic criteria of diabetes used in the large cohorts, the limitation by small numbers of subjects in cross-sectional studies, and the short-term of follow up in clinical trials, we can conclude that currently there is no sufficient accumulated clinical-based evidence for supporting that low serum magnesium intake is associated with the risk for developing type 2 diabetes.

Because the most frequent source of magnesium deficiency is an inadequate diet, healthcare providers should invest more effort in providing an appropriate dietary counsel with the aim of people reach the RDA of magnesium intake. On the other hand, results from long-term trials are needed in order to evaluate the safety and beneficial role of magnesium supplements for reducing the incidence of type 2 diabetes.[27]

References

1. King H, Aubert RE, Herman WH. Global burden of diabetes, 1995–2025: prevalence, numerical estimates, and projections. *Diabetes Care* 1998;21:1414–1431.
2. DeFronzo RA. Lilly Lecture 1987. The triumvirate: B-cell, muscle, liver. A collision responsible for NIDDM. *Diabetes* 1998;37:667–687.

3. Duncan BB, Schmidt MI, Pankow JS, et al. Low-grade systemic inflammation and the development of type 2 diabetes: the atherosclerosis risk in communities study. *Diabetes* 2003;52:1799–1805.
4. Schulze MB, Hoffmann K, Manson JA, et al. Dietary pattern, inflammation, and incidence of type 2 diabetes in women. *Am J Clin Nutr* 2005;82:675–684.
5. Pickup JC. Inflammation and activated innate immunity in the pathogenesis of type 2 diabetes. *Diabetes Care* 2004;27:813–823.
6. Mokdad AH, Bowman BA, Ford ES, et al. The continuing epidemics of obesity and diabetes in the United States. *JAMA* 2001;286:1195–1200.
7. Sullivan PW, Morrato EH, Ghushchyan V, et al. Obesity, inactivity, and the prevalence of diabetes and diabetes-related cardiovascular comorbidities in the U.S., 2000–2002. *Diabetes Care* 2005;28:1599–1503.
8. Guerrero-Romero F, Rodríguez-Morán M. Low serum magnesium levels and metabolic syndrome. *Acta Diabetol* 2002;39:209–213.
9. Mykkanen L, Laakso M, Uusitupa M, et al. Prevalence of diabetes and impaired glucose tolerance in elderly subjects and their association with obesity and family history of diabetes. *Diabetes Care* 1990;13:1099–1105.
10. Lapidus L, Bengtsson C, Lissner L, et al. Family history of diabetes in relation to different types of obesity and change of obesity during 12-yr period. Results from prospective population study of women in Goteborg, Sweden. *Diabetes Care* 1992;15:1455–1458.
11. Wilson FH, Hariri A, Farhi A, et al. A cluster of metabolic defects caused by mutation in a mitochondrial tRNA. *Science* 2004;306:1190–1194.
12. Weyer Ch, Bogardus C, Mott DM, et al. The natural history of insulin secretory dysfunction and insulin resistance in the pathogenesis of type 2 diabetes mellitus. *J Clin Invest* 1999;104:787–794.
13. Durlach J, Bac P, Durlach V, et al. Magnesium status and ageing: an update. *Magnes Res* 1998;11:25–42.
14. Guerrero-Romero F, Rodríguez-Morán M. Relationship between serum magnesium levels and C-reactive protein concentration, in non-diabetic, non-hypertensive obese subjects. *Int J Obes Relat Metab Disord* 2002;26:469–474.
15. Malpuech-Brugere C, Nowacki W, Daveau M, et al. Inflammatory response following acute magnesium deficiency in the rat. *Biochim Biophys Acta* 2000;15;1501:91–98.
16. Song Y, Manson JE, Buring JE, et al. Dietary magnesium intake in relation to plasma insulin levels and risk of type 2 diabetes in women. *Diabetes Care* 2004;27:59–65.
17. Lopez-Ridaura R, Willett WC, Rimm EB, et al. Magnesium intake and risk of type 2 diabetes in men and women. *Diabetes Care* 2004;27:134–140.
18. Colditz GA, Manson JE, Stampfer MJ, et al. Diet and risk of clinical diabetes in women. *Am J Clin Nutr* 1992;55:1018–1023.
19. Meyer KA, Kushi LH, Jacobs DR Jr, et al. Carbohydrates, dietary fiber, and incident type 2 diabetes in older women. *Am J Clin Nutr* 2000;71:921–930.
20. Alzaid AA, Dinneen SF, Moyer TP, et al. Effects of insulin on plasma magnesium in noninsulin-dependent diabetes mellitus: evidence for insulin resistance. *J Clin Endocrinol Metab* 1995;80:1376–1381.
21. White JR Jr, Campbell RK. Magnesium and diabetes: a review. *Ann Pharmacother* 1993;27:775–780.

22. Ma J, Folson AR, Melnick SL, et al. Association of serum and dietary magnesium with cardiovascular disease, hypertension, diabetes, insulin, and carotid arterial wall thickness: the ARIC study. *J Clin Epidemiol* 1995;48:927–940.
23. Guerrero-Romero F, Tamez-Perez HE, González-González G, et al. Oral magnesium supplementation improves insulin sensitivity in non-diabetic subjects with insulin resistance. A double-blind placebo-controlled randomized trial. *Diabetes Metab* 2004;30:253–258.
24. Rodriguez-Moran M, Guerrero-Romero F. Oral magnesium supplementation improves insulin sensitivity and metabolic control in Type 2 diabetic subjects. A randomized, double-blind controlled trial. *Diabetes Care* 2003;26:1147–1152.
25. Gums JG. Magnesium in cardiovascular and other disorders. *Am J Health Syst Pharm* 2004;61:1569–1576.
26. Fox C, Ramsoomair D, Carter C. Magnesium: its proven and potential clinical significance. *South Med J* 2001;94:1195–1201.
27. Guerrero-Romero F, Rodríguez-Morán M. Complementary therapies for diabetes: the case for chromium, magnesium, and antioxidants. *Arch Med Res* 2005;36:250–257.
28. Lukaski HC. Magnesium, zinc, and chromium nutriture and physical activity. *Am J Clin Nutr* 2000;72:585S–593S.
29. Paolisso G, Barbagallo M. Hypertension, diabetes mellitus, and insulin resistance: the role of intracellular magnesium. *Am J Hypertens* 1997;10:346–355.
30. Ebel H, Gunther T. Magnesium metabolism: a review. *J Clin Chem Biochem* 1980;18:257–270.
31. Suarez A, Pulido N, Casla A, et al. Impaired tyrosine-kinase activity of muscle insulin receptors from hypomagnesaemic rats. *Diabetologia* 1995;38:1262–1270.
32. Barbagallo M, Dominguez LJ, Galioto A, et al. Role of magnesium in insulin action, diabetes and cardio-metabolic syndrome X. *Mol Aspects Med* 2003;24:39–52.
33. Rodríguez-Morán M, Guerrero-Romero F. Elevated serum concentration of tumor necrosis factor-alpha is linked to low serum magnesium levels in the obesity-related inflammatory response. *Magnes Res* 2004;17:189–196.
34. Saris NE, Mervaala E, Karppanen H, et al. Magnesium: an update on physiological, clinical and analytical aspects. *Clin Chem Acta* 2000;294:1–26.
35. Jones JE, Manalao R, Flink EB. Magnesium requirements in adults. *Am J Clin Nutr* 1967;20:632–635.
36. Pennington JA, Schoen SA. Total diet study: estimated dietary intakes of nutritional elements 1982–1992. *Int J Vitam Nutr Res* 1996;66:350–362.
37. Galan P, Preziosi P, Durlach V, et al. Dietary magnesium intake in a French adult population. *Magnes Res* 1997;10:321–328.
38. Ford ES. Race, education, and dietary cations: findings from the Third National Health and Nutrition Examination Survey. *Ethn Dis* 1998;8:10–20.
39. Ford ES. Dietary magnesium intake in a National Sample of U.S. adults. *J Nutr* 2003;133:2879–2882.
40. Kawano Y, Matsuoka H, Takishita S, et al. Effects of magnesium supplementation in hypertensive patients: assessment by office, home, and ambulatory blood pressures. *Hypertension* 1998;32:260–265.
41. Lefebvre PJ, Paolisso G, Scheen AJ. Magnesium and glucose metabolism. *Therapie* 1994;49:1–7.
42. Kisters K, Spieker C, Tepel M, et al. New data about the effect of oral physiological magnesium supplementation on several cardiovascular risk factors (lipids and blood pressure). *Magnes Res* 1993;6:355–360.

43. Guerrero-Romero F, Rodríguez-Morán M. Hypomagnesemia is linked to low serum HDL-cholesterol irrespective of serum glucose values. *J Diabetes Complications* 2000;14:272–276.
44. Grifò G, Lo Presti R, Montana M, et al. Plasma, erythrocyte and platelet magnesium in essential hypertension, diabetes mellitus without and with macrovascular complications and atherosclerotic vascular disease. *Recenti Prog Med* 1995; 86:431–436.
45. Grafton G, Bunce CM, Sheppard MC, et al. Effect of Mg^{2+} on Na(+)-dependent inositol transport. Role for Mg^{2+} in etiology of diabetic complications. *Diabetes* 1992;41:35–39.
46. Rodríguez-Morán M, Guerrero-Romero F. Low serum magnesium levels and foot ulcers in subjects with type 2 diabetes. *Arch Med Res* 2001;32:300–303.
47. Kao WHL, Folsom AR, Nieto FJ, et al. Serum and dietary magnesium and the risk for type 2 diabetes mellitus: the Atherosclerosis Risk in Communities Study. *Arch Intern Med* 1999;159:2151–2159.
48. Salmeron J, Ascherio A, Rimm EB, et al. Dietary fiber, glycemic load, and risk of NIDDM in men. *Diabetes Care* 1997;20:545–550.
49. Salmeron J, Manson JE, Stampfer MJ, et al. Dietary fiber, glycemic load, and risk of non-insulin-dependent diabetes mellitus in women. *JAMA* 1997;277:472–477.
50. Hodge AM, English DR, O'Dea K, et al. Glycemic index and dietary fiber and the risk of type 2 diabetes. *Diabetes Care* 2004;27:2701–2706.
51. Balon TW, Gu JL, Tokuyama Y, et al. Magnesium supplementation reduces development of diabetes in a rat model of spontaneous NIDDM. *Am J Physiol* 1995;269: E745–E752.
52. Paolisso G, Sgambato S, Gambardella A, et al. Daily magnesium supplements improve glucose handling in elderly subjects. *Am J Clin Nutr* 1992;55:1161–1167.
53. Yokota K, Kato M, Lister F, et al. Clinical efficacy of magnesium supplementation in patients with type 2 diabetes. *J Am Coll Nutr* 2004;23:506S–509S.
54. Paoliso G, Scheen A, Cozzolino D, et al. Changes in glucose turnover parameters and improvement of glucose oxidation after 4-week magnesium administration in elderly noninsulin-dependent (type 2) diabetic patients. *J Clin Endocrinol Metab* 1994;78:1510–1514.
55. Eibl NL, Koop HP, Nowak HR, et al. Hypomagnesemia in type 2 diabetes: effect of a 3-month replacement therapy. *Diabetes Care* 1995;18:188–192.
56. de Valk HW, Verkaaik R, van Rijn HJM, et al. Oral magnesium supplementation in insulin-requiring type 2 patients. *Diabet Med* 1998;15:503–507.
57. Lima M de L, Cruz T, Pousada JC, et al. The effect of Magnesium supplementation in increasing doses on the control of type 2 diabetes. *Diabetes Care* 1998;21: 682–686.
58. Gullestad L, Jacobsen T, Dolva LO. Effect of magnesium treatment on glycemic control and metabolic parameters in NIDDM patients. *Diabetes Care* 1994;17: 460–461.
59. Paolisso G, Sgambato S, Pizza G, et al. Improved insulin response and action by chronic magnesium administration in aged NIDDM subjects. *Diabetes Care* 1989;12:265–269.
60. Nadler JL, Buchanan T, Natarajan R, et al. Magnesium deficiency produces insulin resistance and increased thromboxane synthesis. *Hypertension* 1993;21:1024–1029.

61. Fung TT, Manson JE, Solomon CG, et al. The association between magnesium intake and fasting insulin concentration in healthy middle-aged women. *J Am Coll Nutr* 2003;22:533–538.
62. Rosolova H, Mayer O, Reaven G. Effect of variations in plasma magnesium concentration on resistance to insulin-mediated glucose disposal in nondiabetic subjects. *J Clin Endocrinol Metab* 1997;82:3783–3785.
63. Rosolova H, Mayer OJ, Reaven GM. Insulin-mediated glucose disposal is decreased in normal subjects with relatively low plasma magnesium concentrations. *Metabolism* 2000;49:418–420.
64. Humphries S, Kushner H, Falkner B. Low dietary magnesium is associated with insulin resistance in a sample of young, nondiabetic black Americans. *Am J Hypertens* 1999;12:747–756.

13
Magnesium Intake and Hepatic Cancer

Andrzej Tukiendorf

Epidemiological studies conducted around the world have primarily focused on the negative correlation of magnesium in drinking water, coupled with differing sample sizes (states, provinces, towns), multiplications of observed morbidity variables (geochemical and geographical factors, analytical data on water), and different causes of mortality (vascular, cardiac, general). The majority of studies carried out have confirmed the existence of this inverse relationship; those on the largest scale have been rather conclusive on this point. However, there are cofactors that prevent us from considering water magnesium as a constant risk. Its effect may not be present when atmospheric and etiological variables are taken into account. We cannot underestimate the influence, especially statistical significance, of such factors when we are telling the water story. That said, the primary goal of this research must be to define and clarify the quantities of magnesium in drinking water that provides nutrition the body needs so that its organs are maintained in a satisfactory condition.[1]

The focus of epidemiological research has expanded from an initial concentration on infectious disease. A wide spectrum of effects, such as mortality and chronic ailments, acute ailments, and psychological factors, as well as environmental ones have been incorporated into study designs and cover what is usually referred to as effect on health.[2]

Regarding magnesium levels in drinking water, the question has been posed as to how a proportionally small intake of magnesium via water can influence the body's total magnesium content.[3] The problem is more complex, and the significance of magnesium intake is in fact both quantitative and qualitative.[1]

Quantitatively speaking, the amount of magnesium required to bridge a deficit in dietary magnesium can be represented by water magnesium. Allowing for differences in water intake via products such as tap water, bottled water, and drinks prepared with magnesium-containing water, water magnesium can represent a crucial contribution that allows for control of a relatively small intake.[1] Therefore a major source of magnesium is foodstuffs.[4]

A qualitative analysis of ingestion and uptake of metals, particularly magnesium, shows that such materials are better absorbed by the intestines when ingested in a waterborne solution as opposed to in food. Recent studies have suggested that tap water intake is a more effective means of maintaining body magnesium levels than dietary supplementation. This results from greater bioavailability: either increased absorption (better biodisposability) and/or increased utilization, probably linked to the biological significance of the hexahydrated form of magnesium.[1]

Metabolic pathology can be influenced by water magnesium intake either in quantitative or qualitative properties. Because cancer cells have high metabolic requirements and the critical quantitative intake of water magnesium (which relieves marginal magnesium deficiency) may explain its importance as a possible protective cancer disease factor in the case of absolute magnesium deficit, the metal has been judged as likely to contribute to human carcinogenesis.[5]

This chapter will initially give a brief description about the epidemiological aspects of magnesium intake in humans. Some basics about liver cancer are given, followed by information about the consequences of a cancerous state on magnesium metabolism and the influence magnesium has on the status of the disease. Reviews of studies on the possible influence of water magnesium on cancer as well as about major factors enhancing hepatic cancer survival are placed in the following sections of the chapter. New personal data on the epidemiology of magnesium intake in the considered aspects will be presented. Finally, some research and public health conclusions will close the chapter.

Magnesium Intake

Magnesium is a widely occurring element, and its primary paths of entry into the organism are through water and food. It is important to note that intestinal absorption of magnesium is very difficult to quantify despite extensive research on the subject.[6] The short half-life of radioactive magnesium, the type of magnesium most commonly used in studies of absorption, causes significant difficulty in research.[4] Due to this there is a paucity of knowledge available on magnesium uptake by the intestines from food.[3]

Apart from the above-mentioned problem, as previously mentioned the most significant proportion of magnesium intake comes from consumables. A great number of foodstuffs contain magnesium, especially vegetables and grains (i.e., leafy green vegetables, whole-grain cereals, beans, nuts, and shellfish). It is estimated that, on average, 30% of magnesium ingested through food is absorbed, but this figure is roughly dependent on the total amount of magnesium present in the diet as well as the amounts of other elements taken in such as calcium, phosphorous, fiber, and phytic acids.[7]

Another deciding factor in the level of magnesium intake is the volume of the element present in cooking water. It has been indirectly proven that an inverse correlation exists between magnesium loss in prepared food and its presence in cooking water; if the level of magnesium in cooking water is high, the magnesium loss in food is lowered.[1]

Approximately 10% to 30% of magnesium intake can be accounted for by drinking water. There are, however, wide variations in the levels of magnesium in drinking water. Ground water in lime and other soft rock contains greater concentrations of magnesium than water from soil and harder rock (granite).[4] Changes in food preparation methods as well as a shift to using ground water with a low magnesium content have caused the relatively new problem of magnesium deficiency. The suggestion has been made that in individuals with a low intake of magnesium through their diet while consuming water with elevated levels of magnesium, the body's total magnesium status is crucially dependent on the contribution of magnesium from water.[8]

Due to a higher bioavailability of water magnesium hydrated ions than magnesium in food, the magnesium concentration in plants bred in regions with magnesium-rich water may be higher, especially in cases where the soil is rich in magnesium along with water of high magnesium content used in irrigation. Residents of such areas eating locally cultivated foodstuffs may, owing to this phenomenon, experience higher magnesium intake. Therefore, food composition tables may not be completely accurate when listing mineral content. It is, however, believed that plant magnesium composition is determined more by genetic factors than environmental ones.[8]

Examining evolutionary changes in intake levels of various nutrients can help us determine, from an ecological point of view, disease risks owing to changes in the diet of modern man. The magnesium intake of a man from the Neolithic era is believed to be greater than 1000 mg per day. In our times the trend is towards decidedly lower levels of magnesium consumption, mainly due to the wholesale removal of magnesium that occurs in the processing and refining of grain; some 70% to 80% of magnesium is lost in this way. From 1910 to 1970, daily magnesium intake fell markedly from 410 mg/daily to 340 mg/daily. In Japan, magnesium intake decreased from 351 mg/daily to 318 mg/daily in the period from 1971 to 1985.[4] Increased levels of industrial treatment and decreasing magnesium levels by as much as 80% to 95% cause most people to ingest lower than the daily recommended allowance of magnesium (6 mg/kg/day). Some sources recommend a magnesium intake level of about 350 mg/daily.[7,9] In drinking water, the level of magnesium content should be approximately 30 mg/L.[1]

Hypomagnesaemia can be attributed to a number of other factors apart from low intake levels. A loss in magnesium can be caused by alcohol consumption. A study of regular consumers of alcohol who were not alcoholics was performed by means of an oral loading test with magnesium. The level of magnesium and calcium expelled in urine was higher among those consuming alcohol and proportional to the amount of alcohol consumed.[10]

In any particular population, deficiencies of magnesium that cause disease can be accounted for by many different risk factors combined with low intake.

Cases such as this are characterized by an accompanying hypocalcaemia, which does not respond to calcium and vitamin D therapy. However, it has been shown that vitamin D plays a role in the regulation of the magnesium uptake.[9] Therefore, it is a question of several weeks required to reconstitute normal intracellular levels of magnesium in deficiency cases. This may be an explanation for the lack of direct success of magnesium-treated hypomagnesaemia-caused disease.[4]

Furthermore, it has been suggested that calcium decreases magnesium uptake, which may itself suggest a decrease in membrane permeability by calcium or a negative influence on the carrier. Optimal uptake probably requires sodium and phosphorous. High levels of fatty acid—magnesium soaps in the intestines, caused by fats consumption, decreases uptake. Protein may decrease or increase uptake.[4,5]

However, the exact proximate causes of differing magnesium intake levels in the body are not fully known. This is particularly the case with meals of mixed ingredients, and total intake cannot be seen as a proxy or relevant measure for uptake. Studies of the body balance of magnesium measured in subjects with controlled levels of magnesium intake in food and water should be done to explore the proportional uptake of magnesium from consumed foodstuffs and water.[4]

There is a sharp decline in the intake of general nutrients by the elderly, and this principle holds true for magnesium. Older persons are also less able to absorb magnesium intestinally. Moreover, older men exhibit lower amounts of magnesium discharged in urine, whereas young women excrete lesser amounts of magnesium than postmenopausal women. This difference was more pronounced in women taking oral contraceptives. Magnesium requirements are, in line with those of other minerals and nutrients, probably raised in the case of the elderly.[11,12]

Magnesium in the Body

Ninety-nine percent of the total amount of magnesium in the body is located intracellularly.[9] More than half of this is found in bone and other tissues (approximately 60%), the greatest amount being located in skeletal muscle and the heart, but only 0.3% is present in the serum.[4]

Magnesium is a cofactor in all enzymes involved in phosphate transfer reactions that utilize adenosine triphosphate (ATP) and other nucleotide triphosphates and substrates. Magnesium plays a crucial role in most of the enzymes processing carbohydrate, lipid, nucleic acid, and protein metabolism (Magnesium is required in over 300 enzymes that aid the metabolic process of these reactions and ion transport.)[13]

Energy-reducing reactions (oxidative phosphorylation) and energy-consuming reactions (contraction) are also dependent on magnesium.[4] It is also involved in the regulation of cellular permeability[5] and neuromuscular excitability.[3]

It is well known that magnesium plays an essential role in a wide range of fundamental cellular reactions and its deficiency in an organism may lead to serious biochemical and symptomatic changes. As regards the liver, it is known that higher collagen concentrations were found in the livers of magnesium-deficient alcohol-fed rats that were observed under control due to increased collagen production in magnesium deficiency.[14] Moreover, the increase in plasma triglycerides and cholesterol observed in magnesium-deficient rats may be the result of increased hepatic synthesis (thus cholesterol output is markedly reduced in deficient animals).[15]

Future investigations should clarify whether these facts have a strong influence on the condition of the liver, including carcinogenesis, that may be caused by the disturbance of magnesium levels in the human organism. Some basics on the possible influence of magnesium on cancer are given in the next section.

Magnesium and Oncogenesis

The relationship between magnesium and cancer is complex and contradictory. Both magnesium load and deficit can produce carcinogenic and anticarcinogenic effects.[16] Carcinogenesis modifies magnesium status, inducing magnesium distribution disturbances which may frequently associate a tumor magnesium load with magnesium depletion in non-neoplastic tissues.[17]

In chemical carcinogenesis, cellular magnesium deficit can be seen in both preneoplastic and neoplastic states. Structural and functional changes of the plasma membrane are caused by magnesium deficit, including a marked reduction of the intracellular cations Mg and K, with a raising of extracellular cations Ca and Na in neoplastic cells.[18] Such extreme changes are similar to those cellular alterations resulting from magnesium deficiency. One basis for the prophylactic use of magnesium in oncology is to avoid the facilitation of the effect of a carcinogenic agent by the triggering of this kind of cellular disturbance.[17]

Disorders in magnesium distribution in carcinogenesis are far more complex than what can be found in simple magnesium deficiency. In the neoplastic state, magnesium contained in intracellular structures is lower, whereas its occurrence in cytosol (as well as mitochondria) rises.[5] Such a variance in magnesium distribution, not found in ordinary magnesium deficiency, points to an alteration of intracellular magnesium regulation. In other words, a depletion of magnesium is starting at the beginning of carcinogenesis.[5]

Later on, variances in the distribution of magnesium are even more complex. At later stages of cancer in humans, the disturbance in magnesium distribution associates an increased magnesium level in the tumor and in blood cells, including erythrocytes and lymphocytes, with some stigmata of magnesium depletion that vary according to the method of treatment.[19] Magnesium depletion occurs more often in patients with solid tumor malignancy than with hemopathic malignancy.[5,17]

Magnesium depletion does not occur equally in all forms of tissue. Experimental and clinical data indicates that the adverse effects of magnesium depletion in non-neoplastic tissue are reversible. It seems, therefore, that an established cancer induces magnesium disturbances that effect magnesium load in tumoral tissue, probably due to magnesium mobilization through blood cells, with magnesium lessening in non-neoplastic tissue.[5,17]

Such a process can be understood in this way—established carcinogenesis induces magnesium disturbances that accentuate tumor magnesium load due to magnesium mobilization through blood cells with magnesium depletion in non-neoplastic tissue. At a premalignant stage and in the case of some hemolymphoreticular malignancies, magnesium deficiency appears carcinogenic, but in the case of solid tumors it exerts a negative influence on tumor growth.[5] These disturbances lead to several logical consequences regarding oncological research and treatment. Indications for cancer therapy of magnesium supplementation or of induced magnesium deficiency are critical. It is possible to hypothesize on the importance of optimal treatment for magnesium disturbances as part of a general cancer treatment regime.[5,16]

There are several studies of magnesium experimental oncogenesis in animal laboratory subjects (mostly rats) that focus on the relation between the magnesium deficiency in diet and lymphomas and leukemias, bone tumors, intestinal tumorlike overgrowth, etc.[20,21] At this time, however, no such outcomes have been reported regarding a relationship between magnesium deficiency and hepatic cancer.

To familiarize the reader with the problem considered, some brief medical information on liver cancer and a concise epidemiological review of water magnesium studies on the disease, as well as hepatic cancer survival, are given in the following sections of the chapter.

Hepatic Cancer

Hepatic cancer is a rare disease and it is many times more common in developing countries, such as Africa and East Asia, than in the United States or Europe.[22] There are four main types of liver cancer that appear in different parts of the organ. The hepatocellular carcinoma starts in the hepatocytes, the main liver cells, and according to the American Cancer Society accounts for 84% of cases. Cholangiocarcinoma appears in the small bile ducts of the liver and accounts for about 13% of liver cancers. Angiosarcoma is a rare cancer that starts in the blood vessels of the liver, accounting for less than 1% of cases according to National Cancer Institute. Finally, hepatoblastoma is the rarest type of liver cancer (in children) affecting usually the right lobe.[23]

It is necessary to stress that no one knows the exact etiology of liver cancer. However, scientists have found that people with certain risk factors are more likely than others to develop liver cancer. Different hypotheses are tested that focus on hepatocarcinogenic mechanisms as well.[23] The epidemiological approach may play an essential role in this scientific activity.

Besides the causes of the disease not being well established, the factors that may prolong survival have also not been recognized so far.[24] That is why the prognoses for patients with liver cancer are generally poor, as most patients have no symptoms until the tumor has grown or spread to other parts of the body. By the time a diagnosis is made, it is often no longer possible to remove all cancerous tissue surgically in order to prolong survival. Late diagnosis may be a reason why the average survival after diagnosis is often quoted as 3 to 6 months.[25] In advanced liver cancer, median survival is usually 2 to 4 months.[24]

Epidemiology of Water Magnesium and Cancer

Although in dozens of epidemiological studies the effects of drinking-water magnesium on morbidity and mortality—predominantly from cardiovascular disease—have been recognized, until now not many reports have confirmed the epidemiology of water magnesium and cancer. Important findings in this field were provided recently by Taiwanese scientists. However, in most of their studies, these authors indicated negative statistical associations of various types of cancer morbidity/mortality but mostly with the hardness of water and calcium rather than magnesium.

In a review of these publications, it is worth noting the results concerning the possible association between the risk of gastric cancer and the levels of calcium and magnesium.[26] The present study suggests that there was a significant protective effect of calcium intake from drinking water on the risk of gastric cancer. Magnesium also exerted a protective effect against gastric cancer, but only for the group with the highest levels of magnesium exposure. In another earlier matched case-control study,[27] the authors found a possible association between the risk of colon cancer and hardness levels in drinking water from municipal supplies in Taiwan (obtained trend analyses showed an increasing odds ratio for the cancer with decreasing hardness in drinking water). Similar epidemiological trends were also achieved for the relations between hardness levels in drinking water and the risk for rectal cancer[28] and pancreatic cancer mortality,[29] however, the researchers did not find any association with magnesium levels (the adjusted odd ratios were not statistically significant for the relationship between magnesium concentrations in drinking water and rectal cancer).[30] One of the strongest epidemiological proofs of a significant protective effect of magnesium intake from drinking water was that given for the risk for esophageal cancer[31] and ovarian cancer.[32] Unfortunately, these authors did not find any results pertaining to the similar trend between drinking-water magnesium and liver cancer.[33] The first strong evidence concerning the possible ecological relation between exposure to water magnesium and hepatic cancer was reported in Eastern Europe by this author.[34,35]

Hepatic Cancer Survival

In patients with hepatic cancer the prediction of survival is difficult. This problem particularly concerns the used therapy.[36] These studies are mostly based on clinical trials and among the favorable prognostic factors that prolong the survival of patients with hepatic cancer. The following treatment methods are suggested: transarterial chemoembolization, partial resection, liver transplantation,[36] percutaneous ethanol injection,[37] cytoreductive surgery,[38] radiotherapy,[39] and a combination of some of these therapies[40] rather than anti-cancer drugs (chemotherapy),[41] although sometimes these theses are questioned.[42] In the cited studies it has been also reported that the survival on liver cancer strongly depends, for example, on an early detection of the disease, stage of disease, tumor size, association with viral infections or alcoholism, etc. Unfortunately, some skeptical opinions in the hepatological community are also claimed. One of these is that most of the cancer hepatic treatments are in fact curative and can palliate only.[43]

New Personal Data on Drinking-Water Magnesium and Hepatic Cancer

Based on extensive ecological data and conducted Bayesian modeling,[44] these results report on epidemiological findings concerning a possible relationship between magnesium in drinking water and risk for liver cancer and its survival. The studies were performed using historical cancer morbidity registers and water quality was assessed by geographical area at the commune level. No information was available on other risk factors for hepatic cancer disease, such as dietary habits, occupational exposure, or smoking. Despite these methodological shortcomings, surprisingly large differences and very strong dose–response relationships were found. Relative estimated contributions of water quality to endemic waterborne disease for magnesium concentrations both for males and females using random effects logistic regression[44] are shown in Figure 13.1.

Both relations presented for males and females accordingly show that for the lowest magnesium concentration in drinking water, the cancer risk was significantly higher than it was for the highest magnesium content (approximately four times in males and three times in females, respectively). Thus, the

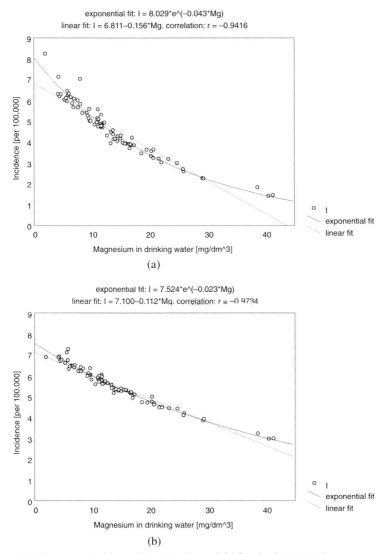

FIGURE 13.1. Liver cancer incidence (I) in (a) males and (b) females (1985–1994) versus magnesium concentration in drinking water (1980–1985) in administrative units of the province of Opole, Poland.[34]

achieved outcomes provide evidence of possible positive health effects of magnesium water supplies on this lethal disease.

The question of how water magnesium exposure might influence survival of liver cancer patients with age using Cox's regression approach[45] is considered in Table 13.1 and Figure 13.2; the results were recently displayed during the scientific conference Magnesium In Biochemical Processes & Medicine

TABLE 13.1. Estimates of hazard ratios (Cox's regression).

Variable	Hazard Ratio	95% Confidence Interval
Gender (M, F)	0.873	(0.711, 1.072)
Age (year)	1.006	(0.997, 1.016)
Magnesium (mg/dm^3)	0.979	(0.962, 0.997)

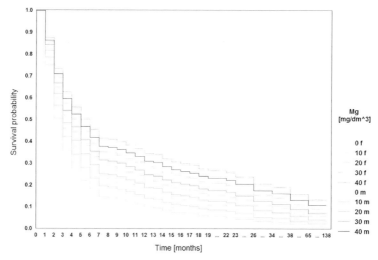

(a) age = 10 years

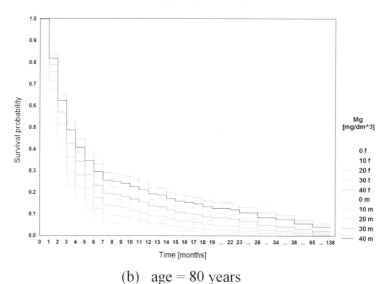

(b) age = 80 years

FIGURE 13.2. Estimated postdiagnosis survival probability in males (m) and females (f) with liver cancer for drinking-water magnesium exposure for patients aged 10 (a) and 80 (b) years old.

organized by the Gordon Research Conferences in Ventura, California (06–11.02.2005).

The values indicate that females have a lower chance of early mortality (over 12%) in comparison to males. Moreover, the mortality risk increased with the age of the patient at diagnosis (around 0.6% with every year of age). Finally, the estimated hazard ratio associated with magnesium supports the hypothesis of a beneficial effect of water magnesium consumption on the chance of early mortality from the disease (the risk of early mortality for the patients decreases about 2% with every unit magnesium concentration increment; see the hazard ratios in Table 13.1). Evidence for the relationship between survival in patients suffering from liver cancer and exposure of the magnesium content of drinking water (for the extreme age bands) can also be identified in the modeled survival curves (Figure 13.2).

The modeled survival curves presented in Figure 13.2 confirm the results set in Table 13.1. For different drinking-water magnesium concentrations for men and women aged 10 and 80, they illustrate the improved survival of patients suffering from liver cancer disease who were exposed to a higher magnesium concentration. Moreover, slightly longer prognoses are noticed for the younger patients.

Conclusions

Speculation on and research into the cause of disease have always been an important focus of interest for man. Researchers in hospitals and medical centers around the world are working to learn more about what causes liver cancer. To the most important etiological factors of the disease, hitherto studies have included the chronic liver infection (hepatitis). Many scientific reports accordingly confirm that there is a strong association between chronic hepatitis B and C viral infection and the development of the disease. Moreover, people with fibrosis and cirrhosis also have an increased risk of liver cancer. Other possible hepatocarcinogens include aflatoxin, nitrosamines, oral estrogen compounds, and numerous other chemicals. The National Cancer Institute mentions risk factors such as family history, being male, and age.[46]

The last two factors seem to be confirmed by the results presented above. From the ecological and medical points of view, a possible impact of water magnesium intake on the risk for hepatic cancer as well as liver cancer survival is attention grabbing. No such relation, except for in the Opole province in Poland, has been reported before in scientific literature.[34,35] Therefore, these results should undergo scientific assessment and verification. Moreover, the creation of this type of epidemiological hypotheses is faced with theoretical problems.

The major problem is the contributions of magnesium in drinking water to the total amount ingested. However, at this stage of knowledge it is suggested that waterborne magnesium is better absorbed than dietary magnesium, and this fact is associated with the general properties of cellular membranes.

Because amnion is a particularly leaky membrane, the effect of water magnesium may be due to the hexahydrated form in connection with linked amnion. Water magnesium may antagonize lead and cadmium on the plasma membrane, particularly during intoxication.[1]

The next question regards the time of latency between the putative cause-specific factor and the follow-up neoplasmatic process. Intuitively, the period of the liver cancer morbidity should have followed the water magnesium consumption. Although the condition seems to be fulfilled in the personal study, no one knows how long this time should be at least approximately assumed.

Nevertheless, the present results and given theoretical background allow for some scientific speculations. One of these could be the reason why alcohol causes abnormalities in magnesium metabolism and results in severe liver diseases (including cirrhosis),[47,48] as well as why the most diseased group (both male and female) consisted of older patients. One of the hypothesized backgrounds for the association between the magnesium exposure and liver cancer could be direct/indirect effects of alcohol consumption. This hypothesis could be justified by the fact that alcohol decreases the magnesium concentration in body and it can be returned into the organism via water ingestion. Then, linking the magnesium higher intakes from the richer water supplies together with its preventive properties in neoplasmatic processes might be a reasonable justification why liver cancer morbidity is diminishing with the increase of drinking water magnesium exposure, and vice versa. If the hypothesis is true, the introduction of social drinking, a habit that is also spreading into the younger age groups, is thus a new phase of our adaptation to a lifestyle that implies severe risks for health effects, including cancer.

If in the light of the given facts concerning drinking-water magnesium and the risk for liver cancer turn more into consistency, the identified epidemiological relationship between survival in patients suffering from hepatic cancer and the element's exposure still remains unclear. The latter effect might be justified by the greater load of magnesium in body in people who were drinking magnesium-rich water. However, any attempts to explain the possible metabolic mechanism need collaborative consultations and discussions of a variety of specialists, that is, biochemists, physiologists, hepatologists, etc.

Further studies of the possible magnesium effects on the hepatic cancer and its survival will open new paths to cancer research. It is believed, however, that present epidemiological findings will be helpful in directing scientific concern towards proper research in the field of a possible relation between magnesium intake and hepatic cancer and to explain the observed evidence.

Accordingly, as concerns public health, it is important that such studies are undertaken. If a risk group can be identified, then public health concerns to restore magnesium deficiency should be targeted exclusively to that group, however if a risk group cannot be identified, the entire population must be protected, probably by adding magnesium to the drinking water in optimal amounts. However, before public health activities are undertaken, it is paramount to determine whether this risk is present but only after more precise clinical or experimental research; thus, possible preventive actions can then be acted upon efficiently.

Acknowledgment. I acknowledge Professor David Spiegelhalter (Fellow of the Royal Society) from the Medical Research Council—Biostatistics Unit, Cambridge, UK, for his statistical support and contribution during computational work.

References

1. Durlach J, Bara M, Guiet-Bara A. Magnesium level in drinking water: its importance in cardiovascular risk. In: Itokawa Y, Durlach J, eds. *Magnesium in Health and Disease*. London: John Libbey; 1989:173–182.
2. Rylander R. Water, health and epidemiology. In: Golding AMB, Noah N, Stanwell-Smith R, eds. *Water and Public Health*. London: Smith-Gordon; 1994:1–14.
3. Rylander R. Magnesium and calcium in drinking water and cardiovascular mortality. *Scand J Work Environ Health* 1991;17:91–94.
4. Rylander R. Environmental magnesium deficiency as a cardiovascular risk factor. *J Cardiovasc Risk* 1996;3:4–10.
5. Durlach J, Bara M, Guiet-Bara A. Magnesium and its relation to oncology. Magnesium and its relationship to oncology. In: Sigel H, Sigel A, eds. *Metal Ions in Biological Systems*. Vol. 26. *Compendium on Magnesium and Its Role in Biology, Nutrition and Physiology*. New York: Marcel Dekker; 1990:549–578.
6. Ebel H, Günther T. Magnesium metabolism: a review. *J Clin Chem Biochem* 1980;18:257–270.
7. Rubenowitz E, Landin K, Rylander R. Magnesium in drinking water and skeletal muscle and sublingual cells. *Magnes Bull* 1994;16:81–86.
8. Rubenowitz E, Molin I, Axelsson G, Rylander R. Magnesium in drinking water in relation to morbidity and mortality from acute myocardial infarction. *Epidemiology* 2000;11:416–421.
9. Rubenowitz E, Axelsson G, Rylander R. Magnesium and calcium in drinking water and death from acute myocardial infarction in women. *Epidemiology* 1999;10:31–36.
10. Rylander R. Mineral/electrolyte-related disease induced by alcohol. In: Watson RR, Preedy VR, eds. *Nutrition and Alcohol: Linking Nutrient Interactions and Dietary Intake*. Boca Raton, FL: CRC Press; 2004:413–424.
11. Seelig MS. Possible roles of magnesium in disorders of the aged: intervention in the ageing process. In: Regelson W, Sinek FM, eds. *Quantitation, Epidemiology, and Clinical Research*. New York: Alan R. Liss; 1983:279–305.
12. Seelig MS. *Role of Magnesium in Disorders of the Aged*. Carlobad, CA: Anti-Aging Nutritional & Education Center; 2002. Available from: http://www.centerforantiaging.com/Role_of_Magnesium_in_Disorders_of_the_Aged.htm (accessed June 2006).
13. Aikawa JK. *Magnesium: Its Biological Significance*. Boca Raton, FL: CRC Press; 1981:21–38.
14. Rayssiguier Y, Chevalier F, Bonnet M, Kopp J, Durlach J. Influence of magnesium deficiency on liver collagen after carbon tetrachloride or ethanol administration to rats. *J Nutr* 1985;115:1656–1662.

15. Rayssiguier Y, Gueux E, Durlach V, Durlach J, Nassir F, Mazur A. Magnesium and the cardiovascular system: I. New experimental data on magnesium and lipoproteins. In: Halpern MJ, ed. *Molecular Biology of Atherosclerosis. Proceedings of the 57th European Atherosclerosis Society Meeting*. London: John Libbey; 1992:507–512.
16. Seelig MS. Magnesium (and trace substance) deficiencies in the pathogenesis of cancer. *Biol Trace Elem Res* 1979;1:273–297.
17. Durlach J, Rinjard P, Bara M, Guiet-Bara A, Collery P. Données nouvelles sur les rapports entre magnésium et cancer. In: Lassere B, ed. *Magnesium—Physiologische Aspekte für die Praxis*. Zurich: Pascentia Verlag; 1987:26–45.
18. Günther T, Averdunk R. Membranes of magnesium deficiency induced neoplastic cells. *Magnes Bull* 1985;7:146–151.
19. Collery P, Anghileri LJ, Coudoux P, Durlach J. Magnesium and cancer: clinical data. *Magnes Bull* 1981;3:11–20.
20. Hass GM, Laing GH, Galt RM, McCreary PA. Role of magnesium deficiency in immunity to neoplasia in the rat. *Magnes Bull* 1981;3:5–11.
21. Hass GM, Laing GH, Galt RM, McCreary PA. Recent advances: immunopathology of magnesium deficiency in rats: induction of tumors; incidence, transmission and prevention of lymphoma-leukemia. *Magnes Bull* 1981;3:217–228.
22. Parkin DM, Muir CS, Whelan SL, Gao Y-T, Ferlay J, Powell J, eds. *Cancer Incidence in Five Continents*. Lyon: IARC Scientific Publications; 1992;120.
23. *Health News & Information*. Buffalo, NY: 2003. Available from: http://www.stayinginshape.com/3chsbuffalo/libv/c15.shtml (accessed June 2006).
24. National Cancer Institute. *MedNews* Catholic Health System; 2003. Available from: http://www.meb.uni-bonn.de/cancer.gov/CDR0000062906.html (accessed June 2006).
25. Asian Liver Center. Stanford, CA: Stanford University; 2003. Available from: http://livercancer.stanford.edu/index2.asp?lang=eng&page=livercancer (accessed June 2006).
26. Yang CY, Cheng MF, Tsai SS, Hsieh YL. Calcium, magnesium, and nitrate in drinking water and gastric cancer mortality. *Jpn J Cancer Res* 1998;89:124–130.
27. Yang CY, Hung CF. Colon cancer mortality and total hardness levels in Taiwan's drinking water. *Arch Environ Contam Toxicol* 1998;35:148–151.
28. Yang CY, Tsai SS, Lai TC, Hung CF, Chiu HF. Rectal cancer mortality and total hardness levels in Taiwan's drinking water. *Environ Res* 1999;80:311–316.
29. Yang CY, Chiu HF, Cheng MF, Tsai SS, Hung CF, Tseng YT. Pancreatic cancer mortality and total hardness levels in Taiwan's drinking water. *J Toxicol Environ Health A* 1999;56:361–369.
30. Yang CY, Chiu HF. Calcium and magnesium in drinking water and risk of death from rectal cancer. *Int J Cancer* 1998;77:528–532.
31. Yang CY, Chiu HF, Tsai SS, Wu TN, Chang CC. Magnesium and calcium in drinking water and the risk of death from esophageal cancer. *Magnes Res* 2002;15:215–222.
32. Chiu HF, Chang CC, Yang CY. Magnesium and calcium in drinking water and risk of death from ovarian cancer. *Magnes Res* 2004;17:28–34.
33. Yang CY, Chiu HF, Tsai SS, Chang CC, Chuang HY. Magnesium in drinking water and the risk of death from liver cancer. *Magnes Res* 2002;15:223–228.
34. Tukiendorf A. Magnesium in drinking water and liver cancer morbidity—a possible relation? *Cent Eur J Public Health* 2002;4:157–162.

35. Tukiendorf A, Rybak Z. New data on ecological analysis of possible relationship between magnesium in drinking water and liver cancer. *Magnes Res* 2004;17:46–52.
36. Sangro B, Herraiz M, Martinez-Gonzalez MA, et al. Prognosis of hepatocellular carcinoma in relation to treatment: a multivariate analysis of 178 patients from a single European institution. *Surgery* 1998;124:1087.
37. Allgaier HP, Deibert P, Olschewski M, et al. Survival benefit of patients with inoperable hepatocellular carcinoma treated by a combination of transarterial chemoembolization and percutaneous ethanol injection—a single-center analysis including 132 patients. *Int J Cancer* 1998;79:601–605.
38. Nagashima J, Okuda K, Tanaka M, Sata M, Aoyagi S. Prognostic benefit in cytoreductive surgery for curatively unresectable hepatocellular carcinoma—comparison to transcatheter arterial chemoembolization. *Int J Oncol* 1999;15:1117–1123.
39. Cheng JC, Chuang VP, Cheng SH, et al. Local radiotherapy with or without transcatheter arterial chemoembolization for patients with unresectable hepatocellular carcinoma. *Int J Radiat Oncol Biol Phys* 2000;47:435–442.
40. Allgaier HP, Deibert P, Olschewski M, et al. Survival benefit of patients with inoperable hepatocellular carcinoma treated by a combination of transarterial chemoembolization and percutaneous ethanol injection—a single-center analysis including 132 patients. *Int J Cancer* 1998;79:601–605.
41. Mok TSK, Leung TWT, Chao S-DLY, et al. A multi-centre randomized phase II study of nolatrexed versus doxorubicin in treatment of Chinese patients with advanced hepatocellular carcinoma. *Cancer Chemother Pharm* 1999;44:307–311.
42. Yao F, Terrault N. Hepatitis C and hepatocellular carcinoma. *Curr Treat Options Oncol* 2001;2:473–483.
43. Lau WY, Leow CK, Li AK. Hepatocellular carcinoma. *Br J Hosp Med* 1997;57:101–104.
44. Spiegelhalter D, Thomas A, Best N, Gilks W. *BUGS. Examples.* Vol. 1. Cambridge: Medical Research Council—Biostatistics Unit; 1996:10–14.
45. Cox DR. Regression models and life-tables. *J Royal Stat Soc B* 1972;34:187–220.
46. *What You Need To Know About Liver Cancer.* Washington, DC: National Cancer Institute; 2003. Available from: http://www.nci.nih.gov/cancerinfo/wyntk/liver (accessed June 2006).
47. Flink EB, Omar M, Shane SR. Alcoholism, liver disease and magnesium. *Magnes Bull* 1981;3:209–216.
48. Duffy JC, Latcham RW. Liver cirrhosis mortality in England and Wales compared to Scotland: an age-period-cohort analysis 1941–1981. *J Royal Stat Soc A* 1986;149:45–59.

Exercise

14
Exercise and Magnesium

Maria José Laires and Cristina Monteiro

Magnesium is essential for the optimal function of a diversity of life-sustaining processes. It is cofactor of more than 300 enzymes, participating in the metabolism of carbohydrates, lipids, proteins, and nucleic acids, in the synthesis of hydrogen transporters, and particularly in all reactions involving the formation and use of adenosine triphosphate (ATP). Magnesium also serves as a regulator of many physiological functions, including neuromuscular, cardiovascular, immunological, and hormonal functions, as well as the maintenance of membrane stability.[1,2]

Magnesium can participate in the reactions involving the formation and use of ATP by two different mechanisms. Kinetic studies of several enzymes requiring both magnesium and ATP showed that enzyme activity depends on the ratio, as well as on the absolute concentrations, of the two cofactors and that magnesium chelates strongly with ATP, forming a Mg(ATP) complex, the active substrate for enzyme action. Similarly, a magnesium complex with adenosine diphosphate (ADP), Mg(ADP), appears to be the active substrate for some enzymes. The second general mechanism of magnesium action is the direct binding of free magnesium (Mg^{2+}) to the enzyme protein and resultant allosteric activation. With some enzymes magnesium has a dual function, not only forming part of the reactive substrate but also activating the enzyme allosterically. However, in some cases Mg(ATP) and Mg^{2+} may have opposite actions. For example, the Mg(ATP) complex stimulates the type-L calcium channels and Mg^{2+} inhibits them.

Magnesium is also known to alter both receptor sites and ion movements across the cell membrane. By making complexes with phospholipids, magnesium stabilizes the membranes, reducing their fluidity and permeability. Thus, magnesium is an important modulator of intracellular ion concentrations. In magnesium deficit, intracellular concentrations of calcium and sodium increase, and concentrations of potassium and phosphorus decrease. Simultaneously, the membrane depolarizes. These alterations may be the result of magnesium's direct effect on sodium, calcium, or potassium channels or the indirect result of its effect on enzymes in the cell membrane that are involved in active transport, for example,

(Na^+K^+)—ATPase. Magnesium also regulates lipid and phosphoinositide-derived second messengers.

Within the cell, magnesium affects the function of organelles such as sarcoplasmatic reticulum, primarily by its ability to alter calcium flux, or mitochondria, by altering their membrane's permeability to protons, which leads to alterations in the coupling of oxidative phosphorylation and electron transport chains, thus affecting the efficiency of ATP production.[1] Magnesium also acts as a calcium antagonist. In the neuromuscular system it reduces the electric excitability of the neurons and inhibits the release of acetylcholine by the nerve endings at the neuromuscular junction, and blocks the effect of N-methyl-D-aspartate, an excitatory neurotransmitter of the central nervous system. In muscle contraction, both stimulation and the activity of the calcium transport system in the sarcoplasmatic reticulum membranes depend on the presence of Mg^{2+}. Troponin contains four calcium-binding sites, two of which have a high affinity for calcium and bind Mg^{2+} competitively. These calcium-magnesium type sites do not seem directly involved in any rapid twitching mechanism, but play a structural role in muscle. Magnesium bound to these sites may maintain the protein permanently in a particular conformational state regardless of the fluctuation in calcium (assuming that both magnesium and calcium-induced structural changes are essentially the same). This conformation may be a prerequisite for calcium activation via binding at the calcium-specific sites.

These and many other functions make it easy to understand why performing exercise is highly dependent on the regulation of magnesium homeostasis. Additionally, there is evidence that exercise performance seems to be impaired under conditions of magnesium deficiency.[3] This is why individuals performing exercise should pay extra attention to magnesium status.

Athletes and Magnesium Status

Frequently, physically active individuals fail to consume a diet that contains adequate amounts of minerals, including magnesium, which leads to marginal nutrient deficiency and results in substandard training and impaired performance.[4-6] Additionally, mineral losses in urine and sweat are more important during exercise than at rest.[7] These conditions may contribute to frequent mineral deficits among athletes.[6,8]

During exercise, compartmental shifts of magnesium have been observed, but data to demonstrate magnesium variations induced by exercise are inconsistent. Such heterogeneity can be partially attributed to differences in experimental designs and work intensity and duration. Moreover, the timing of blood sample and the different analytical protocols have to be taken into account.[9]

With respect to blood extracellular magnesium, various authors have indicated that high-intensity exercise leads to hypermagnesemia as a consequence of the decrease in plasma volume.[10-12] These changes may depend on the relative contribution of anaerobic metabolism to the total energy expended during exercise.[13]

On the other hand, submaximal exercise has been reported to be accompanied by hypomagnesemia.[14–19] Prolonged strenuous exercise, especially under hot conditions, may lead to hypomagnesemia.[14,20–23] Low plasma magnesium levels have been explained by several mechanisms: redistribution of magnesium to red blood cells, adipocytes or myocytes;[24] loss in urine due to increases in aldosterone, anti-diuretic hormone, thyroid hormone, and acidosis that reduce the tubular reabsorption of magnesium;[13] or increased lipolysis due to raises in catecholamines levels induced by exercise.[25–27] Hyperexcretion in sweat only acquire real importance in case of intense activity made in conditions of damp atmosphere and high temperature.[28]

A transient shift of magnesium to the intracellular space during exercise is a probable explanation for a large proportion of hypomagnesaemia. However, regarding red blood cell (RBC) magnesium variations with exercise, dissimilar findings are reported. Red blood cell magnesium levels were reported to be increased after several exercises[8,10,13,29] and were related to the increased metabolic activity during exercise, which would induce a shift of the cation from the plasmatic compartment.

Conversely, RBC magnesium levels were reported to be decreased.[18,30–32] Golf and coworkers[32] postulated that as exercise duration increases, magnesium shifts from the erythrocyte reservoir into the plasma and then into the working muscles. With prolonged exercise (more than 1 h) hypomagnesaemia may occur as a result of the depletion of the erythrocyte reservoir. Several studies suggest that low red blood cell magnesium levels may persists during a season of training.[33–35]

Mg^{2+} concentration is supposed to be a more sensitive parameter than that of total magnesium and so it should give more reliable information about the status and regulation of major, mobilizable magnesium pools in the body. However, only limited information on the exercise effects on the metabolic and regulatory fraction of Mg^{2+} is available.[36]

Recently, Mooren and coworkers[37] showed that, at the end of a treadmill ergometer test, both total blood and serum Mg^{2+} and serum total magnesium decreased. In contrast, Mg^{2+} increased in both thrombocytes and erythrocytes. Intracellular total magnesium was unchanged, making a Mg^{2+} shift between the intra- and extracellular blood compartment unlikely. This study also showed opposite changes of the ratio [Mg^{2+}]/[total Mg] in the intracellular and the extracellular compartment after anaerobic exercise. In in vitro experiments, similar changes of Mg^{2+} in the two blood compartments could be mimicked by application of weak acids like lactic and propionic acid. These authors concluded that changes in the fraction of Mg^{2+} should be enough to influence intracellular signaling and metabolic processes.

Although some explanations have been offered for the compartmental shifts of magnesium, the precise mechanism remains to be clarified. It is important to evaluate whether there is only a transient fall in plasma magnesium concentration, or if the participation in sustained exercise may induce permanent alterations. Several studies indicate that there is a sustained fall in plasma magnesium after strenuous exercise and that hypomagnesaemia persists during a season of training.[34,35,38–40]

Exercise may increase the demand for magnesium and/or increase magnesium loss, potentially leading to magnesium deficit, which can result in muscle weakness, and neuromuscular dysfunction.[20,41]

Overt signs and symptoms of magnesium deficit, for example, hyperirritability, tetany, convulsions, and cardiac arrhythmias, may not be manifested until serum magnesium concentration has decrease below 0.5 mmol/L. Generally, exercise-induced magnesium decrease does not approach the level that would be of concern from a health point of view. Thus, the consequences of exercise-induced magnesium decreases on health status in healthy people with normal magnesium levels seem to be negligible.[42] However, they may be more relevant in people who already have low magnesium. This problem may be exacerbated in those who experience large increases in plasma free fatty acids during exercise, because of the inverse relation between serum magnesium and free fatty acids. It is also important to note that magnesium deficit plays a role in myocardial injury after prolonged sustained exercise, as it may lead to an increased potential for thrombus formation and/or coronary vasospasm.[43]

Reports on magnesium supplementation are also discordant. It has been reported to increase muscle strength and power.[44] Conversely, marathon runners with adequate magnesium status exhibited no improvement in running performance or skeletal muscle function and no increase in muscle magnesium concentrations.[45]

In spite of the fact that most studies lack information about magnesium status of the subjects previous to supplementation, which may justify some of the discordant results, the findings suggest that magnesium supplementation per se does not illicit beneficial effect on physiological function or performance when magnesium status is normal.[5] Doctors and athletes should be aware of the magnesium changes during exercise in order to avoid health risk.[42]

Exercise and Magnesium Deficiency Induce Oxidative Stress

Strenuous physical exercise is capable of inducing oxidative stress, a state where the production of reactive oxygen species (ROS) in the body transcends the anti-oxidant defense capacity.[46,47] Several physiological processes are involved in the ROS generation and propagation during exercise: increased hemoglobin deoxygenation and reoxygenation rates; accelerated mitochondrial electron transport; acidosis and increased release of catecholamines, as during their oxidation to adrenocrome free radicals are formed. Strenuous exercise can also induce trauma, which leads to the extravasation of blood and to the introduction of free iron and copper into tissues, which can catalyze the propagation of ROS and consequently modify a wide range of biomolecules, including the polyunsaturated fatty acids of the cell's membrane.[48] Inflammatory injury, with the infiltration and activation of monocytes, produces a spectrum of free radicals.[49]

If the rise in oxygen free radicals exceeds the protective capacity of cell's anti-oxidant defense systems, it can lead to ROS-mediated injury and consequently to the loss of the cell's membrane integrity and to tissue damage. Several studies have shown that strenuous exercise promotes free-radical formation, resulting in a measurable degree of oxidative modifications to various molecules.[47,50–52]

The efficiency of the anti-oxidant defense systems relays on adequate dietary vitamin and micronutrient intake and on endogenous production of anti-oxidants, such as anti-oxidant enzymes and glutathione.[53]

Like the practice of exercise, magnesium deficiency is also prone to the increase of ROS generation as magnesium has an important role in the inhibition of ROS-induced cell injury. It inhibits catecholamine release and is a cofactor for their methylation that prevents oxidation.[54] It also participates in the de novo synthesis of Reduced glutathione (GSH), which may be important in the restoring of GSH cellular levels during exercise.[55]

In magnesium-deficient rodents, it has been demonstrated that ROS formation and the concentration of thiobarbituric acid reactive substances (TBARS), reflecting lipid peroxidation, are increased.[56,57] These changes were associated with structural damage in skeletal muscles affecting sarcoplasmic reticulum and mitochondria,[56] probably resulting from impairment of intracellular calcium homeostasis by altering the integrity of the sarcoplasmic reticulum membrane.[58] However, the mechanism by which magnesium deficiency potenciates injury remains unclear. Other explanations include increased cytokine concentrations, alterations in iron metabolism, or decreased endogenous anti-oxidant capacity.[59] The lack of magnesium was also associated with a depletion of selenium and reduced glutathione peroxidase activity.[60] Chugh and coworkers[61] supplemented patients with oxidative stress diseases with magnesium and suggested its anti-peroxidant effect when observing a decrease in lipid peroxidation.

In view of the above, during the practice of exercise, the conditions are prone to both the increase of reactive oxygen species generation and the decrease of magnesium status. As a consequence, individuals practicing exercise are more susceptible to free radical—mediated injury. We emphasize the importance of evaluating magnesium status in athletes, not only because its deficit may compromise performance, but also because the practice of exercise with a magnesium deficit may render the individual more susceptible to the occurrence of cellular damage, especially those associated with oxidative stress.

Immune Function Related to Exercise: Possible Roles of Magnesium

Magnesium has a strong relation with the immune system in both nonspecific and specific immune response.[62]

The attempt of an infectious agent to enter the body immediately activates the innate system. The nonspecific line comprises the action of macrophages or natural killer (NK) lymphocytes. Natural killer cells are rapidly activated and capable of policing the host until the more specific immune competent cells are activated. It has become clear that NK cells are highly influenced by various types of stress, including acute physical exercise.[63–65]

Failure of the innate system activates the acquired system. Monocytes or macrophages process and present foreign material (antigens) to lymphocytes. This is followed by clonal proliferation of T and B lymphocytes that possess receptors able to recognize the antigen. Action and function of both branches of the immune system are governed by the action of cytokines, which include interleukin 1 (IL-1), interleukin-6 (IL-6), and tumor necrosis factor (TNF). Their excessive or insufficient production may contribute to many infections or immunoinflammatory diseases.[66] While TNF and IL-1 have pro-inflammatory effects, IL-6 has more restorative effects by being the main inducer of the acute phase response by the liver.[67]

Exercise, both high intensity and prolonged, is a stress to the body that is proportional to the intensity and duration of the exercise, relative to the maximal capacity of the athlete. Exercise stress leads to a proportional increase in stress hormone levels and concomitant changes in several aspects of immunity, including high cortisol; neutrophilia; lymphopenia; decreases in granulocyte oxidative burst, NK cell activity, lymphocyte proliferation, production of cytokines in response to mitogens, and in nasal and salivary immunoglobulin A (IgA) levels; and increases in blood granulocyte and monocyte phagocytosis, and pro- and anti-inflammatory cytokines.[68]

The acute effects of exercise on lymphocytes are mediated by adrenaline and noradrenaline, whereas neutrophyls are mediated by cortisol.[69]

Many aspects of immune function can be depressed temporarily by either a single bout of very severe exercise or a longer period of excessive training.[68,70]

Although the disturbance is usually quite transient, it can be sufficient to allow a clinical episode of infection, particularly upper respiratory tract infections.[71,72] However, regular and moderate exercise has been reported to improve the ability of the immune system to protect the host from infection.[73,74] Resting levels of NK cells are enhanced as a result of training.[69] Leucocyte number is clinically normal and remains unchanged with training.[73]

The response of the immune system to exercise is varied, with different behaviors for each cell type and dependent on the intensity and duration of the exercise test and on the training of the subject. Besides these factors, nutritional deficiencies alter immunocompetence and increase the risk of infection. Both heavy exercise and nutrition exert separate influences on immune function, appearing to be greater when acting synergistically.[62] Exercise training increases the body requirement for most nutrients. However, some athletes adopt an unbalanced dietary regimen predisposing them to immunosuppression.[75]

Several elements are known to exert modulatory effects on immune function, including zinc, iron, selenium, calcium, copper, and magnesium.[68,72,76–80]

Several groups leading investigations in nutrition and immunology have shown evidence that magnesium plays a key role in the immune response: cofactor for immunoglobulin synthesis, C'3 convertase, immune cell adherence, antibody-dependent cytolysis, IgM lymphocyte binding, macrophage response to lymphokines, T helper—B cell adherence, binding of substance P to lymphoblasts, and antigen binding to macrophage RNA.[81,82] Most of these studies have been designed in animal models, focusing on what happens in magnesium-depleted animals.

The reason for assuming an association between magnesium and immune function was based on findings that magnesium deficiency leads to increased inflammation.[83,84] The appearance of clinical signs of inflammation is one of the early symptoms of magnesium deficiency in the rat.

The activation of immune cells, such as monocytes, macrophages, and polymorphonuclear neutrophils, which synthesize a variety of biological substances, some of which are powerful inducers of inflammatory events (cytokines, free radicals, eicosanoids), was also reported.[85] High levels of circulating cytokines, such as IL-6, could be detected early after initiating the magnesium-deficient diet, leading to the release of the acute phase proteins by the liver.[59,84]

In vitro assays showed that macrophages and neutrophils from magnesium – deficient rats were metabolically stimulated with a higher production of superoxide anion when compared with the control group. When increasing concentrations of magnesium in the diet, the rate of superoxide anion formation was decreased.[85]

There is increasing evidence that these mechanisms could be mediated by an increase in substance P during magnesium deficiency.[59,84,86] However, the magnitude of the induced increases of substance P was not similar between studies, probably due to different inflammatory responses.[87] The increase in substance P induces a neurogenic inflammatory response, including the activation of mast cells, macrophages, and both B and T lymphocytes. In addition, T lymphocytes from magnesium-deficient animals exhibit increased expression of substance P receptors associated with a marked increase in the release of histamine, PgE_2, IL-1, IL-6, and TNF. When magnesium-deficient rats were treated with a specific substance P receptor blocker, a marked attenuation of the elevations of histamine, PgE_2 and TNF and of the oxidative-induced loss of blood glutathione was observed. The authors hypothesized that substance P may initiate a cascade of inflammatory/pro-oxidant events leading to the cardiomyopatic characteristics associated with severe dietary magnesium restriction.[59] It was also observed that substance P plays a direct role in promoting activation of neutrophils and endothelium as well as the induction of NO production.[88] These processes may participate in the oxidative stress that contributes to the depletion of blood glutathione and cardiac pathology. The association between the early release of substance P and the subsequent inflammatory response is unclear. Studies conducted in human populations are less extended than those using animal models. Bussiere and coworkers[89] observed in vitro that high magnesium concentrations of the medium decrease human leukocyte activation and suggested that extracellular magnesium can diminish leukocyte activation by its calcium antagonism, as magnesium counteracts calcium in many physiological and pathological processes.

Mooren and coworkers[90] observed that, after a 2-month period of magnesium supplementation, an exercise test until exhaustion induced an activation of the immune system as indicated by an increase in granulocyte count and a postexercise lymphopenia. However, magnesium supplementation seems to have been unable to prevent any exercise-associated alterations in immune cell function in athletes with balanced magnesium status.

The similarities between exercise-induced alterations in immune function and the changes in immunity caused by the deficiency in magnesium are noteworthy.

Prolonged, exhaustive exercise has been shown to be associated with temporary immunosuppression. However, there is accumulating evidence that this type of exercise also leads to considerable magnesium decreases. Therefore, particularly if dietary intake is low, athletes may be prone to magnesium deficit. Magnesium deficit has been shown to be related to impaired cellular and humoral immune function.

Although considerably more work needs to be done in this area, there is evidence that immunoregulation during and after intense physical exercise is influenced by transient or manifest deficiencies in magnesium.

Further research is needed and special efforts should be made to establish the most adequate dose in nutritional supplements in athletes to reach benefits on heath and performance.

Acknowledgments. The authors thank Maria de Fátima Raposo for her proficient technical assistance.

References

1. Durlach J, Bara M. Le magnésium en biologie et en médecine. Cachami Edition Medicales: Internationales, 2nd ed., EM, ed. Cachan, 2000.
2. Seelig MS. *Magnesium Deficiency in the Pathogenesis of Disease*. New York: Plenum Press; 1980.
3. Clarkson PM. Micronutrients and exercise: anti-oxidants and minerals. *J Sports Sci* 1995;13:S11–S24.
4. Bohl CH, Volpe SL. Magnesium and exercise. *Crit Rev Food Sci Nutr* 2002;42: 533–563.
5. Lukaski HC. Vitamin and mineral status: effects on physical performance. *Nutrition* 2004;20:632–644.
6. Seelig MS. Consequences of magnesium deficiency on the enhancement of stress reactions; preventive and therapeutic implications (a review). *J Am Coll Nutr* 1994;13:429–446.

7. Córdova A, Navas FJ, Gómez-Carramiñana M, et al. Evaluation of magnesium intake in elite sportsmen. *Magnes Bull* 1994;16:59–63.
8. Casoni I, Guglielmini C, Graziano L, et al. Changes of magnesium concentrations in endurance athletes. *Int J Sports Med* 1990;11:234–237.
9. Rayssiguier Y, Guezennec CY, Durlach J. New experimental and clinical data on the relationship between magnesium and sport. *Magnes Res* 1990;3:93–102.
10. Cordova A. Changes on plasmatic and erythrocytic magnesium levels after high-intensity exercises in men. *Physiol Behav* 1992;52:819–821.
11. Joborn H, Akerstrom G, Ljunghall S. Effects of exogenous catecholamines and exercise on plasma magnesium concentrations. *Clin Endocrinol (Oxf)* 1985;23:219–226.
12. Monteiro CP. *Equilíbrio Oxirredutor: Um Estudo Em Nadadores E Em Não Atletas, Em Repouso E Em Resposta Ao Exercício. Ciências Da Motricidade*. Lisboa: Faculdade de Motricidade Humana, Universidade Técnica de Lisboa; 2005:209.
13. Deuster PA, Dolev E, Kyle SB, et al. Magnesium homeostasis during high-intensity anaerobic exercise in men. *J Appl Physiol* 1987;62:545–550.
14. Franz KB, Ruddel H, Todd GL, et al. Physiologic changes during a marathon, with special reference to magnesium. *J Am Coll Nutr* 1985;4:187–194.
15. Guerra M, Monje A, Perez-Beriain R, et al. Ionic magnesium and selenium in serum after a cycle-ergometric test in football-players. In: Centeno JA, Collery P, Vernet G, Finkelman RB, Gibb H, Etienne JC, eds. *Metal Ions in Biology and Medicine*. Vol. 6. Paris: John Libbey and Company Ltd. 2000:501–504.
16. Laires MJ, Alves F, Halpern MJ. Changes in serum and erythrocyte magnesium and blood lipids after distance swimming. *Magnes Res* 1988;1:219–222.
17. Laires MJ, Alves F. Changes in plasma, erythrocyte, and urinary magnesium with prolonged swimming exercise. *Magnes Res* 1991;4:119–122.
18. Laires MJ, Madeira F, Sergio J, et al. Preliminary study of the relationship between plasma and erythrocyte magnesium variations and some circulating pro-oxidant and antioxidant indices in a standardized physical effort. *Magnes Res* 1993;6:233–238.
19. Rose LI, Caroll DR, Lowe SL, et al. Serum electrolyte changes after marathon running. *J Appl Physiol* 1970;29:449–451.
20. Lijnen P, Hespel P, Fagard R, et al. Erythrocyte, plasma and urinary magnesium in men before and after a marathon. *Eur J Appl Physiol Occup Physiol* 1988;58:252–256.
21. Welsh RC, Warburton DER, Haykowksy MJ, et al. Hematological response to the half ironman triathlon. *Med Sci Sports Exerc* 1999;31:s63.
22. Refsum HE, Meen HD, Stromme SB. Whole blood, serum and erythrocyte magnesium concentrations after repeated heavy exercise of long duration. *Scand J Clin Lab Invest* 1973;32:123–127.
23. Stendig-Lindberg G, Shapiro Y, Epstein Y, et al. Changes in serum magnesium concentration after strenuous exercise. *J Am Coll Nutr* 1987;6:35–40.
24. Haymes EM. Vitamin and mineral supplementation to athletes. *Int J Sport Nutr* 1991;1:146–169.
25. Rayssiguier Y, Larvor P. Hypomagnesemia following stimulation of lipolysis in ewes: effects of cold exposure and fasting. *Magnes Health Dis* 1980;9:68–72.
26. Rayssiguier Y. Hypomagnesemia resulting from adrenaline infusion in ewes: its relation to lipolysis. *Horm Metab Res* 1977;9:309–314.

27. Elliot DA, Rizack MA. Epinephrine and adrenocorticotropic hormone-stimulated magnesium accumulation in adipocytes and their plasma membranes. *J Biol Chem* 1974;249:3985–3990.
28. Speich M, Pineau A, Ballereau F. Minerals, trace elements and related biological variables in athletes and during physical activity. *Clin Chim Acta* 2001;312:1–11.
29. Stromme SB, Stenwold IE, Meen HD, et al. Magnesium metabolism during prolonged heavy exercise. In: Howald H, Poortmans JR, eds. *Metabolic Adaptation to Prolonged Physical Exercise*. Basel: Birkhauser; 1975:361–366.
30. Lijnen P, Hespel P, Fagard R, et al. Erythrocyte 2,3-diphosphoglycerate concentration before and after a marathon in men. *Eur J Appl Physiol Occup Physiol* 1988;57:452–455.
31. Pereira D, Laires MJ, Monteiro CP, et al. Oral magnesium supplementation in heavily trained football players. Impact on exercise capacity and lipoperoxidation. In: Halpern MJ, Durlach J, eds. *Current Research on Magnesium*. London: John Libbey; 1996:237–241.
32. Golf SW, Happel O, Graef V. Plasma aldosterone, cortisol and electrolyte concentration in physical exercise after magnesium supplementation. *J Clin Chem Biochem* 1984;22:717–721.
33. Laires MJ, Sainhas J, Fernandes JS, et al. *Algunos Efectos Del Esfuerzo Sobre Los Parámetros De Magnesio Sanguíneo*. Vol. 24. Congresso Científico Olímpico; Málaqa 1995:96–100.
34. Monteiro CP, Palmeira A, Felisberto GM, et al. Magnesium, calcium, trace elements and lipid profile in trained volleyball players: influence of training. In: Halpern MJ, Durlach J, eds. *Current Research in Magnesium*. London: John Libbey; 1996:231–235.
35. Resina A, Brettoni M, Gatteschi L, et al. Changes in the concentrations of plasma and erythrocyte magnesium and of 2,3-diphosphoglycerate during a period of aerobic training. *Eur J Appl Physiol Occup Physiol* 1994;68:390–394.
36. Murphy E. Mysteries of magnesium homeostasis. *Circ Res* 2000;86:245–248.
37. Mooren FC, Golf SW, Lechtermann A, et al. Alterations of ionized Mg^{2+} in human blood after exercise. *Life Sci* 2005;77:1211–1225.
38. Golf VS, Graef V, Gerlach JJ, et al. Veänderungen der serum-CK- und serum-CK-MB aktivatäten in abhängigkeit von einer magnesium-substitution bein leistungs sportlerinnen. *Magnes Bull* 1983;2:43–46.
39. Huet F, Keppling J, Marajo J, et al. Comportement nutritionnel du coureur de demi-fond. Aspects qualitatifs et quantitatifs. Rapports avec la depense energétique de l'entraînement. *Sci Sports* 1988;3:17–28.
40. Stendig-Lindberg G, Wacker WEC, Shapiro Y. Long term effects of peak strenuous effort on serum magnesium, lipids, and blood sugar in apparently healthy young men. *Magnes Res* 1991;4:59–65
41. Resina A, Gatteschi L, Castellani W, et al. Effect of aerobic training and exercise on plasma and erythrocyte magnesium concentration. In: Kies CV, Driskell JA, eds. *Sports Nutrition: Minerals and Electrolytes*. London: CRC Press; 1995:189–203.
42. Warburton DE, Welsh RC, Haykowsky MJ, et al. Biochemical changes as a result of prolonged strenuous exercise. *Br J Sports Med* 2002;36:301–303.
43. Rowe WJ. Endurance exercise and injury to the heart. *Sports Med* 1993;16:73–79.
44. Brilla LR, Haley TF. Effect of magnesium supplementation on strength training in humans. *J Am Coll Nutr* 1992;11:326–329.

45. Terblanche S, Noakes TD, Dennis SC, et al. Failure of magnesium supplementation to influence marathon running performance or recovery in magnesium-replete subjects. *Int J Sport Nutr* 1992;2:154–164.
46. Halliwell B, Gutteridge JMC. *Oxygen Free Radicals in Biology and Medicine.* Oxford: Clarendon Press; 1989.
47. Sen CK. Oxidants and antioxidants in exercise. *J Appl Physiol* 1995;79:675–686.
48. Duthie GG, Robertson JD, Maughan RJ, et al. Blood antioxidant status and erythrocyte lipid peroxidation following distance running. *Arch Biochem Biophys* 1990;282:78–83.
49. Conn CA, Kozak WE, Tooten PC, et al. Effect of exercise and food restriction on selected markers of the acute phase response in hamsters. *J Appl Physiol* 1995;78:458–465.
50. Sureda A, Tauler P, Aguilo A, et al. Relation between oxidative stress markers and antioxidant endogenous defences during exhaustive exercise. *Free Radic Res* 2005;39:1317–1324.
51. Tauler P, Sureda A, Cases N, et al. Increased lymphocyte antioxidant defences in response to exhaustive exercise do not prevent oxidative damage. *J Nutr Biochem* 2005. In press, available online 28 Nov 2005.
52. Vollaard NB, Shearman JP, Cooper CE. Exercise-induced oxidative stress:myths, realities and physiological relevance. *Sports Med* 2005;35:1045–1062.
53. Packer L, Singh VN. Nutrition and exercise introduction and overview. *J Nutr* 1992;122:758–759.
54. Langer SZ. Presynaptic regulation of the release of catecholamines. *Pharmacol Rev* 1981;32:337.
55. Minnich V, Smith MB, Brauner MJ, et al. Glutathione biosynthesis in human erythrocytes—I. Identification of the enzymes of glutathione synthesis in hemolysates. *J Clin Invest* 1971;50:507–513.
56. Rock E, Astier C, Lab C, et al. Dietary magnesium deficiency in rats enhances free radical production in skeletal muscle. *J Nutr* 1995;125:1205–1210.
57. Rayssiguier Y, Gueux E, Bussiere L, et al. Dietary magnesium affects susceptibility of lipoproteins and tissues to peroxidation in rats. *J Am Coll Nutr* 1993;12:133–137.
58. Malpuech-Brugère C, Rock E, Astier C, et al. Exacerbated immune stress response during experimental magnesium deficiency results from abnormal cell calcium homeostasis. *Life Sci* 1998;63:1815–1822.
59. Weglicki WB, Kramer JH, Mak IT, et al. Proinflammatory neuropeptides in magnesium deficiency. In: Centeno PC, Vernet G, Finkelman RB, Gibb H, Etienne JC, eds. *Metal Ions in Biology and Medicine.* Vol. 6. Paris: John Libbey; 2000:472–474.
60. Zhu Z, Kimura M, Itokawa Y. Selenium concentration and glutathione peroxidase activity in selenium and magnesium deficient rats. *Biol Trace Elem Res* 1993;37:209–217.
61. Chugh SN, Kolley T, Kakkar R, et al. A critical evaluation of anti-peroxidant effect of intravenous magnesium in acute aluminium phosphide poisoning. *Magnes Res* 1997;10:225–230.
62. Tam M, Gomez S, Gonzalez-Gross M, et al. Possible roles of magnesium on the immune system. *Eur J Clin Nutr* 2003;57:1193–1197.
63. Pedersen BK. *Exercise Immunology.* New York: Springer; 1997.

64. Pedersen BK. Influence of physical activity on the cellular immune system: mechanisms of action. *Int J Sports Med* 1991;12:S23–S29.
65. Pedersen BK, Ullum H. NK cell response to physical activity: possible mechanisms of action. *Med Sci Sports Exerc* 1994;26:140–146.
66. Bendtzen K. Immune hormones (cytokines); pathogenic role in autoimmune rheumatic and endocrine diseases. A review. *Autoimmunity* 1989;2:177–189.
67. Cannon JG, Evans WJ, Hughes VA, et al. Physiological mechanisms contributing to increased interleukin-1 secretion. *J Appl Physiol* 1986;61:1869–1874.
68. Venkatraman JT, Pendergast DR. Effect of dietary intake on immune function in athletes. *Sports Med* 2002;32:323–337.
69. Pedersen BK, Toft AD. Effects of exercise on lymphocytes and cytokines. *Br J Sports Med* 2000;34:246–251.
70. Bishop NC, Blannin AK, Walsh NP, et al. Nutritional aspects of immunosuppression in athletes. *Sports Med* 1999;28:151–176.
71. Gleeson M, Nieman DC, Pedersen BK. Exercise, nutrition and immune function. *J Sports Sci* 2004;22:115–125.
72. Shephard RJ, Shek PN. Heavy exercise, nutrition and immune function: is there a connection? *Int J Sports Med* 1995;16:491–497.
73. Mackinnon LT. Chronic exercise training effects on immune function. *Med Sci Sports Exerc* 2000;32:S369–S376.
74. Peters EM. Exercise, immunology and upper respiratory tract infections. *Int J Sports Med* 1997;18(suppl 1):S69–S77.
75. Gleeson M, Bishop NC. Elite athlete immunology: importance of nutrition. *Int J Sports Med* 2000;21(suppl 1):S44–S50.
76. Mooren FC, Lechtermann A, Fromme A, et al. Alterations in intracellular calcium signaling of lymphocytes after exhaustive exercise. *Med Sci Sports Exerc* 2001;33:242–248.
77. Beisel WR, Edelman R, Nauss K, et al. Single-nutrient effects on immunologic functions. Report of a workshop sponsored by the Department of Food and Nutrition and its nutrition advisory group of the American Medical Association. *J Am Med Assoc* 1981;245:53–58.
78. Galan P, Thibault H, Preziosi P, et al. Interleukin 2 production in iron-deficient children. *Biol Trace Elem Res* 1992;32:421–426.
79. Singh A, Failla ML, Deuster PA. Exercise-induced changes in immune function: effects of zinc supplementation. *J Appl Physiol* 1994;76:2298–2303.
80. Laires MJ, Monteiro CP. Magnesium status: Influence on the regulation of exercise induced oxidative stress and immune function in athletes. In: Rayssiguier Y, Mazur A, Durlach J, eds. *Advances in Magnesium Research: Nutrition and Health.* London: John Libbey; 2001:433–441.
81. Galland L. Magnesium and immune function: an overview. *Magnesium* 1988;7:290–299.
82. Perraud AL, Knowles HM, Schmitz C. Novel aspects of signaling and ion-homeostasis regulation in immunocytes. The TRPM ion channels and their potential role in modulating the immune response. *Mol Immunol* 2004;41:657–673.
83. Malpuech-Brugère C, Nowacki W, Rock E, et al. Enhanced tumour necrosis factor-alpha production following endotoxin challenge in rats in an early event during magnesium deficiency. *Biochim Biophys Acta* 1999;1453:35–40.
84. Weglicki WB, Phillips TM. Pathobiology of magnesium deficiency: a cytokine/neurogenic inflammation hypothesis. *Am J Physiol* 1992;263:R734–R737.

85. Rayssiguier Y, Brussière F, Malpuech-Brugère C, et al. Activation of phagocytic cell and inflammatory response during experimental magnesium deficiency. In: Centeno PC, Vernet G, Finkelman RB, Gibb H, Etienne JC, eds. Metal Ions in Biology and Medicine. Vol. 6. Paris: John Libbey; 2000:534–536.
86. Kabashima H, Nagata K, Maeda K, et al. Involvement of substance P, mast cells, TNF-alpha and ICAM-1 in the infiltration of inflammatory cells in human periapical granulomas. *J Oral Pathol Med* 2002;31:175–180.
87. Malpuech-Brugere C, Nowacki W, Daveau M, et al. Inflammatory response following acute magnesium deficiency in the rat. *Biochim Biophys Acta* 2000;1501:91–98.
88. Mak IT, Kramer JH, Weglicki WB. Suppression of neutrophil and endothelial activation by substance P receptor blockade in the Mg-deficient rat. *Magnes Res* 2003;16:91–97.
89. Bussiere FI, Mazur A, Fauquert JL, et al. High magnesium concentration in vitro decreases human leukocyte activation. *Magnes Res* 2002;15:43–48.
90. Mooren FC, Golf SW, Volker K. Effect of magnesium on granulocyte function and on the exercise induced inflammatory response. *Magnes Res* 2003;16:49–58.

Metabolic Syndrome

15
Review of Magnesium and Metabolic Syndrome

Yasuro Kumeda and Masaaki Inaba

With the westernization of lifestyle in people all over the world, especially in advanced countries in recent years, dietary intake of magnesium via grain, barley, seaweed, vegetable, and nuts has remarkably diminished. As a result, people may more easily develop hypomagnesemia. Likewise, metabolic syndrome has been an increasing problem for health, probably resulting from various causes such as increased intake of animal fat, exercise insufficiency, and accumulation of various stresses. Metabolic syndrome is often complicated with obesity, hypertension, hyperglycemia, and hyperlipidemia, and thus people with the syndrome may be susceptible to cardiovascular events. Hypomagnesemia may cause an increase of vascular tonus by intracellular magnesium depletion, resulting in an increase of blood pressure. Furthermore, it might cause impaired insulin secretion, insulin resistance, and hyperlipidemia, and finally lead to the development of metabolic syndrome. Therefore, the importance of magnesium intake for the maintenance of health should be recognized.

Magnesium has been proven to be a very important for more than 300 kinds of enzyme activation in vivo. It is easy for the intake of cereals, including barley, beans, seaweed, and green vegetables to decrease along with the change in recent years to European and American lifestyles, causing hypomagnesia. Similarly, metabolic syndrome is recognized as an important problem along with the change of lifestyle. The increase of animal fat intake, the lack of physical activity, and the accumulation of life stress in recent years, perhaps causes an increase of metabolic syndrome frequency, and this brings about the diseases that cause blood vessel problems such as obesity, high blood pressure, hyperglycemia, and hyperlipidemia. As a result, the risk of the onset of cerebral infarction and cardiac infarction is very high. The relationship between magnesium intake shortage and the onset of metabolic syndrome has been recognized in recent years, and the relevance is outlined in this chapter.

What Is Metabolic Syndrome?

Metabolic syndrome is a group of diseases with increasing tendency in recent years. Hyperlipidemia, diabetes (including the boundary type), high blood pressure, and smoking are indicated as the four largest risk factors of an arteriosclerotic disease. It is shown that the risk of ischemic heart disease increases rapidly, and takes the appearance of a cardiovascular disease when these risk factors overlap even when the extent is light, and this kind of recognition develops into the concept of metabolic syndrome. In metabolic syndrome, hyperinsulinemia continues chronically because the insulin level of the blood could not decrease easily due to insulin resistance, and if the hyperinsulinemia continues, it becomes easy to develop hypertension, hypertriglycemia, hypo high-density lipoproteinemia (HDL), the storage of visceral fat, and so on. The compound lifestyle disease that had been called metabolic syndrome, syndrome X^1, deadly quartet[2], insulin resistance syndrome[3], and visceral fat syndrome up to now is thought to be deeply related to lifestyle, including obesity, eating habits, and exercise, in addition to hereditary predisposition and aging. Metabolic syndrome might develop more once obesity, hypertension, high blood sugar, and hyperlipemia are present, that is, the risk factors of arteriosclerosis, and develops easily into cardiac infarction and cerebral infarction. For obesity, visceral fat–type obesity is more harmful for health than subcutaneous fat–type obesity, and often complicates hyperinsulinemia.

In the investigation of 120,000 Japanese workers, it appeared that the person who has one risk factor of obesity, hypertension, high blood sugar, hypertriglycemia, or hypercholesterolemia, even if it is a slight illness, increases their risk factor for heart disease 5 times; the person who has two risk factors increases the risk of heart disease 10 times; and the person who has three to four factors increases 31 times the risk of heart disease. The diagnosis standard of metabolic syndrome is shown by the United States Hyperlipemia treatment guideline[4] and WHO.[5] In Japan, eight societies, such as the Japanese Society of Internal Medicine, defined the disease concept of metabolic syndrome and the diagnosis standard was set at the Japanese Internal Medicine Department Society general meeting in April 2005 (see Table 15.1). At that time, it was agreed that metabolic syndrome is the concept of screening for a high risk of an arteriosclerosis, and has aimed to improve diagnosis of the disease that presents as hyperlipidemia, hypertension, and hyperglycemia.

Recently, a report of the importance of the relationship between metabolic syndrome and magnesium was presented. In a cross-sectional population-based study comparing 192 individuals with metabolic syndrome and 384 disorder-free control subjects, matched by age and gender, low serum magnesium levels were identified in 126 (65.6%) and 19 (4.9%) individuals with and without metabolic syndrome ($p < 0.00001$). The mean serum magnesium level among subjects with metabolic syndrome was 1.8 ± 0.3 mg/dL, and among control subjects 2.2 ± 0.2 mg/dL ($p < 0.00001$). There was a strong independent

TABLE 15.1. Diagnosis standards of metabolic syndrome.

The United States hyperlipemia treatment guideline (ATPIII: Adult Treatment Panel III, NCEP National Cholesterol Education Program)

More than three of the five following items:
1. The waist (abdominal circumference) is 102 cm or more (85 cm or more in the Japanese) in men and 88 cm or more in women (90 cm or more in Japanese).
2. Triglycerides are 150 mg/dL or more.
3. HDL cholesterol is less than 40 mg/dL in men and less than 50 mg/dL in women.
4. 130 mm Hg or more in the systolic blood pressure or 85 mm Hg or more in the diastolic blood pressure.
5. Fasting blood sugar level is 110 mg/dL or more.

Diagnosis standard by WHO

Two or more of the following conditions, in addition to hyperinsulinemia (high rank 25% of nondiabetic) or fasting blood sugar is 110 mg/dL or more.
1. Visceral fat obesity waist/hip ratio is >0.9 (man), >0.85 (woman), or BMI is 30 or more, or the abdominal circumference is 94 cm or more.
2. Lipid metabolic disorder: Triglycerides are 150 mg/dL or more, or HDL cholesterol is less than 35 mg/dL in men and less than 39 mg/dL in women.
3. Blood pressure is 140/90 mm Hg or more, or a hypotensive drug is being taken.
4. Microalbuminuria (urinary albumin excretion rate is 20 μg/min or more), or urinary albumin/creatinine ratio is 30 mg/g.Cr or more.

Diagnosis standard by eight-society Joint Committee in Japan[a]

Item 1, plus two or more of items 2 through 4.
1. A waist diameter is 85 cm or more in men, 90 cm or more in women. (These waist diameters correspond to visceral fat area 100 cm^2 in both men and women.)
2. Hyperlipidemia (triglycerides are 150 mg/dL or more, and/or the HDL cholesterol is less than 40 mg/dL).
3. Hypertension: Systolic blood pressure is 130 mm Hg or more, and/or diastolic blood pressure is 85 mm Hg or more.
4. Hyperglycemia: 110 mg/dL or more.

[a]Japan Atherosclerosis Society, Japan Diabetes Society, Japanese Society of Hypertension, Japan Society for the Study of Obesity, Japanese Circulation Society, Japanese Society of Nephrology, The Japanese Society on Thrombosis and Hemostasis, and Japanese Society of Internal Medicine.

relationship between low serum magnesium levels and metabolic syndrome [odds ratio (OR) = 6.8; 95% confidence interval (95% CI), 4.2–10.9]. Among the components of metabolic syndrome, dyslipidemia (OR 2.8; 95% CI, 1.3–2.9) and high blood pressure (OR 1.9; 95% CI, 1.4–2.8) were strongly related to low serum magnesium levels. This study revealed a strong relationship between decreased serum magnesium and metabolic syndrome.[6]

However, when we pay attention to each of the diseases that compose metabolic syndrome, it is revealed that the pathology in connection with magnesium deficiency is important. Moreover, it is recognized that the lifestyle that causes magnesium deficiency plays a major role in the appearance of the diseases of metabolic syndrome.

Next, we make a detailed explanation of the relation between each disease that composes metabolic syndrome and magnesium.

Magnesium and Diabetes Mellitus

Urinary excretion of magnesium is increased in either type 1 or type 2 diabetes mellitus patients, resulting in the reduction of the levels of magnesium in both serum and intracellular components.[7] In patients with cerebrovascular attack, hypertension, and diabetes, serum magnesium level is decreased. Serum magnesium level correlated in a negative manner with fasting plasma glucose and insulin in the patients with cerebrovascular stroke,[8] indicating that magnesium deficiency causes the occurrence of insulin resistance in those patients.

The magnesium ion (Mg^{2+}) plays a pivotal role in insulin secretion from pancreatic β cells. The first step to stimulate physiological insulin secretion is the intracellular uptake of glucose into pancreatic β cells via glucose transporter in plasma membrane. Glucose is metabolized via the Tricarboxylic acid (TCA) cycle to produce adenosine triphosphate (ATP). ATP then binds to Mg^{2+} to close ATP-sensitive K channels, resulting in the depolarization of plasma membrane. The depolarization opens up voltage-dependent calcium ion (Ca^{2+}) channel to stimulate intracellular entry of Ca^{2+}. The resultant increase of intracellular Ca^{2+} level stimulates insulin secretary vesicle to secrete insulin outside the cells. Mg^{2+} deficiency inhibits glucose metabolism via the TCA cycle to suppress ATP production; Mg^{2+} deficiency and reduced ATP production both inhibit the synthesis of Mg-ATP, and, thus, the activation of ATP-sensitive K channel, resulting in the suppression of insulin secretion.

Although mice fed on Mg^{2+} deficiency showed an attenuated insulin secretary response to glucose, magnesium supplementation restored, although partially, its response.[9]

For intracellular uptake of glucose by insulin-sensitive cells, the translocation of glucose transporter (GLUT) into plasma membrane is a vital step, which insulin induces by autophosphorylating β subunits of insulin receptor after its specific binding.

Mg^{2+} acts in a stimulative manner in the step of insulin binding to its receptor and autophosphorylation of β subunits. Furthermore, intracellular magnesium stimulates intracellular glucose transport and its oxidation step.[10] Because insulin by itself stimulates intracellular magnesium uptake in insulin-sensitive cells, magnesium deficiency forms a vicious cycle, inhibiting insulin action mainly by inhibiting insulin binding to its receptor and intracellular glucose metabolism. These mechanisms cause the occurrence of insulin resistance and glucose intolerance, and possibly leading to the development of diabetes.

Very recently, the evidence that hypomagnesemia and/or intracellular decrease of Mg^{2+} inhibit insulin receptor–associated tyrosine kinase activity may explain the development of insulin resistance in magnesium deficiency.[11]

In summary, magnesium deficiency inhibits insulin secretion from pancreatic β cells and insulin sensitivity in insulin-targeted cells, and possibly worsens glucose intolerance to cause the development of diabetes.

Magnesium and Hyperlipidemia and Atherosclerosis

Magnesium deficiency inhibits lipoprotein lipase (LPL) activity to induce hypertriglyceridemia. Magnesium deficiency also impairs the metabolism of saturated fatty acid to polyunsaturated fatty acid by inhibiting desaturate activity. The resultant increase of saturated fatty acids in plasma membrane may accelerate atherosclerosis.[12]

Because the activation step of lecithin-cholesterol acyltransferase, which, as well as LPL, plays important role in increasing HDL cholesterol and degrading triglycerides, requires magnesium, magnesium deficiency leads to increases of very low low-density lipoprotein (VLDL), low-density lipoprotein (LDL), and triglycerides, and a decrease of HDL cholesterol.[13]

In rabbits fed on a magnesium-deficient and cholesterol-enriched diet, arterial thickening in the intimal layer of aorta was accelerated by increased accumulation of lipid in arterial wall. Magnesium supplementation to those rabbits lowered serum cholesterol and triglycerides, and attenuated the progression of atherosclerosis in magnesium-deficient rabbits.[14]

Lal and colleagues reported the effect of magnesium supplementation on lipid metabolism in diabetes patients.[15] They found a significant decrease of serum magnesium levels in 40 type 2 diabetes patients, compared to 50 control subjects (diabetes, 1.44 ± 0.48 mg/dL, vs. controls, 2.29 ± 0.33 mg/dL; $p < 0.001$). They administered magnesium oxide at the daily dose of 600 mg for 12 weeks to type 2 diabetes patients. Serum total, LDL, and triglycerides decreased and HDL cholesterol increased significantly by 4 weeks, without any change in fasting or postprandial plasma glucose levels, concluding that magnesium administration is effective in improving lipid metabolism.

Magnesium and Hypertension

Magnesium supplementation to patients with essential hypertension or alcohol-loaded hypertensive rats is known to suppress blood pressure, the mechanism by which is explained in relation to intracellular magnesium and calcium level.[16] The intimate mechanism is magnesium-induced inhibition of intracellular Ca^{2+} entry via L-type calcium channel of vascular smooth muscle cells.[17]

Magnesium deficiency suppresses intracellular Mg^{2+} levels in vascular smooth muscle cells, resulting in an increase of Ca/Mg ratio, thereby causing hypertension by constricting vascular smooth muscle cells.[18]

Mg^{2+}, in addition to its direct effect on vascular smooth muscle cells, suppresses blood pressure by its effect via peripheral sympathetic nerves. Shimozawa and colleagues[19] reported that in the perfused mesenteric arteries system, norepinephrine release was significantly attenuated (51% ± 2%,

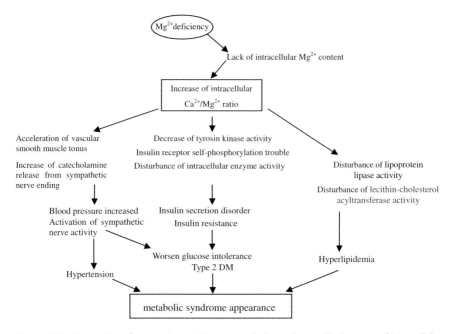

FIGURE 15.1. The outline of magnesium action on metabolic syndrome. The increase of intracellular Ca^{2+}/Mg^{2+} ratio due to magnesium shortage accelerates the vascular smooth muscle tonus and increases the catecholamine release from sympathetic nerve endings, consequentially becoming high blood pressure. Moreover, high intracellular Ca^{2+}/Mg^{2+} ratio causes the decrease of tyrosine kinase activity, self-phosphorylation disorder of the insulin receptor, and the disturbance of intracellular enzyme activity; as a result, insulin secretion disorder and insulin resistance is caused. As for the lipid metabolism, magnesium shortage causes the disturbance of both lipoprotein lipase activity and lecithin-cholesterol acyltransferase activity, and as a result, VLDL, LDL, and trigylcerides increase. Therefore, hyperlipemia is caused. Metabolic syndrome comes to develop as these diseases come in succession.

$p < 0.01$) by high Mg^{2+} concentration solution (4.8 mmol/L) compared with normal Mg^{2+} solution (1.2 mmol/L). They proved that Mg^{2+} blocked voltage-gated Ca^{2+} currents in a concentration-dependent manner by the perforated whole-cell patch clamp method to nerve growth factor–treated PC12 cells. The majority of the voltage-gated Ca^{2+} currents were carried through N-type channels, followed by L-type channels. Mg^{2+} blocked both of these channels. From these experimental results, they suggested that Mg^{2+} blocks mainly N-type Ca^{2+} channels at nerve endings, and thus inhibits norepinephrine release, which decreases blood pressure independent of its direct vasodilating action.

The action of magnesium on the appearance of disease of above-mentioned metabolic syndrome is shown in Figure 15.1.

Magnesium and Obesity

The obese population of Europe and America has increased in recent years, and it is said that an increase of childhood obesity is a hotbed for the future increase of lifestyle diseases. Milagros and colleagues[20] made comparative study, for 24 obese nondiabetic children [body mass index (BMI) ⩾ 85th percentile] and 24 sex- and puberty-matched lean control subjects (BMI < 85th percentile), of magnesium intake and body composition of serum magnesium, the insulin receptivity index, and meal. The following became clear as a result that serum magnesium was significantly lower in obese children (0.748 ± 0.015 mmol/L, mean $\pm$ SE) compared with lean children (0.801 ± 0.012 mmol/L; $p = 0.009$). Serum magnesium was inversely correlated with fasting insulin ($r_s = -0.36$; 95% CI, -0.59 to -0.08; $p = 0.011$) and positively correlated with quantitative insulin sensitivity ($r_s = 0.35$; 95% CI, 0.06–0.58; $p = 0.015$). Dietary magnesium intake was significantly lower in obese children (obese, 0.12 ± 0.004 vs. lean, 0.14 ± 0.004 mg/kcal; $p = 0.003$). Dietary magnesium intake was inversely associated with fasting insulin ($r_s = -0.43$; 95% CI, -0.64 to -0.16; $p = 0.002$) and directly correlated with insulin sensitivity ($r_s = 0.43$; 95% CI, 0.16–0.64; $p = 0.002$). From this, it is clear that the association between magnesium deficiency and insulin resistance (IR) is present during childhood, and that serum magnesium deficiency in obese children may be secondary to decreased dietary magnesium intake. Therefore, magnesium supplementation or increased intake of magnesium-rich foods may be an important tool in the prevention of type 2 diabetes in obese children.

Conclusion

In recent years, the diseases of modern people has changed remarkably due to the dramatic change in lifestyle, including dining habits. It is clear that intake shortage of dietary magnesium is related to the appearance of metabolic syndrome; this is a social issue. In addition, it is known that metabolic syndrome may take part in the honeshiti deterioration due to osteoporosis, one of the lifestyle diseases not mentioned in this text. It is assumed that the importance of the magnesium intake via food is acknowledged more widely, and a part of the prevention of lifestyle diseases, including metabolic syndrome, becomes possible by increasing dietary magnesium intake. Detailed clarification of magnesium participation in the process of metabolic syndrome is needed.

References

1. Reaven GM. Role of insulin resistance in human disease. *Diabetes* 1988;37:1595–1607.
2. Kaplan NM. The deadly quartet. *Arch Int Med* 1989;149:1514–1520.

3. DeFronzo RA, Ferrannini E. Insulin resistance: a multifaceted syndrome responsible for NIDDM, obesity, hypertension, dyslipidemia, and atherosclerotic cardiovascular disease. *Diabetes Care* 1991;14:173–194.
4. Expert Panel on Detection, Evaluation, and Treatment of High Blood Cholesterol in Adults. Executive Summary of The Third Report of The National Cholesterol Education Program (NCEP) Expert Panel on Detection, Evaluation, And Treatment of High Blood Cholesterol In Adults (Adult Treatment Panel III). *JAMA* 2001;285:2486–2497.
5. Zimmet P, Alberti KG, Shaw J. Global and societal implications of the diabetes epidemic. *Nature* 2001;414:782–787.
6. Guerrero-Romero F, Rodriguez-Moran M. Low serum magnesium levels and metabolic syndrome. *Acta Diabetol* 2002;39:209–213.
7. Altura BM, Altura BT. Magnesium and cardiovascular biology: an important link between cardiovascular risk factors and atherogenesis. *Cell Mol Biol Res* 1995;41:347–359.
8. Ma J, Folsom AR, Melnick SL, et al. Associations of serum and dietary magnesium with cardiovascular disease, hypertension, diabetes, insulin, and carotid arterial wall thickness: the ARIC study. Atherosclerosis Risk in Communities Study. *J Clin Epidemiol* 1995;48:927–940.
9. Legrand C, Okitolonda W, Pottier AM, et al. Glucose homeostasis in magnesium-deficient rats. *Metabolism* 1987;36:160–164.
10. Polisso G, Barbagallo M. Hypertension, diabetes mellitus and insulin resistance. The role of intracellular magnesium. *Am J Hypertens* 1997;10:346–355.
11. McKeown NM. Whole grain intake and insulin sensitivity: evidence from observational studies. *Nutr Rev* 2004;62:286–291.
12. Morrill GA, Gupta RK, Kostellow AB, et al. Mg^{2+} modulates membrane lipids in vascular smooth muscle: a link to atherogenesis. *FEBS Lett* 1997;408:191–194.
13. Rayssiguier Y, Gueux E. Magnesium and lipids in cardiovascular disease. *J Am Coll Nutr* 1986;5:507–519.
14. Altura BT, Brust M, Bloom S, et al. Magnesium dietary intake modulates blood lipid levels and atherosclerosis. *Proc Natl Acad Sci U S A* 1990;87:1840–1844.
15. Lal J, Vasudev K, Kela AK, et al. Effect of oral magnesium supplementation on the lipid profile and blood glucose of patients with type 2 diabetes mellitus. *J Assoc Physicians India* 2003;51:37–42.
16. Hsieh ST, Sano H, Saito K, et al. Magnesium supplementation prevents the development of alcohol-induced hypertension. *Hypertension* 1992;19:175–182.
17. Delpiano MA, Altura BM. Modulatory effect of extracellular Mg^{2+} ions on K+ and Ca2+ currents of capillary endothelial cells from rat brain. *FEBS Lett* 1996;394:335–339.
18. Paolisso G, Barbagallo M. Hypertensions, diabetes mellitus and insulin resistance. The role of intracellular magnesium. *Am J Hypertens* 1997;10:346–355.
19. Shimosawa T, Takano K, Ando K, et al. Magnesium inhibits norepinephrine release by blocking N-type calcium channels at peripheral sympathetic nerve endings. *Hypertension* 2004;44:897–902.
20. Huerta MG, Roemmich JN, Kington ML, et al. Magnesium deficiency is associated with insulin resistance in obese children. *Diabetes Care* 2005;28:1175–1181.

16
Diabetes Mellitus and Magnesium

Masanori Emoto and Yoshiki Nishizawa

The number of individuals with diabetes is now steadily increasing worldwide.[1] Research over 50 years has demonstrated that abnormal magnesium homeostasis exists in both type 1 and type 2 diabetes. In the 1980s and 1990s, the association between hyperglycemia and magnesium metabolism and the mechanisms of impairment of insulin sensitivity and secretion in magnesium deficiency were vigorously investigated. Later in the 1990s, epidemiological studies including large numbers of subjects established the close association of low dietary magnesium intake with development of type 2 diabetes. It is now of great interest to determine whether research on magnesium can provide new means of preventing diabetes. In this chapter, we review historical to contemporary findings concerning magnesium and diabetes in an attempt to obtain new insights in this field.

Hypomagnesemia in Diabetes

More than 50 years ago, reduced blood levels of magnesium in diabetic individuals were documented in the literature.[2] In 1979, Mather and colleagues reported plasma magnesium levels measured by atomic absorption spectrophotometry in 582 unselected diabetic outpatients and 140 controls.[3] They found that mean plasma magnesium level in diabetic patients was lower than that in controls, and that 25% of diabetic patients had magnesium levels below those of all controls. Hypomagnesemia is frequently observed in both type 1 and type 2 diabetes.[3-8] The frequency of hypomagnesemia in type 2 diabetes is reported to vary from 25% to 48%.[3,5,8-11] Because magnesium is predominantly an intracellular ion, its plasma and serum levels do not necessarily reflect intracellular and tissue magnesium status. Serum and intracellular levels measured by nuclear magnetic resonance of erythrocytes are also reduced in stable type 2 diabetic patients, even when total serum magnesium level is comparable to that in control subjects.[10] Intracellular magnesium deficiency may occur even if plasma or serum magnesium level is normal. However, once decrease in plasma magnesium level becomes evident, it closely reflects deficiency of intracellular magnesium.[12,13] Only a few studies have reported normal or even high levels of magnesium in serum in diabetic patients.[14] These variations in plasma and serum magnesium levels in individuals with diabetes may be attributed to differences in state of glycemic control, diabetic complications, and oral intake of magnesium, as well as other factors.

Possible Mechanisms of Hypomagnesemia in Diabetes

Magnesium is abundant in whole grains, nuts, legumes, green leafy vegetables, and reduced-fat dairy products. Magnesium homeostasis is maintained mainly by intestinal absorption and renal excretion.[15] Loss of magnesium in sweat is negligible except in special conditions such as marked sudation. Normally, 30% to 50% of dietary magnesium intake is absorbed by the small intestine. Net magnesium absorption is determined by the magnesium content of ingested foods and the retention of magnesium in digestive fluids. Approximately 70% to 80% of serum magnesium is filtered through the glomeruli, but only 20% to 30% of filtered magnesium is re-absorbed along the proximal tubules. On average, one third of dietary intake of magnesium is eliminated in the urine, as determined by analysis of magnesium intake and urinary losses.

Two possible mechanisms have been proposed for low levels of magnesium in plasma and blood cells in diabetic patients: (1) low magnesium intake from diet, and (2) increased urinary loss of magnesium. A few studies in the United States and Denmark found that dietary intake of magnesium was low in individuals with type 1 and type 2 diabetes.[16] The westernization of dietary habits has sometimes resulted in daily intake of magnesium below the recommended daily allowance. In fact, as described below, recent epidemiological studies have demonstrated a close association of low dietary magnesium intake with the development of type 2 diabetes.[17-21]

Most previous studies have demonstrated that urinary excretion of magnesium is increased in patients with diabetes[6,22,23] and that poor glycemic control is inversely correlated with hypomagnesemia.[8,22,24-26] Sjogren and colleagues found that glycosylated HbA1 level was inversely correlated with magnesium levels in plasma, mononuclear cells, and skeletal muscle, in a study of muscle biopsies obtained from type 1 diabetic patients.[24] Their subsequent study of 18 type 2 diabetic patients revealed lower levels of magnesium in skeletal muscle and higher excretion of urine of magnesium than in control subjects, but no correlation of plasma magnesium level with magnesium levels in skeletal muscle or circulating blood cells.[22] Another study of 12 type 1 diabetic patients reported that urinary magnesium excretion exhibited an inverse correlation with magnesium level in erythrocytes, but not with fasting plasma glucose or glycosylated hemoglobin levels.[23] In a large cohort study, the Atherosclerosis Risk in Communities (ARIC) Study including 15,248 participants aged 45 to 64 years, serum magnesium levels were found to be inversely associated with fasting serum insulin and glucose levels.[26] On the other hand, no correlations of plasma magnesium level with intracellular magnesium level or glycemic control were found in 128 type 2 diabetic patients reported by the Lima Mde group.[11]

Several studies have attempted to determine whether improvement of glycemic control can improve magnesium deficiency in diabetes. Schnack and coworkers found in a study of 50 type 2 diabetic patients that neither initiation of insulin therapy nor intensification oral hypoglycemic agent treatment improved plasma magnesium level during a 3-month follow-up period despite marked improvement of glycemic control.[27] Interestingly, Djurhuus and colleagues, in an attempt to clarify the effect on magnesium deficiency of improvement of glycemic control in type 1 diabetic patients,[28] found that their intervention, which involved a mean increase in insulin dose of 10%, with resulting a 20% decrease in fasting plasma glucose, induced a 14% decrease in urinary excretion of magnesium concomitant with improvement of serum lipid profile, increase in high-density lipoprotein (HDL) cholesterol level, and decrease in triglyceride level. Recently, Walti and coworkers found that neither absorption nor retention of dietary magnesium was impaired in well-controlled type 2 diabetic patients compared with healthy controls, in a study using meals test-labeled with a magnesium isotope.[29]

Magnesium and Insulin Secretion

In Vitro and In Vivo Studies

The role of intracellular magnesium in insulin secretion by pancreatic β cells remains unclear, and there have been few in vivo or in vitro studies of this subject. Magnesium is involved in more than 300 enzymatic reactions including those involving adenosine triphosphate (ATP), and also acts as an intracellular calcium blocker. It is thus likely that intracellular magnesium deficiency affects the insulin secretory capacity of pancreatic β cells, although the interaction of intracellular magnesium with insulin secretion signaling appears to be complex. Legrand and colleagues investigated the effect of a low-magnesium diet on insulin secretory capacity in the rat.[30] In their study, insulin responses during oral and intravenous and oral glucose tolerance tests were markedly reduced in rats fed a magnesium-deficient diet for 4 weeks, and these changes could be reversed by a high-magnesium diet. Recently, Palanivel and colleagues clarified the role of regulation of acetyl-CoA carboxylase by a magnesium-activated protein phosphatase in pancreatic β cells derived from the GK rat, a model of type 2 diabetes.[31] These findings suggest that magnesium plays a role in insulin secretion.

Clinical Evidence Concerning the Relationship of Magnesium and Insulin Secretion

Several studies have investigated the effects of oral supplementation of magnesium on insulin secretion in diabetic, elderly, and healthy subjects. Mentzel and colleagues failed to detect an effect of serum magnesium level on residual insulin secretion assessed by an oral glucose tolerance test with glucagon stimulation in 33 type 1 diabetic and 10 control subjects.[32] Paolisso and coworkers evaluated insulin secretion carefully using an intravenous glucose test and the glucose clamp technique, the gold standard for measurement of insulin secretion and sensitivity in humans.[33-35] They reported that in eight elderly type 2 diabetic patients, oral supplementation of 3 g magnesium per day significantly increased acute insulin response after an intravenous glucose (0.33 g/kg for 3 min) and arginine (5 g) test, and that net increase in this response was correlated with increase in magnesium content in erythrocytes.[33] Subsequently, they also showed that chronic supplementation of 2 g magnesium per day increased both acute insulin response after glucose loading and the rate of disappearance of glucose during euglycemic hyperinsulinemic clamp, with a positive correlation of insulin response with net increase in magnesium level in erythrocytes in elderly type 2 diabetic patients[34] and elderly subjects.[35] Oral magnesium supplementation thus appears to stimulate insulin secretion in certain specific conditions, although clinical use of it for the purpose of glycemic control in diabetes will require further investigation.

Magnesium and Insulin Resistance

In Vitro Studies of Magnesium and Insulin Signaling

Magnesium is a cofactor for multiple enzymes involved in carbohydrate metabolism.[36] The reason for the strong associations of magnesium deficiency with decrease in insulin sensitivity and insulin resistance has yet to be clearly determined. In vivo animal and in vitro studies have revealed that magnesium deficiency strongly reduces insulin-mediated glucose uptake in muscle and adipocytes.

In rat adipocytes cultured in low-magnesium media, insulin-mediated glucose uptake, oxidation, and incorporation into triglycerides were significantly reduced, but could be restored to normal levels by re-incubation in high-magnesium medium.[37] They found a significant decrease of 60% in intracellular magnesium level in adipocytes cultured in low-magnesium medium compared with that in adipoctyes cultured in medium with physiological magnesium level.[37]

In rats made hypomagnesemic by feeding of a low-magnesium diet for 4 days, 40% reduction of rate of disappearance of glucose after glucose intravenous administration and 45% reduction of glucose-stimulated insulin secretion were found compared with control rats.[38] Furthermore, muscle perfusion experiments and receptor analyses revealed a 50% reduction in insulin sensitivity (ED50) without changes in basal or maximum glucose uptake, 50% reduction of autophosphorylation of the β-subunit of the insulin receptor, and reduction of tyrosine kinase activity of insulin receptors without changes in insulin binding to insulin receptors or abundance of glucose transporter protein (GLUT 4). Similar findings have been reported in fructose-fed insulin-resistant rats.[39] This impairment of glucose uptake in magnesium deficiency is at least partly reversible by improvement of intracellular magnesium level in vitro[37] and of serum magnesium level in vivo.[39] Magnesium supplementation has also been reported to delay the development of diabetes in spontaneously obese diabetic rats.[40] These findings indicate that intracellular magnesium deficiency causes up to 50% decrease in insulin-mediated glucose disposal in muscle and adipocytes, mainly due to impairment of autophosphorylation and tyrosine kinase activity of insulin receptors at the postreceptor level.

Other intracellular relationships between the insulin signaling and magnesium homeostasis are complex and not yet fully elucidated.[16,41] Intracellular ionic magnesium exerts various effects by interacting with intracellular ionic calcium as a natural physiological calcium blocker.[15,41] It is known that magnesium regulates the influx and efflux of ionic calcium in cells. Protein kinase C (PKC), a constitutive negative regulator of the insulin receptor, is activated by increase in intracellular ionic calcium secondary to a deficiency of intracellular ionic magnesium.[41,42] Another possible mechanism of insulin resistance is an interaction of magnesium with PPAR-γ receptor, a member of the steroid/thyroid hormone nuclear receptor superfamily of transcription factors. This receptor plays an important role in glucose/lipid metabolism and adipocyte differentiation. Intracellular ionic magnesium acts as an important cofactor in the phosphorylation of PPAR-γ and its coactivator, PGC-1α, which regulate enzymes participating in liver gluconeogenesis such as glucose-6-phosphtase and phosphoenolpyruvate carboxylase.[43] Interestingly, Guerrero-Romero and colleagues recently found that treatment with pioglitazone 30 mg per day for 12 weeks increased serum magnesium level by 112% above pretreatment level, in a randomized, controlled trial with 60 subjects.[44] This finding suggests the possibility of a new link between PPARs and magnesium.

Epidemiological and Clinical Evidence Concerning the Relationship between Magnesium and Insulin Resistance

Epidemiological Evidence

Noninvasive indirect simple indexes of insulin sensitivity can be used in a larger number of subjects in epidemiological studies, such as fasting insulin, homeostasis model assessment (HOMA-IR),[45,46] and the quantitative insulin check index (QUICKI).[47,48] In the Atherosclerosis Risk in Communities (ARIC) Study including 15,248 participants aged 45 to 64 years, serum magnesium level and dietary magnesium intake were each inversely associated with fasting serum insulin level, suggesting an association of magnesium with fasting hyperinsulinemia and insulin resistance.[26] In 349 subjects randomly selected from 39,345 American women in the Women's Health Study (WHS), mean fasting insulin level was significantly decreased in the highest quartile group of total dietary magnesium intake.[21] Huerta and colleagues were the first to report, in a study of 48 children, that dietary magnesium intake (mg/kcal) was inversely correlated with HOMA-IR ($r_s = -0.43$, $p = 0.002$) and positively correlated with QUICKI ($r_s = 0.43$, $p = 0.002$).[49]

Clinical Evidence Obtained Using Glucose Clamp Technique, Minimal Model Analysis, and Steady-State Plasma Insulin and Plasma Glucose (SSPG)

In human studies, the euglycemic hyperinsulinemic glucose clamp is the gold standard for assessing whole-body insulin sensitivity and separately estimating rates of oxidative and nonoxidative glucose metabolism in combination with indirect calorimetry. Using this technique, Paolisso and coworkers successfully demonstrated the effect of chronic magnesium supplementation on insulin sensitivity in a double-blind crossover study of elderly type 2 diabetic patients[34] and healthy elderly subjects.[35] Chronic supplementation of 2 g magnesium per day for 4 weeks increased total body glucose disposal by 128% and glucose oxidation rate during hyperinsulinemia by 152%, compared to the placebo period.[35] These improvements in insulin sensitivity were positively correlated with net change in erythrocyte magnesium level (Pearson's $r = 0.67$, $p < 0.01$). On the other hand, neither basal hepatic glucose production nor rate of nonoxidative glucose metabolism during hyperinsulinemia was significantly altered compared with the placebo period. Alzaid and coworkers subsequently reported that physiological hyperinsulinemia and hyperglycemia induced by the glucose clamp technique decreased plasma magnesium level in both type 2 diabetic and control subjects, and that this decrease was correlated with the whole-body rate of disappearance of glucose during euglycemic clamp ($r = 0.55$, $p < 0.01$). This finding indicates that magnesium uptake to insulin-sensitive tissue by insulin is also impaired, as well as glucose uptake in type 2 diabetes.[50] In 179 young adults who had been followed longitudinally, total dietary magnesium intake in milligrams/kilograms of fat-free mass was found to be weakly but significantly correlated with whole-body rate of glucose disposal during euglycemic hyperinsulinemic clamp ($r = 0.15$, $p < 0.05$ overall; in men, $r = 0.25$, $p < 0.02$) and with sum of plasma insulin levels during the oral glucose tolerance test.[51]

We have examined insulin resistance in Japanese individuals with type 2 diabetes in detail in various clinical conditions by means of euglycemic hyperinsulinemic glucose clamp with an artificial pancreas (STG-22, Nikkiso Co. Ltd, Tokyo). To the best of our knowledge, no previous studies have examined the association of magnesium with insulin resistance determined using glucose clamp technique in Japanese type 2 diabetic patients. Total serum magnesium level and insulin resistance assessed by euglycemic hyperinsulinemic clamp were retrospectively investigated in a cross-sectional study of 104 type 2 diabetic patients attending the Diabetes Center of Osaka City University Hospital,

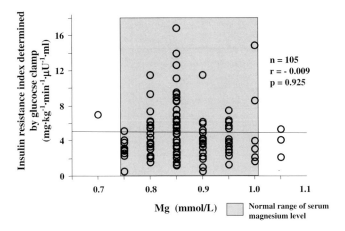

FIGURE 16.1. The relationship between total serum magnesium level and insulin resistance index assessed by euglycemic hyperinsulinemic clamp in 104 type 2 diabetic patients at the Diabetes Center of Osaka City University Hospital. Most of our Japanese type 2 diabetic patients had normal levels of serum magnesium, with a mean of 0.9 ± 0.1 (SD) mmol/L, as indicated by the dotted area. No significant correlation was found between serum magnesium level and insulin resistance index ($r = -0.009$, $p = 0.925$). Possible explanations for this finding are provided in the text.

as previously described elsewhere.[46,48] As shown in Figure 16.1, we failed to find any significant association of serum magnesium level with insulin resistance in our diabetic patients, who regularly visited our hospital and were treated by diabetologists (unpublished data). This discrepancy between our findings and those of previous studies is probably due to our evaluation of total serum magnesium level, rather than ionic serum or intracellular level of magnesium, in well-controlled diabetic patients. Total serum magnesium levels were within normal range in most of our subjects. As described above, total serum magnesium level does not precisely reflect the state of intracellular magnesium deficiency. Furthermore, it is likely that the in-sulin resistance in our subjects was more strongly affected by confounding factors, such as obesity, fat deposition, and various drugs, than by magnesium per se.

Another technique for assessing insulin sensitivity is the intravenous glucose tolerance test with minimal model analysis and steady-state plasma insulin and plasma glucose (SSPG) technique. Using minimal model analysis, Nadler and colleagues found a significant decrease by 75% in insulin sensitivity after a magnesium-deficient diet, with reduction of both serum and intracellular free magnesium levels in six subjects.[52] In 98 healthy subjects, Rosolova and colleagues found the highest fasting insulin and SSPG levels during SSPG in the group with the lowest tertile of plasma magnesium levels and an inverse correlation of plasma magnesium level with SSPG ($r = -0.27$, $p < 0.01$).[53,54]

Magnesium and Development of Diabetes

Magnesium Intake and Development of Type 2 Diabetes

There is increasing clinical and epidemiological evidence that low magnesium intake is associated with development of type 2 diabetes (Table 16.1). In 1992, the Nurse Healthy Study, which included 84,360 women with a 6-year follow-up period, first demonstrated that high dietary magnesium intake was associated with reduction of risk of development of diabetes, with a relative risk of 0.68 compared with the group with the lowest quintile of magnesium intake.[55] Various prospective epidemiological studies have subsequently been performed, as shown in Table 16.1. The Iowa Women's Healthy Study, a prospective cohort study including 35,988 older women with 6-year follow-up, found a strong inverse association of total grain, whole grain, total dietary fiber, cereal fiber, and dietary magnesium intakes with incidence of diabetes after adjustment for potential nondietary confounding variables.[17] The lowest multivariate-adjusted relative risk of diabetes was 0.67 compared to the highest quintile of dietary magnesium intake. Similarly, the Health Professionals Follow-up Study, which included 42,898 men with 12 years follow up, found a relative risk of development of diabetes of 0.58 compared with the lowest quintile of whole-grain intake.[56] Subsequently, the Nurse Health Study, which had a 16-year follow-up period, investigated the beneficial effects of intake of nuts and peanut butter, principal food sources of magnesium, on the development of diabetes. The relative risk of development of diabetes in women consuming peanut butter five times or more per week was 0.79 compared with those never or almost never consuming peanut butter.[57] The ARIC study found that the incidence of type 2 diabetes during 6-year follow up increased approximately twofold across the highest-to-lowest serum magnesium level in Caucasian participants.[58] The relative risk of diabetes was 1.76 with a 95% confidence interval of 1.18–2.61, compared with the group with lowest serum magnesium levels. However, in that study, no significant association was found between dietary magnesium intake and the incidence of diabetes.

Cross-sectional studies have also supported the association of magnesium intake with metabolic disorders related to development of diabetes such as fasting insulin level, waist, and body mass index (BMI).[18,26,51] For example, in the Framingham Offspring Study, including 2941 subjects, intake of whole grains, which are major food sources of magnesium, was closely associated with fasting insulin level as well as body mass index and waist-to-hip ratio.[18] These associations were less strong after adjusting for dietary magnesium.

Recently, two prospective studies by Lopez-Ridaura and colleagues and Song and colleagues with large numbers of subjects and long follow-up periods have yielded definite findings regarding magnesium intake and development of diabetes.[20,21] In a study involving follow up of 85,060 women and 42,872 men, a food-frequency questionnaire was validated every 2 to 4 years over 18 years in women and 12 years in men. The relative risk (RR) of type 2 diabetes after

TABLE 16.1. Principal prospective epidemiological studies concerning associations of dietary magnesium (Mg) intake, serum magnesium level, and the development of type 2 diabetes.

Report	Study cohort	Subjects (N)	Follow-up period (years)	Relative risk (RR[a]) of diabetes
Colditz, 1992[55]	Nurse Healthy Study	84,360 American women	5 years	0.63 (p^a = 0.02) for magnesium compared with in the lowest quintile of magnesium intake.
Kao, 1999[58]	Atherosclerosis Risk in Communities Study	12,128 nondiabetic middle-aged adults	6 years	1.76 (95% CI, 1.18–2.61; P^a = 0.01) compared with lowest serum Mg. No association with magnesium intake
Meyer, 2002[17]	The Iowa Women's Healthy Study	35,988 older women	6 years	0.67 (p^a = 0.0003) across quintiles of magnesium intake
Fung[56]	Health Professionals Follow-up Study	42,898 men	12 years	0.58 (95% CI, 0.47–0.70; p < 0.0001) compared with the lowest quintile of whole-grain intake
Jiang, 2002[57]	Nurse Healthy Study	83,818 women	15 years	0.79 (95% CI, 0.68–0.91; p^a < 0.001) for women consuming peanut butter 5 times or more per week[b]
Lopez-Ridaura, 2004[20]	—	85,060 women and 42,872 men	8 years in women and 12 years in men	0.66 (95% CI, 0.60–0.73; p < 0.001) for women and 0.67 (95% CI 0.56–0.80; p < 0.001) for men.
Song, 2004[21]	The Women's Health Study	39,345 American women	6 years	0.89 (p = 0.05) for women and 0.78 (p = 0.02) for overweight or obese women

Abbreviations: RR, multivariate-adjusted relative risk of diabetes by confounding factor or by comparison with the group with the lowest risk of diabetes; 95% CI, 95% confidence interval.
[a]The details about the relative risk (RR) of diabetes are described in the text. p for trend across quintile groups
[b]Compared with those who never/almost never ate peanut butter.

adjustment for various confounding factors was found to be 0.66 [95% confidential interval (95% CI), 0.60–0.73; $p < 0.001$) for women and 0.67 (95% CI, 0.56–0.80; $p < 0.001$) for men, in a comparison of highest and lowest quintiles of total magnesium intake.[20] In the Women's Health Study, with a cohort of 39,345 American women and an average follow-up period of 6 years, the multivariate-adjusted RR of diabetes of the lowest quintile of magnesium intake was 0.89 compared to the highest ($p = 0.05$), while that for overweight or obese women with BMI greater than 25kg/m^2 was 0.78 ($p = 0.02$).[21] These findings strongly suggest the usefulness of increased consumption of major food sources of magnesium, such as whole grains, nuts, and green leafy vegetables, in preventing type 2 diabetes.

Interventional Trial of Magnesium Supplementation in Type 2 Diabetes

Controversy exists concerning whether oral magnesium supplementation improves metabolic profile, and especially glycemic control, in type 2 diabetic patients (Table 16.2). Table 16.2 shows the findings of well-conducted clinical trials concerning this issue. Some studies of subjects with type 2 diabetes[35,59–61] have found decreases in fasting plasma glucose and insulin level as well as HbA1c with improvement of insulin sensitivity, as represented by index of homeostasis model assessment (HOMA-R) and euglycemic hyperinsulinemic clamp, while others have not.[11,62–64] These discrepancies in findings probably arise from the differences in the magnesium salts and doses used, the degree of magnesium deficiency, and the period of intervention.

In a recent randomized, double-blind, placebo-controlled trial conducted by Rodriguez-Moran and coworkers,[60] a notable effect of oral supplementation of $MgCl_2$ solution (2.5 g per day) for 16 weeks was found in a total of 63 type 2 diabetic patients treated with glibenclamide who had low serum magnesium levels (below 0.74 mmol/L). It is surprising that after 16-week treatment, the group with magnesium supplementation had lower HOMA-R (3.8 ± 1.1 vs. 5.0 ± 1.3, $p = 0.005$), fasting glucose level (8.0 ± 2.4 mmol/L vs. 10.3 ± 2.1 mmol/L, $p = 0.01$), and HbA1c (8.0% ± 2.4% vs. 10.1% ± 3.3%, $p = 0.04$) with higher plasma magnesium (0.74 ± 0.10 mmol/L vs. 0.65 ± 0.07 mmol/L, $p = 0.02$) compared with the placebo group. Subsequently, they also demonstrated improvement of the HOMA-R index to 57% below baseline mean value after 3-month supplementation, with increase in serum magnesium in nondiabetic subjects with insulin resistance.[65] The magnesium chloride used in their studies is believed to have excellent bioavailability and thus to be useful for improving serum magnesium level within the first month of treatment.[60,65]

In addition to better glycemic control, other beneficial effects of oral magnesium supplementation can be expected in type 2 diabetic patients, such as improvement of lipid profile, decrease in triglyceride level, and improvement of hypertension.[64] The major side effects of oral magnesium supplementation are headache, nausea, hypotension, and mild abdominal and bone pain,

TABLE 16.2. Principal previous reports concerning the clinical benefits of oral magnesium supplementation on metabolic profile.

Reports	Subjects	N	Supplementation (g/day, period)	Increase in Mg	Results
Paolisso, 1992[35]	T2DM	12	Mg pidolate 4.5 g, 4 weeks	Erythrocytes	Increase in insulin secretion and [activity?] Decreases in FPG and PPG
Paolisso, 1994[59]	T2DM	9	Mg pidolate, 3 weeks	Plasma Erythrocytes	Increases in insulin sensitivity and glucose oxidation No change in glucose level
Eibl, 1995[63]	T2DM	40	Mg citrate 30 mmol, 3 months	Plasma	No changes in HbA1c
De Valk, 1998[64]	T2DM	50	Mg aspartate15 mmol, 3 months	Plasma Urinary [excretion?]	No changes in FPG, HbA1c, lipid profile, or blood pressure
Lima Mde, 1998[11]	T2DM	128	Mg oxide 20.7 mmol, 30 days	No changes in plasma or monocytes	No changes of FPG, Fructosamine, HbA1
Rodriguez-Moran, 2003[60]	T2DM	63	Mg chloride 2.5 g, 16 weeks	Serum	Decrease of HOMA-R Decrease of FPG, HbA1c
Yokota, 2004[61]	T2DM	9	Natural Mg 0.3 g, 30 days	—	Decrease of FIRI, BP, TG, HOMA-R
Guerrero-Romero, 2004[65]	Insulin-resistant	60	Mg chloride 2.5 g, 3 months	Serum	Decrease HOMA-R index

Abbreviations: FPG or PPG, fasting or postprandial plasma glucose; HOMA-R, insulin resistance index of homeostasis model assessment; Mg, magnesium; T2DM, type 2 diabetes mellitus.

which usually require neither specific treatment nor cessation of magnesium supplementation.

Perspectives on Research in Magnesium and Diabetes

Magnesium, the second most abundant cation in the body, acts as an important cofactor for many enzymes and is a natural calcium blocker, and consequently is involved in many biological reactions in various cells and tissues. This makes the physiological and pathological functions and critical roles played by magnesium difficult to clearly determine. Many problems remain to be solved, as stated above. The following issues are of great importance to the clinical treatment of diabetes: (1) the intracellular mechanisms of insulin resistance in skeletal muscle, fat, and liver, and insulin secretion by pancreatic β cells in magnesium deficiency; (2) the clinical effects of oral supplementation of magnesium on metabolic profiles, especially in individuals with glucose intolerance, diabetes, and related disorders such as hypertension and hyperlipidemia; and (3) the beneficial effects of consumption of major food sources, such as whole grains, nuts, and green leafy vegetables, in preventing the development of diabetes and related disorders in high-risk populations.[66] Further randomized prospective clinical trials can be anticipated to address the third of these issues.

References

1. Zimmet P, Alberti KG, Shaw J. Global and societal implications of the diabetes epidemic. *Nature* 2001;414:782–787.
2. Stutzman FL, Amatuzio DS. Blood serum magnesium in portal cirrhosis and diabetes mellitus. *J Lab Clin Med* 1953;41:215–219.
3. Mather HM, Levin GE. Magnesium status in diabetes. *Lancet* 1979;1:924.
4. McNair P, Christiansen C, Madsbad S, et al. Hypomagnesemia, a risk factor in diabetic retinopathy. *Diabetes* 1978;27:1075–1077.
5. Levin GE, Mather HM, Pilkington TR. Tissue magnesium status in diabetes mellitus. *Diabetologia* 1981;21:131–134.
6. Fujii S, Takemura T, Wada M, Akai T, Okuda K. Magnesium levels of plasma, erythrocyte and urine in patients with diabetes mellitus. *Horm Metab Res* 1982;14:161–162.
7. Durlach J, Collery P. Magnesium and potassium in diabetes and carbohydrate metabolism. Review of the present status and recent results. *Magnesium* 1984;3:315–323.
8. Vanroelen WF, Van Gaal LF, Van Rooy PE, De Leeuw IH. Serum and erythrocyte magnesium levels in type I and type II diabetics. *Acta Diabetol* 1985;22:185–190.
9. McNair P, Christensen MS, Christiansen C, Madsbad S, Transbol I. Renal hypomagnesaemia in human diabetes mellitus: its relation to glucose homeostasis. *Eur J Clin Invest* 1982;12:81–85.
10. Resnick LM, Altura BT, Gupta RK, Laragh JH, Alderman MH, Altura BM. Intracellular and extracellular magnesium depletion in type 2 (non-insulin-dependent) diabetes mellitus. *Diabetologia* 1993;36:767–770.

11. Lima Mde L, Cruz T, Pousada JC, Rodrigues LE, Barbosa K, Cangucu V. The effect of magnesium supplementation in increasing doses on the control of type 2 diabetes. *Diabetes Care* 1998;21:682–686.
12. Ryzen E, Servis KL, DeRusso P, Kershaw A, Stephen T, Rude RK. Determination of intracellular free magnesium by nuclear magnetic resonance in human magnesium deficiency. *J Am Coll Nutr* 1989;8:580–587.
13. Guerrero-Romero F, Rodriguez-Moran M. Complementary therapies for diabetes: the case for chromium, magnesium, and antioxidants. *Arch Med Res* 2005;36: 250–257.
14. Yajnik CS, Smith RF, Hockaday TD, Ward NI. Fasting plasma magnesium concentrations and glucose disposal in diabetes. *BMJ* 1984;288:1032–1034.
15. Laires MJ, Monteiro CP, Bicho M. Role of cellular magnesium in health and human disease. *Front Biosci* 2004;9:262–276.
16. Paolisso G, Barbagallo M. Hypertension, diabetes mellitus, and insulin resistance: the role of intracellular magnesium. *Am J Hypertens* 1997;10:346–355.
17. Meyer KA, Kushi LH, Jacobs DR Jr, Slavin J, Sellers TA, Folsom AR. Carbohydrates, dietary fiber, and incident type 2 diabetes in older women. *Am J Clin Nutr* 2000;71: 921–930.
18. McKeown NM, Meigs JB, Liu S, Wilson PW, Jacques PF. Whole-grain intake is favorably associated with metabolic risk factors for type 2 diabetes and cardiovascular disease in the Framingham Offspring Study. *Am J Clin Nutr* 2002;76:390–398.
19. Abbott RD, Ando F, Masaki KH, et al. Dietary magnesium intake and the future risk of coronary heart disease (the Honolulu Heart Program). *Am J Cardiol* 2003; 92:665–669.
20. Lopez-Ridaura R, Willett WC, Rimm EB, et al. Magnesium intake and risk of type 2 diabetes in men and women. *Diabetes Care* 2004;27:134–140.
21. Song Y, Manson JE, Buring JE, Liu S. Dietary magnesium intake in relation to plasma insulin levels and risk of type 2 diabetes in women. *Diabetes Care* 2004;27: 59–65.
22. Sjogren A, Floren CH, Nilsson A. Magnesium, potassium and zinc deficiency in subjects with type II diabetes mellitus. *Acta Med Scand* 1988;224:461–466.
23. Gurlek A, Bayraktar M, Ozaltin N. Intracellular magnesium depletion relates to increased urinary magnesium loss in type I diabetes. *Horm Metab Res* 1998;30:99–102.
24. Sjogren A, Floren CH, Nilsson A. Magnesium deficiency in IDDM related to level of glycosylated hemoglobin. *Diabetes* 1986;35:459–463.
25. Pun KK, Ho PW. Subclinical hyponatremia, hyperkalemia and hypomagnesemia in patients with poorly controlled diabetes mellitus. *Diabetes Res Clin Pract* 1989;7: 163–167.
26. Ma J, Folsom AR, Melnick SL, et al. Associations of serum and dietary magnesium with cardiovascular disease, hypertension, diabetes, insulin, and carotid arterial wall thickness: the ARIC study. Atherosclerosis Risk in Communities Study. *J Clin Epidemiol* 1995;48:927–940.
27. Schnack C, Bauer I, Pregant P, Hopmeier P, Schernthaner G. Hypomagnesaemia in type 2 (non-insulin-dependent) diabetes mellitus is not corrected by improvement of long-term metabolic control. *Diabetologia* 1992;35:77–79.
28. Djurhuus MS, Henriksen JE, Klitgaard NA, et al. Effect of moderate improvement in metabolic control on magnesium and lipid concentrations in patients with type 1 diabetes. *Diabetes Care* 1999;22:546–554.

29. Walti MK, Zimmermann MB, Walczyk T, Spinas GA, Hurrell RF. Measurement of magnesium absorption and retention in type 2 diabetic patients with the use of stable isotopes. *Am J Clin Nutr* 2003;78:448–453.
30. Legrand C, Okitolonda W, Pottier AM, Lederer J, Henquin JC. Glucose homeostasis in magnesium-deficient rats. *Metabolism* 1987;36:160–164.
31. Palanivel R, Veluthakal R, McDonald P, Kowluru A. Further evidence for the regulation of acetyl-CoA carboxylase activity by a glutamate- and magnesium-activated protein phosphatase in the pancreatic beta cell: defective regulation in the diabetic GK rat islet. *Endocrine* 2005;26:71–77.
32. Menzel R, Pusch H, Ratzmann GW, et al. Serum magnesium in insulin-dependent diabetics and healthy subjects in relation to insulin secretion and glycemia during glucose-glucagon test. *Exp Clin Endocrinol* 1985;85:81–88.
33. Paolisso G, Passariello N, Pizza G, et al. Dietary magnesium supplements improve B-cell response to glucose and arginine in elderly non-insulin dependent diabetic subjects. *Acta Endocrinol* 1989;121:16–20.
34. Paolisso G, Sgambato S, Pizza G, Passariello N, Varricchio M, D'Onofrio F. Improved insulin response and action by chronic magnesium administration in aged NIDDM subjects. *Diabetes Care* 1989;12:265–269.
35. Paolisso G, Sgambato S, Gambardella A, et al. Daily magnesium supplements improve glucose handling in elderly subjects. *Am J Clin Nutr* 1992;55:1161–1167.
36. Paolisso G, Scheen A, D'Onofrio F, Lefebvre P. Magnesium and glucose homeostasis. *Diabetologia* 1990;33:511–514.
37. Kandeel FR, Balon E, Scott S, Nadler JL. Magnesium deficiency and glucose metabolism in rat adipocytes. *Metabolism* 1996;45:838–843.
38. Suarez A, Pulido N, Casla A, Casanova B, Arrieta FJ, Rovira A. Impaired tyrosine-kinase activity of muscle insulin receptors from hypomagnesaemic rats. *Diabetologia* 1995;38:1262–1270.
39. Balon TW, Jasman A, Scott S, Meehan WP, Rude RK, Nadler JL. Dietary magnesium prevents fructose-induced insulin insensitivity in rats. *Hypertension* 1994;23:1036–1039.
40. Balon TW, Gu JL, Tokuyama Y, Jasman AP, Nadler JL. Magnesium supplementation reduces development of diabetes in a rat model of spontaneous NIDDM. *Am J Physiol* 1995;269:E745–E752.
41. Takaya J, Higashino H, Kobayashi Y. Intracellular magnesium and insulin resistance. *Magnes Res* 2004;17:126–136.
42. Resnick LM, Gupta RK, Bhargava KK, Gruenspan H, Alderman MH, Laragh JH. Cellular ions in hypertension, diabetes, and obesity. A nuclear magnetic resonance spectroscopic study. *Hypertension* 1991;17:951–957.
43. Puigserver P, Spiegelman BM. Peroxisome proliferator-activated receptor-gamma coactivator 1 alpha (PGC-1 alpha): transcriptional coactivator and metabolic regulator. *Endocr Rev* 2003;24:78–90.
44. Guerrero-Romero F, Rodriguez-Moran M. Pioglitazone increases serum magnesium levels in glucose-intolerant subjects. A randomized, controlled trial. *Exp Clin Endocrinol Diabetes* 2003;111:91–96.
45. Matthews DR, Hosker JP, Rudenski AS, Naylor BA, Treacher DF, Turner RC. Homeostasis model assessment: insulin resistance and beta-cell function from fasting plasma glucose and insulin concentrations in man. *Diabetologia* 1985;28:412–419.

46. Emoto M, Nishizawa Y, Maekawa K, et al. Homeostasis model assessment as a clinical index of insulin resistance in type 2 diabetic patients treated with sulfonylureas. *Diabetes Care* 1999;22:818–822.
47. Katz A, Nambi SS, Mather K, et al. Quantitative insulin sensitivity check index: a simple, accurate method for assessing insulin sensitivity in humans. *J Clin Endocrinol Metab* 2000;85:2402–2410.
48. Yokoyama H, Emoto M, Fujiwara S, et al. Quantitative insulin sensitivity check index and the reciprocal index of homeostasis model assessment are useful indexes of insulin resistance in type 2 diabetic patients with wide range of fasting plasma glucose. *J Clin Endocrinol Metab* 2004;89:1481–1484.
49. Huerta MG, Roemmich JN, Kington ML, et al. Magnesium deficiency is associated with insulin resistance in obese children. *Diabetes Care* 2005;28:1175–1181.
50. Alzaid AA, Dinneen SF, Moyer TP, Rizza RA. Effects of insulin on plasma magnesium in noninsulin-dependent diabetes mellitus: evidence for insulin resistance. *J Clin Endocrinol Metab* 1995;80:1376–1381.
51. Humphries S, Kushner H, Falkner B. Low dietary magnesium is associated with insulin resistance in a sample of young, nondiabetic Black Americans. *Am J Hypertens* 1999;12:747–756.
52. Nadler JL, Buchanan T, Natarajan R, Antonipillai I, Bergman R, Rude R. Magnesium deficiency produces insulin resistance and increased thromboxane synthesis. *Hypertension* 1993;21:1024–1029.
53. Rosolova H, Mayer O Jr, Reaven G. Effect of variations in plasma magnesium concentration on resistance to insulin-mediated glucose disposal in nondiabetic subjects. *J Clin Endocrinol Metab* 1997;82:3783–3785.
54. Rosolova H, Mayer O Jr, Reaven GM. Insulin-mediated glucose disposal is decreased in normal subjects with relatively low plasma magnesium concentrations. *Metabolism* 2000;49:418–420.
55. Colditz GA, Manson JE, Stampfer MJ, Rosner B, Willett WC, Speizer FE. Diet and risk of clinical diabetes in women. *Am J Clin Nutr* 1992;55:1018–1023.
56. Fung TT, Hu FB, Pereira MA, et al. Whole-grain intake and the risk of type 2 diabetes: a prospective study in men. *Am J Clin Nutr* 2002;76:535–540.
57. Jiang R, Manson JE, Stampfer MJ, Liu S, Willett WC, Hu FB. Nut and peanut butter consumption and risk of type 2 diabetes in women. *JAMA* 2002;288:2554–2560.
58. Kao WH, Folsom AR, Nieto FJ, Mo JP, Watson RL, Brancati FL. Serum and dietary magnesium and the risk for type 2 diabetes mellitus: the Atherosclerosis Risk in Communities Study. *Arch Intern Med* 1999;159:2151–2159.
59. Paolisso G, Scheen A, Cozzolino D, et al. Changes in glucose turnover parameters and improvement of glucose oxidation after 4-week magnesium administration in elderly noninsulin-dependent (type II) diabetic patients. *J Clin Endocrinol Metab* 1994;78:1510–1514.
60. Rodriguez-Moran M, Guerrero-Romero F. Oral magnesium supplementation improves insulin sensitivity and metabolic control in type 2 diabetic subjects: a randomized double-blind controlled trial. *Diabetes Care* 2003;26:1147–1152.
61. Yokota K, Kato M, Lister F, et al. Clinical efficacy of magnesium supplementation in patients with type 2 diabetes. *J Am Coll Nutr* 2004;23:506S–509S.
62. Gullestad L, Jacobsen T, Dolva LO. Effect of magnesium treatment on glycemic control and metabolic parameters in NIDDM patients. *Diabetes Care* 1994;17:460–461.

63. Eibl NL, Kopp HP, Nowak HR, Schnack CJ, Hopmeier PG, Schernthaner G. Hypomagnesemia in type II diabetes: effect of a 3-month replacement therapy. *Diabetes Care* 1995;18:188–192.
64. de Valk HW, Verkaaik R, van Rijn HJ, Geerdink RA, Struyvenberg A. Oral magnesium supplementation in insulin-requiring Type 2 diabetic patients. *Diabetes Med* 1998;15:503–507.
65. Guerrero-Romero F, Tamez-Perez HE, Gonzalez-Gonzalez G, et al. Oral magnesium supplementation improves insulin sensitivity in non-diabetic subjects with insulin resistance. A double-blind placebo-controlled randomized trial. *Diabetes Metab* 2004;30:253–258.
66. Nadler JL. A new dietary approach to reduce the risk of type 2 diabetes? *Diabetes Care* 2004;27:270–271.

17
Magnesium Metabolism in Insulin Resistance, Metabolic Syndrome, and Type 2 Diabetes Mellitus

Mario Barbagallo, Ligia J. Dominguez, Virna Brucato, Antonio Galioto, Antonella Pineo, Anna Ferlisi, Ernesto Tranchina, Mario Belvedere, Ernesto Putignano, and Giuseppe Costanza

Magnesium plays a key role in regulating insulin action, insulin-mediated glucose uptake, and vascular tone. Intracellular magnesium depletion may result in a defective tyrosine–kinase activity at the insulin receptor level, in a postreceptorial impairment in insulin action, and clinically in a worsening of insulin resistance. Intra- and extracellular alterations of magnesium metabolism have been identified in clinical states characterized by insulin resistance, such as metabolic syndrome, hypertension, altered glucose tolerance, type 2 diabetes, and aging. Several studies, from our and other's groups, have confirmed the clinical relevance of alterations of magnesium homeostasis in these conditions and have highlighted the importance of an accurate definition of the magnesium status. While measurements of total serum magnesium levels have been proven inadequate for this purpose because important magnesium depletions are required before total serum level decreases, two technologies, ^{31}P nuclear magnetic resonance (^{31}P-NMR) spectroscopy and magnesium-specific ion-selective electrodes, that, respectively, measure intracellular and extracellular free levels of magnesium, have a higher sensitivity in detecting magnesium deficits. A number of evidences have confirmed that magnesium supplementation is indicated in conditions associated with magnesium deficit, although well-designed therapeutic trials with oral magnesium supplements to study the beneficial effects in metabolic syndrome and in type 2 diabetes are needed.

An increasing number of evidences have suggested a clinical relevance for the altered magnesium (Mg) metabolism present in states of increased peripheral insulin resistance. As discussed below, we have suggested a role for Mg deficit as a possible unifying mechanism of the insulin resistance of hypertension and conditions associated with altered glucose tolerance, including metabolic syndrome and type 2 diabetes mellitus.[1] Our group has used a ion-based approach to demonstrate: (1) the critical importance of Mg metabolism in regulating insulin sensitivity, as well as vascular tone and blood pressure homeostasis; (2) that Mg deficiency, defined on the basis of intracellular free magnesium levels (Mgi), and or serum ionized magnesium (MgI) is a common feature of insulin resistant-states, as well as various cardiovascular and metabolic processes and aging; (3) the ability of environmental factors such as dietary nutrient sugar and mineral content to alter the set point of steady-state cell ion activity.[2,3]

Magnesium and Glucose Metabolism

Magnesium is the second most abundant intracellular cation (after potassium) present in living cells and its plasma concentration is remarkably constant in healthy subjects. Ninety-nine percent of Mg is distributed in the intracellular fluid, and 1% is distributed in the extracellular fluid. The levels of Mg in the plasma of healthy people are remarkably constant, being on average 0.85 mmol/L and varying less than 15% from this value. Circulating Mg exists in three forms: a protein-bound fraction (25% bound to albumin and 8% bound to globulins), a chelated fraction (12%), and the physiologically active ionized fraction (MgI, 55%).[4] Because Mg is predominantly an intracellular ion, and in the serum only the ionized active form is metabolically available, its total serum concentrations (MgT) may not reflect the Mg status or intracellular pool, and intracellular Mg depletion can be seen with normal MgT concentrations.[5–8]

Magnesium is directly involved in numerous important biochemical reactions, and particularly is a necessary cofactor in over 300 enzymatic reactions and specifically in all those processes that involve the utilization and transfer of adenosine triphosphate (ATP). Thus, intracellular Mg is a critical cofactor for several enzymes in carbohydrate metabolism, and because of its role as part of the activated Mg-ATP complex required for all of the rate-limiting enzymes of glycolysis, regulates the activity of all enzymes involved in phosphorylation reactions. Intracellular free magnesium concentration is critical in the phosphorylation of the tyrosine–kinase of the insulin receptor, as well as all other protein kinases, and all ATP and phosphate transfer-associated enzymes, such as the CaATPases in plasma membrane and endoplasmic reticulum. Magnesium deficiency may result in disorders of tyrosine–kinase activity on the insulin receptor, an event related to the development of a of postreceptorial insulin resistance and decreased cellular glucose utilization[9]; that is, the lower the basal Mgi, the greater the amount of insulin required to metabolize the same glucose load, indicating decreased insulin sensitivity. Specifically, in skeletal muscle and fat tissue, insulin resistance would be the expected outcome in the presence of suppressed cellular Mg. Significant decrements in these enzyme activities can already be observed at the range of Mg values seen in disease states such as type 2 diabetes and hypertension.[10]

Diagnostic Tools Available to Define Magnesium Status of Magnesium-Deficient Subjects

One of the principle reasons Mg metabolism has not become more the focus of routine attention in clinical practice has been the absence of an easily available, accurate, and reproducible measure of Mg status. Measurements of MgT levels, which includes the protein-bound, chelated, and ionized fractions, that are the clinical measurement routinely used for assessing circulating Mg are not useful, and have been proven inadequate for this purpose because important Mg depletion is required before its total serum levels decrease.[5,11] Two technologies, ^{31}P-NMR spectroscopy and Mg-specific ion-selective electrodes (ISE), that measure Mgi and extracellular MgI, respectively, are a major advance in this regard, and have a higher sensitivity in detecting Mg deficits. This is because MgT levels are a late marker of a depletion of Mg stores (the complexed serum MgT is the last store to be depleted when an already important Mg depletion has occurred), while the declines in ionized unbound MgI may occur significantly earlier. While ^{31}P-NMR techniques are a research-based test because of the expenses in setting up and maintaining the NMR equipment, the development of a Mg-selective electrode apparatus may be particularly useful in routine clinical use. Using these techniques, a deficiency of intra- and ionized extracellular Mg levels have been consistently demonstrated in diabetes, often when MgT levels were within normal limits,[5] and significant relationship have been found with blood pressure, and cardiovascular and metabolic parameters.[3] From the intracellular point of view, these relations appear to be continuous, and do not display a threshold value within the range of clinically observable cellular free Mg levels, that is, the lower the free Mgi, the stiffer the blood vessels, the higher the blood pressure, the greater the insulin resistance, etc.[1,3,12]

Magnesium Deficiency in Type 2 Diabetes Mellitus and Metabolic Syndrome

The presence of a Mg deficit in diabetic patients has long been recognized.[13,14] Epidemiological studies have found a high prevalence of hypomagnesaemia in subjects with diabetes, especially in those with poor glycemic control.[15,16] Because of the lack of sensitivity of serum total MgT, a suppressed level of intracellular Mgi and serum ionized MgI can be found in many subjects with total serum MgT still in the normal range (see above). Using the Nova-8 Mg ISE to measure serum ionized Mg in a preliminary sample of 50 subjects with type 2 diabetes, we have recently found significantly lower MgI levels compared to normal controls (0.49 + 0.01 mmol/L vs. 0.52 + 0.01 mmol/L, $p < 0.05$), without significant changes in MgT (Figure 17.1). When MgI and Mgi levels were measured concurrently in the same subjects using ^{31}P-NMR and the ISE Mg-selective electrode, Resnick and colleagues found that both Mgi and MgI (but not serum total MgT) were significantly reduced in type 2 diabetes subjects, and that a close direct relationship was present between the ionized extra- and the intracellular Mg measurement.[5]

Low dietary Mg intake and increased Mg urinary losses[17–19] are the main causes of the Mg deficit in diabetic subjects. The use of loop and thiazide

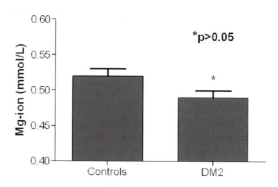

Figure 17.1. Ionized Mg levels (MgI, mmol/L) in type 2 diabetic subjects versus normal controls.

diuretics, which promote Mg wasting, may worsen Mg depletions. A Mg-deficient diet has been found to be associated with a significant impairment of insulin-mediated glucose uptake.[20] Hyperglycemia and hyperinsulinemia may have a role in the increased urinary Mg excretion, thus contributing to Mg depletion, and the reduced sensitivity to insulin may affect Mg transport.

Magnesium deficiency, which may take the form of a chronic latent Mg deficit rather than clinical hypomagnesemia, may have clinical importance because of the crucial role of Mg as a cofactor in many enzymatic reactions regulating glucose metabolism. A deficient Mg status may not just be a secondary consequence of diabetes, but experimental and epidemiological data suggest that it may precede and cause insulin resistance and altered glucose tolerance, and even type 2 diabetes.[21] However, independent to the cause of poor plasma and intracellular Mg content, a depletion of Mg seems to be a cofactor for a further derangement of insulin resistance.[1,22]

At the cellular level, using gold-standard NMR techniques, our group have shown lower steady-state Mgi and reciprocally increased Cai levels in subjects with type 2 diabetes mellitus, compared with young nondiabetic subjects.[23-25] We have recently extended our experience with NMR technique to study cytosolic free Mg directly in situ, in vivo, and in human living tissues, such as muscle and brain, and have shown that in living tissues Mgi values are quantitatively and inversely related to both systolic and diastolic blood pressures.[26]

Intracellular free magnesium levels quantitatively predict the fasting and postglucose levels of hyperinsulinemia, as well as peripheral insulin sensitivity. Specifically, (1) fasting insulin levels[27]; (2) the integrated insulinemic response to a standard oral glucose tolerance test[28]; (3) the steady-state plasma glucose response to insulin infusion; and (4) indices for peripheral insulin sensitivity derived from euglycemic hyperinsulinemic clamps, are all inversely related to Mg levels, whether measured as Mgi in situ in brain, free or total Mg in peripheral red cells,[21,29,30] or even as circulating Mg.[31,32] Furthermore, direct and inverse relations, respectively, are observed between Mgi and Cai levels and HbA1c, fasting blood glucose, and the glycemic response to oral glucose loading in normal, hypertensive, and diabetic subjects.[33]

Aging, Insulin Resistance, and Magnesium

Old age is frequently associated with insulin resistance and glucose intolerance. We have specifically studied the behavior of ion content with age and have shown a continuous age-dependent fall of Mgi levels in peripheral blood cells of healthy elderly subjects,[25] these alterations being indistinguishable from those occurring, independently of age, in essential hypertension or diabetes.[25] In other terms, essential hypertension and/or type 2 diabetes appear to determine an acceleration of natural age-dependent Mg depletion, suggesting that these ionic changes may be clinically significant, underlying the predisposition in elderly subjects to cardiovascular and metabolic diseases, and might therefore help to explain the age-related increased incidence of these diseases. At the same time, the naive concept of both hypertension and type 2 diabetes as diseases of accelerated vascular aging may be more literally true then previously thought, because both of these diseases display these same ionic changes at all ages. Thus, having these conditions is ionically the same as getting older, for example, a 48-year-old diabetic having cellular ionic alterations indistinguishable from a healthy 84-year-old, suggesting a role for Mg deficit in the increased incidence of hypertension and glucose intolerance with age.[2,25]

Altogether, this accumulating evidence of the relevance of altered cellular Mg metabolism to tissual insulin sensitivity suggests a critical role of Mg metabolism in contributing to the clinical coincidence of Mg depletion to states of insulin resistance, such as hypertension, metabolic syndrome, type 2 diabetes, as well as the increased incidence of each of these conditions with age, a condition itself characterized by a tendency to intracellular Mg depletion. Thus, pathophysiologically, Mgi depletion can directly promote tissual insulin resistance and altered vascular tone, thus helping to understand the mechanisms underlying the clinical association among these apparently different conditions (Figure 17.2).

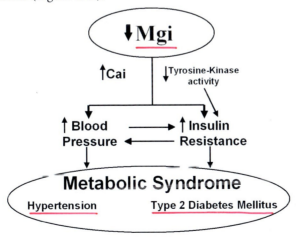

FIGURE 17.2. Overall hypothesis in which intracellular Mg deficiency may mediate the relationship between insulin resistance, hypertension, and type 2 diabetes mellitus.

Insulin and Glucose Acute Effects on Dynamic Changes of Cellular Magnesium

Although these data suggest that cellular free ion content may determine clinically relevant biologic outcomes such as blood pressure, cardiac mass, and cellular insulin responsiveness,[2] it is also concomitantly true that metabolic and dietary factors have a role in contributing to insulin sensitivity and in regulating intracellular ion metabolism. Thus, dietary salt[34] or sugar loading allowed us to assess the role of different dietary circumstances in regulating blood pressure and cellular ion metabolism. Specifically, the transient hyperglycemia of oral glucose loading reproduces in normal subjects the same altered ionic profile of depleted Mgi/increased Cai levels that occurs chronically in diabetic subjects,[27] which dynamically appears to be equally and inversely true for serum ionized Mg. The contribution of hyperglycemia and hyperinsulinemia to the intracellular Mg depletion of diabetes has been confirmed by in vitro studies from our group showing that both glucose and insulin may, in turn, alter Mgi levels. Thus, glucose in a specific, concentration- and time-dependent manner, at concentrations achieved clinically, and independent of insulin lowered Mgi and reciprocally elevated Cai.[35] Barbagallo and colleagues reported that hyperglycemia also alter ionic content in cultured vascular smooth muscle cells, suggesting an ionic mechanism for the increased vasoconstriction present in chronic diabetic states.[36] This glucose-mediated effects are independent of insulin action, because hyperglycemia induces these changes both in vascular smooth muscle cells, and in erythrocytes, where glucose transport is unaffected by insulin.[35,37,38] Among its many actions, insulin has specific ionic effects to stimulate the transport of Mg from the extracellular to the intracellular compartment, thus increasing Mgi content.[39] Using NMR techniques to measure Mgi, we have shown that the ionic action of insulin is specific, dose-related, and independent of cellular glucose uptake. Insulin in the incubation medium was able to induce an accumulation of Mgi which shifted from a basal values of 177 ± 11 mmol/L to 209 ± 19 mmol/L.[39] The ionic effects of insulin were time and dose dependent. In the dose-response study, the dose of insulin at which we observed these ionic effects started at 10 µU/mL and peaked at 200 µU/mL, this dose corresponding roughly to the maximal physiological response in humans, adding to the physiological relevance of this effect. Such results further suggest that insulin is an important modulator of intracellular Mg content; furthermore, there are indications that, as in other energy producing systems, an ATPase-dependent pump is involved in the mechanism by which insulin regulates the erythrocyte Mgi content.

The overall hypothesis that Mgi content is a crucial determinant of cellular responsiveness is supported by other data from our group showing that the ability of insulin at physiologically maximal concentrations to stimulate Mgi is impaired in cells from hypertensive individuals, in which the basal Mgi content is reduced,[21] confirming that insulin action is strictly dependent from the cellular Mgi content; for all subjects, independently of their designation as normotensive or hypertensive, cellular Mgi responsiveness to insulin was closely and directly related to basal cellular Mgi levels; that is, the lower the basal Mgi, the less responsive was the cell to insulin. Furthermore, a blunting of Mgi responses to insulin could be reproduced in normal cells that were Mg depleted by prior treatment either with A23187 in a calcium-free medium or with high glucose concentrations (15 mmol/L).[40] Once again, insulin responsiveness followed basal Mgi levels ($r = 0.637$). Altogether, these data demonstrating ionic aspects of cellular insulin resistance has led us to hypothesize that the insulin resistance, currently defined by measurements of tissue glucose uptake, might equally well be defined on the basis of altered cellular ion responsiveness. Similar cellular behavior was found in cell responsiveness to glucose action. As was the case for insulin the lower the Mgi, the less responsive is the cell to glucose.[37] Thus, this ionic insulin resistance is probably not specific for insulin, but may rather be one tissue manifestation of a general property of cells in which steady-state Mgi and/or Cai level may determine cell responsiveness to insulin, glucose, or to other external stimuli. It is possible to hypothesize therefore, that the insulin resistance of hypertension and diabetes, currently defined by measurements of tissue glucose uptake, might equally well be defined on the basis of altered cellular ion responsiveness (Figure 17.2).

The link between Mg deficiency and the development of insulin resistance and type 2 diabetes is strengthened by the observation that several treatments for diabetes appear to increase Mg levels. Metformin, for example, raises Mg levels in the liver. Pioglitazone, a thiazolidinedione anti-diabetic agent that increases insulin sensitivity, increases free Mg concentration in adipocytes.[41] Other research of our group have demonstrated that the action of the insulin mimeting substances vanadate and IGF-1 are both associated to a direct effect to stimulate intracellular free Mg levels,[42] and that the effects of antioxidants glutathione and vitamin E to improve glucose and insulin metabolism may derive at least in part from their action on Mg metabolism. Barbagallo and coworkers demonstrated that the action of glutathione and vitamin E to increase insulin sensitivity in hypertensive subjects is associated with a concurrent increase in Mgi, and that a significant direct relationship between glucose disposal increase and Mgi levels was present.[29,30]

Is Providing Oral Magnesium Supplementation Effective in Reversing the Magnesium and Clinical Abnormalities?

The effects of Mg supplements on the metabolic profile of type 2 diabetic subjects are still controversial,[43] benefits having been found in some,[44–46] but not all, clinical studies.[47–49] Differences in baseline Mg status and metabolic control may explain the differences among these studies. Thus, a recent clinical trial specifically conducted among diabetic subjects with low total serum Mg levels (index of an already advanced Mg deficit) found a beneficial effect of oral Mg supplementation on fasting and postprandial glucose levels and insulin sensitivity[43] and we have shown an improvement in insulin-mediated glucose uptake measured by euglycemic insulin clamp in diabetic subjects after oral Mg supplementations.[21] A significant relationship was found between the increase in plasma and erythrocyte Mg concentration and the parallel progressive increase in the insulin sensitivity in diabetic patients that were supplemented with increased dosage of Mg supplements (5.1–11.5 mmol of elemental Mg).[21] Among nondiabetic, apparently healthy subjects, there are also some evidences of a relatively small but significant beneficial effects of Mg supplements on insulin sensitivity.[32,50] Recent epidemiological data have showed a significant inverse association between Mg intake and diabetes risk supporting the priority of Mg deficit in the development of glucose intolerance and diabetes.[51] Thus, taking into account that both dietary Mg and serum plasma Mg content have been associated with an increased risk to develop glucose intolerance and diabetes,[13] the use of Mg supplements could be an alternative tool for the prevention of type 2 diabetes,[52] a hypothesis that needs to be confirmed by specific and well-designed trials with Mg that are needed in the near future.[21,53]

Altogether, with the recent advances in the accurate measurement of intracellular and extracellular Mg levels, we have now the tools to translate what is known of the critical importance of Mg in glucose and insulin metabolism into clinical practice, both in routinely monitoring Mg status and in the therapeutic use of Mg supplementation in those conditions and those subjects in whom Mg deficiency can be demonstrated.

Acknowledgment. We are grateful to Mr. Castrense Giordano for his helpful technical assistance in the measurement of ionized free magnesium.

References

1. Barbagallo M, Dominguez LJ, Galioto A, et al. Role of magnesium in insulin action, diabetes and cardio-metabolic syndrome X. *Mol Aspects Med* 2003;24: 39–52.
2. Barbagallo M, Resnick LM. Calcium and magnesium in the regulation of smooth muscle function and blood pressure: the ionic hypothesis of cardiovascular and metabolic diseases and vascular aging. In: Sowers JR, ed. *Endocrinology of the Vasculature*. 1st ed. Totowa, NJ: Humana Press; 1996:283–300.
3. Resnick LM. Ionic basis of hypertension, insulin resistance, vascular disease, and related disorders. The mechanism of "syndrome X". *Am J Hypertens* 1993;6: 123S–134S.
4. Saris NE, Mervaala E, Karppanen H, Khawaja JA, Lewenstam A. Magnesium. An update on physiological, clinical and analytical aspects. *Clin Chim Acta* 2000;294: 1–26.

5. Resnick LM, Altura BT, Gupta RK, Laragh JH, Alderman MH, Altura BM. Intracellular and extracellular magnesium depletion in type 2 (non-insulin-dependent) diabetes mellitus. *Diabetologia* 1993;36:767–770.
6. Altura BT, Burack JL, Cracco RQ, et al. Clinical studies with the NOVA ISE for IMg2+. *Scand J Clin Lab Invest Suppl* 1994;217:53–67.
7. Altura BT, Shirey TL, Young CC, et al. A new method for the rapid determination of ionized Mg2+ in whole blood, serum and plasma. *Methods Find Exp Clin Pharmacol* 1992;14:297–304.
8. Altura BT, Shirey TL, Young CC, et al. Characterization of a new ion selective electrode for ionized magnesium in whole blood, plasma, serum, and aqueous samples. *Scand J Clin Lab Invest Suppl* 1994;217:21–36.
9. Kolterman OG, Gray RS, Griffin J, et al. Receptor and postreceptor defects contribute to the insulin resistance in noninsulin-dependent diabetes mellitus. *J Clin Invest* 1981;68:957–969.
10. Laughlin MR, Thompson D. The regulatory role for magnesium in glycolytic flux of the human erythrocyte. *J Biol Chem* 1996;271:28977–28983.
11. Barbagallo M, Resnick LM, Dominguez LJ, Licata G. Diabetes mellitus, hypertension and ageing: the ionic hypothesis of ageing and cardiovascular-metabolic diseases. *Diabetes Metab* 1997;23:281–294.
12. Resnick LM. The cellular ionic basis of hypertension and allied clinical conditions. *Prog Cardiovasc Dis* 1999;42:1–22.
13. Kao WH, Folsom AR, Nieto FJ, Mo JP, Watson RL, Brancati FL. Serum and dietary magnesium and the risk for type 2 diabetes mellitus: the Atherosclerosis Risk in Communities Study. *Arch Intern Med* 1999;159:2151–2159.
14. Mather HM, Nisbet JA, Burton GH, et al. Hypomagnesaemia in diabetes. *Clin Chim Acta* 1979;95:235–242.
15. Schnack C, Bauer I, Pregant P, Hopmeier P, Schernthaner G. Hypomagnesaemia in type 2 (non-insulin-dependent) diabetes mellitus is not corrected by improvement of long-term metabolic control. *Diabetologia* 1992;35:77–79.
16. Sjogren A, Floren CH, Nilsson A. Magnesium, potassium and zinc deficiency in subjects with type II diabetes mellitus. *Acta Med Scand* 1988;224:461–466.
17. McNair P, Christensen MS, Christiansen C, Madsbad S, Transbol I. Renal hypomagnesaemia in human diabetes mellitus: its relation to glucose homeostasis. *Eur J Clin Invest* 1982;12:81–85.
18. Djurhuus MS, Klitgaard NA, Pedersen KK, et al. Magnesium reduces insulin-stimulated glucose uptake and serum lipid concentrations in type 1 diabetes. *Metabolism* 2001;50:1409–1417.
19. Djurhuus MS. New data on the mechanisms of hypermagnesuria in type I diabetes mellitus. *Magnes Res* 2001;14:217–223.
20. Matsunobu S, Terashima Y, Senshu T, Sano H, Itoh H. Insulin secretion and glucose uptake in hypomagnesemic sheep fed a low magnesium, high potassium diet. *J Nutr Biochem* 1990;1:167–171.
21. Paolisso G, Barbagallo M. Hypertension, diabetes mellitus, and insulin resistance: the role of intracellular magnesium. *Am J Hypertens* 1997;10:346–355.
22. Takaya J, Higashino H, Kobayashi Y. Intracellular magnesium and insulin resistance. *Magnes Res* 2004;17:126–136.
23. Resnick LM, Gupta RK, Bhargava KK, Gruenspan H, Alderman MH, Laragh JH. Cellular ions in hypertension, diabetes, and obesity. A nuclear magnetic resonance spectroscopic study. *Hypertension* 1991;17:951–957.

24. Resnick LM. Hypertension and abnormal glucose homeostasis. Possible role of divalent ion metabolism. *Am J Med* 1989;87:17S–22S.
25. Barbagallo M, Gupta RK, Dominguez LJ, Resnick LM. Cellular ionic alterations with age: relation to hypertension and diabetes. *J Am Geriatr Soc* 2000;48: 1111–1116.
26. Resnick LM, Barbagallo M, Bardicef M, et al. Cellular-free magnesium depletion in brain and muscle of normal and preeclamptic pregnancy: a nuclear magnetic resonance spectroscopic study. *Hypertension* 2004;44:322–326.
27. Resnick LM, Bardicef O, Barbagallo M, et al. 31PNMR spectroscopic studies of oral glucose loading and in situ skeletal muscle ion content in essential hypertension. *Hypertension* 1995;26:552.
28. Resnick LM, Gupta RK, Gruenspan H, Alderman MH, Laragh JH. Hypertension and peripheral insulin resistance. Possible mediating role of intracellular free magnesium. *Am J Hypertens* 1990;3:373–379.
29. Barbagallo M, Dominguez LJ, Tagliamonte MR, Resnick LM, Paolisso G. Effects of glutathione on red blood cell intracellular magnesium: relation to glucose metabolism. *Hypertension* 1999;34:76–82.
30. Barbagallo M, Dominguez LJ, Tagliamonte MR, Resnick LM, Paolisso G. Effects of vitamin E and glutathione on glucose metabolism: role of magnesium. *Hypertension* 1999;34:1002–1006.
31. Rosolova H, Mayer O Jr, Reaven GM. Insulin-mediated glucose disposal is decreased in normal subjects with relatively low plasma magnesium concentrations. *Metabolism* 2000;49:418–420.
32. Rosolova H, Mayer O Jr, Reaven G. Effect of variations in plasma magnesium concentration on resistance to insulin-mediated glucose disposal in nondiabetic subjects. *J Clin Endocrinol Metab* 1997;82:3783–3785.
33. Resnick LM. Cellular ions in hypertension, insulin resistance, obesity, and diabetes: a unifying theme. *J Am Soc Nephrol* 1992;3:S78–S85.
34. Barbagallo M, Resnick LM. The role of glucose in diabetic hypertension: effects on intracellular cation metabolism. *Am J Med Sci* 1994;307(suppl 1):S60–S65.
35. Resnick LM, Barbagallo M, Gupta RK, Laragh JH. Ionic basis of hypertension in diabetes mellitus. Role of hyperglycemia. *Am J Hypertens* 1993;6:413–417.
36. Barbagallo M, Shan J, Pang PK, Resnick LM. Glucose-induced alterations of cytosolic free calcium in cultured rat tail artery vascular smooth muscle cells. *J Clin Invest* 1995;95:763–767.
37. Barbagallo M, Dominguez LJ, Bardicef O, Resnick LM. Altered cellular magnesium responsiveness to hyperglycemia in hypertensive subjects. *Hypertension* 2001;38: 612–615.
38. Barbagallo M, Gupta RK, Resnick LM. Independent effects of hyperinsulinemia and hyperglycemia on intracellular sodium in normal human red cells. *Am J Hypertens* 1993;6:264–267.
39. Barbagallo M, Gupta RK, Resnick LM. Cellular ionic effects of insulin in normal human erythrocytes: a nuclear magnetic resonance study. *Diabetologia* 1993;36: 146–149.
40. Barbagallo M, Gupta RK, Bardicef O, Bardicef M, Resnick LM. Altered ionic effects of insulin in hypertension: role of basal ion levels in determining cellular responsiveness. *J Clin Endocrinol Metab* 1997;82:1761–1765.
41. Nadler J, Scott S. Evidence that pioglitazone increases intracellular free magnesium concentration in freshly isolated rat adipocytes. *Biochem Biophys Res Commun* 1994;202:416–421.

42. Dominguez LJ, Barbagallo M, Sowers JR, Resnick LM. Magnesium responsiveness to insulin and insulin-like growth factor I in erythrocytes from normotensive and hypertensive subjects. *J Clin Endocrinol Metab* 1998;83:4402–4407.
43. Guerrero-Romero F, Rodriguez-Moran M. Complementary therapies for diabetes: the case for chromium, magnesium, and antioxidants. *Arch Med Res* 2005;36:250–257.
44. Rodriguez-Moran M, Guerrero-Romero F. Oral magnesium supplementation improves insulin sensitivity and metabolic control in type 2 diabetic subjects: a randomized double-blind controlled trial. *Diabetes Care* 2003;26:1147–1152.
45. Yokota K, Kato M, Lister F, et al. Clinical efficacy of magnesium supplementation in patients with type 2 diabetes. *J Am Coll Nutr* 2004;23:506S–509S.
46. Paolisso G, Scheen A, Cozzolino D, et al. Changes in glucose turnover parameters and improvement of glucose oxidation after 4-week magnesium administration in elderly noninsulin-dependent (type II) diabetic patients. *J Clin Endocrinol Metab* 1994;78:1510–1514.
47. de Valk HW, Verkaaik R, van Rijn HJ, Geerdink RA, Struyvenberg A. Oral magnesium supplementation in insulin-requiring Type 2 diabetic patients. *Diabetes Med* 1998;15:503–507.
48. Lima Mde L, Cruz T, Pousada JC, Rodrigues LE, Barbosa K, Cangucu V. The effect of magnesium supplementation in increasing doses on the control of type 2 diabetes. *Diabetes Care* 1998;21:682–686.
49. Gullestad L, Jacobsen T, Dolva LO. Effect of magnesium treatment on glycemic control and metabolic parameters in NIDDM patients. *Diabetes Care* 1994;17:460–461.
50. Guerrero-Romero F, Tamez-Perez HE, Gonzalez-Gonzalez G, et al. Oral magnesium supplementation improves insulin sensitivity in non-diabetic subjects with insulin resistance. A double-blind placebo-controlled randomized trial. *Diabetes Metab* 2004;30:253–258.
51. Lopez-Ridaura R, Willett WC, Rimm EB, et al. Magnesium intake and risk of type 2 diabetes in men and women. *Diabetes Care* 2004;27:134–140.
52. McCarty MF. Nutraceutical resources for diabetes prevention—an update. *Med Hypotheses* 2005;64:151–158.
53. Resnick LM. Magnesium in the pathophysiology and treatment of hypertension and diabetes mellitus: where are we in 1997? *Am J Hypertens* 1997;10:368–370.

Cardiovascular Disease

18
Cardiovascular Disease and Magnesium

Naoyuki Hasebe and Kenjiro Kikuchi

The effects of magnesium on cardiovascular function and disease have received increasing attention during the last decade. Mg^{2+} is the second most abundant intracellular cation after potassium (K^+), and a critical cofactor in more than 300 enzymatic reactions critically involving in energy metabolism, glucose utilization, protein synthesis, fatty acid synthesis and breakdown, adenosine triphosphatase (ATPase) functions, and virtually all hormonal reactions in the cardiovascular system. The mechanisms of cardiovascular protective effects of magnesium associated with depression embrace calcium antagonistic action on L-type and N-type calcium (Ca) channels, suppression of cathecholamine release from sympathetic nervous terminal, and suppression of aldosteron secretion from adrenal gland (Figure 18.1).

Blood Pressure Control and Hypertension

It is now clear that the magnesium ion, although not directly involved in the biochemical process of contraction, modulates vascular smooth muscle tone and contractility by affecting calcium ion concentrations and its availability at critical sites.[1] Magnesium ions actively promote relaxation, offset calcium-related excitation–contraction coupling, and decrease cellular responsiveness to depolarizing stimuli, by stimulating Ca^{2+}-dependent K^+ channels, which serve to offset the potential depolarizing influence of cellular calcium accumulation by activating the membrane sodium (Na)-K-ATPase pump, which is critical for the maintenance of the resting cellular membrane potential, by competitively inhibiting Ca binding to calmodulin, and by stimulating both plasma membrane and sarcoplasmic reticulum Ca ATPases.

Magnesium status has a direct effect upon the relaxation capability of vascular smooth muscle cells and the regulation of the cellular cations important to blood pressure—cellular sodium:potassium (Na:K) ratio and intracellular calcium (iCa^{2+}). Mg^{2+} (intra- and extracellular) exists in three states: (1) free, ionized fraction (the physiologically active form); (2) complexed to anions (citrate, phosphate, bicarbonate); and (3) protein bound. In extracellular fluid,

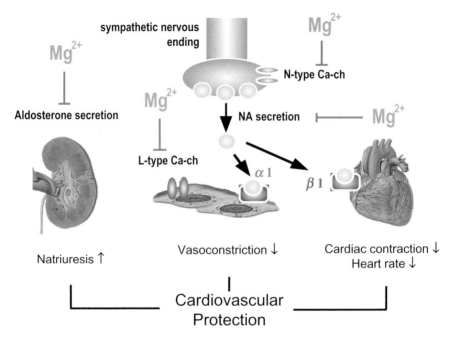

FIGURE 18.1. Cardiovascular protective effects of magnesium. Ca-ch, calcium channel; NA, noradrenaline.

free Mg^{2+} composes 61% of total Mg^{2+}, 6% is complexed, and 33% is protein bound.[2] Both cellular and whole body Na:K ratios are crucial to the maintenance of normal blood pressure; when sodium becomes too high and potassium too low, high blood pressure is one result. At the cellular level, proper function of the sodium–potassium pump maintains potassium at a high intracellular concentration and sodium at a high extracellular concentration. Magnesium is required for the proper function of the sodium–potassium pump, which requires Mg-ATP as a source of energy and is responsible for maintaining the separation of sodium and potassium across cellular membranes. Physiologically, when magnesium status is low enough to cause hypomagnesemia, serum potassium level drops even when potassium intake is adequate or above adequate. Thus, when magnesium status becomes low or deficient, the Mg-ATP–driven sodium–potassium pump can be hindered, proper potassium metabolism is disturbed, potassium flows out of the cell, sodium flows into the cell, and even if adequate potassium is nutritionally maintained under these circumstances, an effective potassium deficit and abnormal Na:K ratio is expressed until the magnesium status is repaired. Hypertension, in such cases, can only be cured by adequate nutritional magnesium intake but is often treated with anti-hypertensive medications, some of which deplete the body of K and/or Mg.

As a result, nutritional magnesium has both direct and indirect impacts on the regulation of blood pressure and therefore on the occurrence of hypertension. Hypertension occurs when cellular Na:K ratios become too high, a consequence of a high-sodium, low-potassium diet or, indirectly, through a magnesium-deficient state that causes a pseudopotassium deficit.

Likewise, magnesium deficiency impairs proper calcium metabolism, affecting blood pressure. When cells' natural calcium channel blocker magnesium becomes deficient, calcium will rush abnormally into cells, creating high iCa^{2+}, low serum calcium, and low urinary calcium states even when calcium intake is adequate. Many aspects of the rise in intracellular calcium ion concentration and its impact on hypertensive elements have been elucidated at the cellular level.[3,4] What has not been studied nor considered in many of these studies is magnesium's impact on calcium ion metabolism; when magnesium is low, there is a concomitant rise in intracellular calcium ion due, primarily, to a magnesium deficiency, creating a magnesium:calcium (Mg:Ca) imbalance. In the case of vascular smooth muscle cells, such an imbalance instills a perpetually high iCa^{2+}, causing a perpetual vasoconstriction state and the condition we know as hypertension.[5] In the case of neurons, healthy intracellular magnesium ion concentrations block N-type Ca^{2+} channels at nerve endings, thus inhibiting norepinephrine release and maintaining normal blood pressure.[6] Many other hypertensive effects of raised iCa^{2+} can be set in motion by low intracellular magnesium. Such effects become added to the loss of magnesium ion's direct vasodilator effect. Thus, we see that long-term normal blood pressure requires maintenance of healthy iCa^{2+} levels which, in turn, require adequate cellular magnesium. High iCa^{2+} and high cellular Na:K ratio both occur when cellular magnesium becomes too low and the Mg-ATP–driven sodium–potassium pump and calcium pump become functionally impaired. High iCa^{2+} has several vasoconstrictive effects which lead to hypertension, an indirect result of low magnesium status. Dietary calcium is directly proportional to dietary magnesium. Serum magnesium does not reflect true magnesium status as do intracellular magnesium measurements.

At the whole body level, when magnesium status is low enough to exhibit hypomagnesemia, serum calcium drops below normal regardless of the intake of calcium, and only the ingestion of magnesium in adequate amounts can dispel the hypocalcemia. Low magnesium status causing a low serum calcium also causes a low urinary calcium. Resnick and coworkers showed intracellular magnesium-to-calcium ion ratio to be the crucial factor in the etiology of hypertension and other aspects of metabolic syndrome X.[7] Resnick's work has shown that Mg:Ca cellular ratios below a certain level result in hyperinsulinemia, insulin resistance, platelet aggregation, cardiac hypertrophy, and hypertension. Confirming this view, Kisters and colleagues reported aortic smooth muscle cells of the spontaneously hypertensive rat to have a Ca:Mg ratio of 3.5 compared with normal rats' ratio of only 2.2.[8]

Another mechanistic consideration is the linkage of Mg^{2+} deficiency with formation of reactive oxygen species (ROS) and potential effects on redox-sensitive growth factor signaling pathways. The knowledge in the field of ROS and vascular function and pathology is currently growing at an exponential rate, and it is becoming clear that these highly reactive molecules are probably not only involved in vascular pathology, but also in normal cell signaling.[9] Anti-oxidants are compounds that hinder the oxidative processes and thereby delay or prevent oxidative stress. In this sense, magnesium is one of the physiological, efficient anti-oxidant. Mg^{2+} deficiency leads to an increase in thiobarbituric acid-reactive substances (TBARS), which indicates an increase in ROS and which is reversed by tempol, a superoxide dismutase mimetic. A broader question that begs answering is how Mg^{2+} deficiency might induce these effects. The traditional belief is that Mg^{2+} exerts many of its cellular actions by serving as a Ca^{2+} antagonist and there is no doubt that Mg^{2+} can compete with Ca^{2+} for binding sites on a variety of cellular targets. It is also thought that Mg^{2+} deficiency may affect cell function by causing the concentration of free intracellular Mg^{2+} to fall below its KD for key enzymes and hence decrease their activity (i.e., the Mg^{2+}-dependent Na^+ and K^+ activated ATPase that establishes the Na^+ and K^+ gradients across the cell membrane and the Mg^{2+}-dependent myosin ATPase which is critical for muscle contraction). It is well known that platelet Ca^{2+} mobilization is elevated in humans and animals with elevated blood pressure. This finding has been interpreted by many to indicate that there is a basic defect Ca^{2+} metabolism in platelets that might extend to other reactive cells, including vascular smooth muscle. However, the increase in platelet Ca^{2+} might also result from an increase in shear stress. This basic question needs to be addressed for the effect of Mg^{2+} on blood pressure and generation of ROS. In view of the continued interest regarding the role of Mg^{2+} in blood pressure regulation and vascular pathology, particularly in light of the hugely successful Dietary Approaches to Stop Hypertension (DASH) trials, additional work needs to be conducted in this area. Focus should continue to be directed towards understanding mechanisms of action, with a concerted effort being made towards making the animal models more closely match the human condition, and with an increased focus on molecular mechanisms, linking between an essential dietary nutrient and the key molecular pathways involved in regulating vascular smooth muscle growth and structure.

Acute Myocardial Infarction

Over the past decade, several reviews have focused on the relevance of magnesium in cardiac disease. A special interest was developed on the importance of magnesium as a pharmacological agent in the treatment of acute myocardial infarction (AMI). Evidence exists that magnesium is the biological element that could diminish or even prevent reperfusion injury in AMI. Coronary artery disease is characterized pathomorphologically by arteriosclerosis and is expressed clinically by reduced coronary blood flow with consecutive myocardial ischemia and hypoxemia. Magnesium depletion and tissue ischemia accelerate myocardial hypoperfusion, myocardial Ca overload, coronary arteriosclerosis, and platelet aggregation, suggesting the role of magnesium in the pathogenesis of ischemic heart disease and sudden death (Figure 18.2).

Clinical application of magnesium is still controversial in acute myocardial infarction because of the conflicting results of previous clinical trials. Magnesium therapy has been reported to reduce mortality in acute myocardial infarction in Leicester Intravenous Magnesium Intervention Trial 2 (LIMIT-2).[10] In LIMIT-2, magnesium administration demonstrated the 24% reduction in the 28-day mortality rate of acute myocardial infarction. However, it could not demonstrate significant effect on mortality in International Study of Infarct Survival 4 (ISIS-4).[11] One of the reasons suggested was a lower mortality rate of the control group in ISIS-4. The ISIS-4 investigators enrolled 58,050 patients: 29,011 to magnesium and 29,030 to control. The control group mortality was 7.2% in ISIS-4 compared with 7.6% in the magnesium group. The Magnesium in Coronaries (MAGIC) trial investigated the benefits of early administration of intravenous magnesium to high-risk patients with acute ST-segment elevation myocardial infarction (STEMI).[12] At 30 days, the mortality rate was 15.3% in the magnesium group and 15.2% in the placebo group.

It is possible that the effects of magnesium are not sufficiently powerful to demonstrate significant reduction in mortality rate in mild-to-moderate acute myocardial infarction, which has already received interventional therapy and full medication with other effective agents.

Despite these conflicting results of clinical trials, several previous reports of experimental myocardial infarction have suggested that magnesium treat-

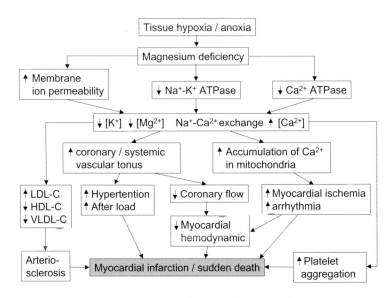

FIGURE 18.2. Myocardial ischemia and magnesium kinetics.

ment is potentially effective to reduce ventricular arrhythmias and reduce infarct size. However, the mechanism of its efficacy is still not fully understood. We have demonstrated that magnesium administration has an infarct-size–limiting effect independent of hemodynamic changes and incidence of arrhythmias in myocardial infarction in rabbits.[13] A high dose of magnesium significantly reduced the myocardial infarct size accompanied by a significant reduction of double products. The negative inotropic and chronotropic actions of a high dose of magnesium potentially produces energy sparing effects and reduces the myocardial infarct size. However, a lower dose of magnesium demonstrated a similar infarct-size–limiting effects independent of hemodynamic changes. We focused on the effects of magnesium on the reduction of myocardial infarct size, which is one of the major determinant of the mortality rate in acute myocardial infarction. The infarct-size–limiting effect of magnesium is attributable, at least in part, to attenuation of calcium overload and augmentation of adenosine mechanism (Figure 18.3).[13] Adenosine is a major cardioprotective substance in ischemia. Ischemic myocardium is salvaged by administration of adenosine or by inhibition of adenosine breakdown. The main pathway of adenosine synthesis in ischemic myocardium is decomposition of adenosine monophosphate by ecto 5′-nucleotidase (5′-ND), the main enzyme of producing adenosine in ischemic myocardium.[14] Magnesium is an important cofactor of 5′-ND. We confirmed that the activity of ecto 5′-ND extracted from rabbit myocardium was enhanced by magnesium in a dose-dependent fashion. It is conceivable that the infarct-size–limiting effect of magnesium is mediated, at least in part, by adenosine through the enhancement of activity of ecto 5′-ND (Figure 18.3).[13]

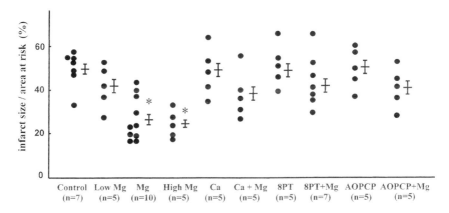

FIGURE 18.3. Myocardial infarct size limiting effects of magnesium. $*p < 0.05$ versus control.[13]

Calcium overload in myocardium is one of the most important mechanisms of myocardial reperfusion injury.[15] It has been reported that the pretreatment with calcium channel antagonists has infarct-size–limiting effects. We have demonstrated that calcium administration abolished the infarct-size–limiting effect of magnesium, suggesting that there exists a competitive interaction between magnesium and calcium in ischemia–reperfusion myocardium (Figure 18.3).[13] Magnesium is known as a natural calcium antagonist. Its calcium antagonistic action is multifunctional, including inhibition of voltage-dependent L-type calcium channel, suppression of Na^+/Ca^{2+} exchanging system, inhibition of calcium release from sarcoplasmic reticulum (SR), potentiation of calcium sequestration into SR, and suppression of calcium binding to specific site of troponin C. Although it was not a study on the myocardial infarct size, Ferrari and coworkers reported that magnesium reduces mitochondrial calcium overload and protects myocardial injury in isolated rabbit hearts. Although calcium administration potentially augments calcium overload mechanism, the elevated extra- and intracellular calcium levels are not the sole determinant of the myocardial infarct size because calcium administration itself did not increase the infarct size. In either case, the infarct-size–limiting effect of magnesium was abolished by calcium administration. It is conceivable that magnesium reduces myocardial infarct size, at least in part, via the magnesium–calcium interaction mechanism in rabbits.

Magnesium depletion has been implicated in the pathogenesis of ischemic heart disease as a result of altering blood lipid composition and accelerating atherogenesis, and causing coronary artery spasm. However, the evidence supporting the role of magnesium in the pathogenesis of ischemic heart disease, at this time, is not very convincing. Magnesium deficiency has been shown to increase the infarct size.[16] Several beneficial effects of magnesium other than infarct-size–limiting effect have been proposed in myocardial infarction: coronary vasodilation increasing coronary blood flow, inhibition of catecholamine release,[17] and inhibition of platelet aggregation protecting against thrombosis. It is noteworthy that magnesium supplementation is inexpensive, easy to administer, and relatively free from side effects. In conclusion, magnesium administration has an infarct-size–limiting effect independent of hemodynamic changes and incidence of arrhythmias in myocardial infarction in rabbits. The infarct-size–limiting effect of magnesium is attributable, at least in part, to attenuation of calcium overload and augmentation of adenosine mechanism.

Ischemic–reperfusion injury of the microvasculature results in progressive diminution of perfusion to previously ischemic tissues despite restoration of flow in the conduit arteries supplying these tissues, that is, the no-reflow phenomenon.[18] The no-reflow phenomenon is a significant clinical problem. It occurs in coronary interventions and in patients receiving thrombolytic therapy during AMI. No reflow is associated with a higher incidence of early and prolonged congestive heart failure (CHF) compared with the absence of no reflow. The primary insult in no reflow is probably reperfusion-induced (and oxygen free radical–mediated) injury to endothelium.[19] Circumstantial evidence suggests that Mg^{2+} reduces endothelial injury. First, deficiency of Mg^{2+} potentiates oxygen free radical–induced postischemic injury in working isolated rat hearts. Second, agents that attenuate the initial ischemic injury, namely Ca^{2+} antagonists administered before reperfusion, also reduce the severity of no reflow and preserve endothelial function. Finally, experimental areas of no reflow are decreased, and vascular endothelial and smooth muscle function are preserved after administration of Mg^{2+} cardioplegia (16 mM).[20]

Catecholamine

Myocardial dysfunction is accompanied by increased catecholamine concentrations in blood and urine. The serum catecholamine level is an indicator of the clinical severity of heart failure.[21] Although sympathetic activation is a crucial compensatory mechanism in the initial stage of heart failure, excess catecholamine itself induces cardiac cell injury, leading to cardiac dysfunction. Therefore, ß-adrenergic receptor blockade improves prognosis as well as cardiac function in chronic heart failure. Excess isoproterenol (ISO), a selective ß-adrenergic receptor agonist, induces myocardial necrosis and apoptosis, interstitialfibrosis, and left ventricular (LV) hypertrophy. The mechanism of cardiac dysfunction induced by excess ISO may be explained by calcium overload[22] and free radical generation.[23] Moreover, excess ISO induces ß-adrenergic desensitization, one of the hallmarks of heart failure, which potentially exacerbates cardiac dysfunction. We have demonstrated that magnesium supplement prevents ISO-induced cardiac dysfunction and ß-adrenergic desensitization in dogs.[24] The mechanism of cardiac dysfunction induced by excess ISO remains not fully understood. The most plausible explanation of ISO cardiotoxicity is intracellular calcium overload through ß-adrenergic receptor stimulation. Cardiomyocytes are injured by excessive Ca^{2+} influx,[25] leading to irreversible cell injury, apoptosis, and necrosis. Therefore, ISO-induced cardiac dysfunction is effectively prevented by calcium channel blockades as well as ß-adrenergic receptor blockades. Mg^{2+} inhibits extracellular Ca^{2+} influx via a voltage-sensitive Ca^{2+} channel (L-type Ca^{2+} channel), Ca^{2+} release from sarcoplasmic reticulum (SR), that is, Ca^{2+} induced Ca^{2+} release, and potentiate sequestration of released Ca^{2+} by SR in myocardial cells. These findings suggest that magnesium supplementation potentially suppresses Ca^{2+} overload in myocardium.

Prolonged exposure of hearts to ß-adrenergic agonist results in decreased inotropic responsiveness to agonists, that is, ß-adrenergic desensitization. It represents the mechanism that contributes to the progression of heart failure. The process of ß-adrenergic desensitization involves decreases in ß-adrenergic receptor density, which is triggered by phosphorylation of receptors, a rapid uncoupling of receptor from Gs, and decreases in the basal adenylate cyclase activities and calcium channel density. Although the mechanism of ISO-induced ß-adrenergic receptor desensitization has not been entirely clarified, the time-dependent decrease in inotropic response to ISO was significantly prevented by magnesium supplement (Figure 18.4).[24] Magnesium is known to modulate the receptor-G protein–catalytic interactions at several sites. Magnesium acts as a cofactor with ATP at the catalytic site, and is required for GTPase activity and Gs activation as well as GTP binding. Feldman reported that ß-adrenergic desensitization is obscured in a high-magnesium condition by maintaining the activity of adenylate cyclase and cyclic adenosine monophosphate (cAMP)-dependent protein kinase.[26] These findings suggest that magnesium supplement may inhibit ß-adrenergic desensitization by the regulation of the receptor-G protein–catalytic interactions.

Another possible effect of magnesium on cardiac dysfunction and ß-adrenergic desensitization is an antioxidative action. Excess ISO can generate oxygen free radicals, suppress the activity of anti-oxidative enzyme, consequently augment oxidative stress. Excess catecholamine itself can be oxidized and generate oxidative products. An oxidation product of catecholamines may be responsible for impaired inotropic responses to adrenergic stimulation, as well as myocardial necrosis and contractile failure. The time-dependent increase in lipid peroxides, radical products of polyunsatulated fatty acid, was significantly diminished by magnesium supplement (Figure 18.4).[24] A chain reaction of lipid peroxidation by ROS impairs myocardial cell membrane, and, consequently, causes cardiac dysfunction. Magnesium has been reported to reduce radical production in ischemia reperfusion of myocardium. The anti-oxidative effects of magnesium may contribute to prevention of cardiac dysfunction and ß-adrenergic desensitization induced by excess ISO.

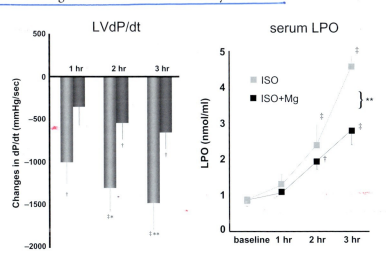

FIGURE 18.4. Magnesium suppresses isoproterenol-induced cardiac dysfunction and oxidative stress. ISO, isoproterenol; LPO, lipid peroxides. †$p < 0.05$ versus baseline; ‡$p < 0.01$ versus baseline; *$p < 0.05$ versus ISO; **$p < 0.01$ versus ISO.[24]

Arrhythmia

Extracellular and/or intracellular magnesium depletion has been implicated in a variety of cardiovascular disturbances, including causation of ventricular arrhythmias, predisposing to digitalis intoxication, modulating vascular tone, and promoting atherogenesis. Administration of MgSO$_4$ during electrophysiological evaluation of patients has demonstrated two effects of Mg^{2+} relevant to the treatment of supraventricular tachyarrhythmias: (1) prolongation of atrioventricular nodal conduction time (anterograde and retrograde) and refractory period[27]; and (2) suppression of conduction in accessory pathways with and without atrioventricular nodelike properties,[27] although conflicting results have been reported. Prolongation of atrioventricular nodal conduction by Mg^{2+} is most likely attributable to inhibition of Ca^{2+} current, the primary mode of impulse conduction through the atrioventricular node, but also may result from Mg^{2+}-induced attenuation of sympathetic activity at the atrioventricular node. Cardiac arrhythmia is an important complication of magnesium depletion that has been associated with a prolonged QT$_c$ interval. In addition, magnesium supplementation has been shown to reduce the QT$_c$ intervals even in patients with normal serum magnesium levels. Torsades de pointes is a repetitive polymorphous ventricular tachycardia that occurs in the presence of QT prolongation, and is usually induced by drugs that prolong the QT interval. Because of its effect on prolonging the QT interval, magnesium depletion has also been implicated in the pathogenesis of torsades de pointes, although such a relation has rarely been shown. Because of its ability to shorten the QT interval, magnesium supplementation has been used with some success in treating torsades de points. In regard to cardiac function, there is a close association between magnesium and potassium. Magnesium has been shown to attenuate the electrophysiological effects of hyperkalemia. Furthermore, in view of the relationship between magnesium and intracellular and extracellular potassium depletion, magnesium depletion has been implicated as a potential cause of digitals intoxication. This is supported by the finding that an acute induction of hypomagnesemia in dogs with dialysis facilitates the development of digitalis intoxication and arrhythmias. Moreover, ventricular arrhythmias, including those induced by digitalis, are sensitive to magnesium therapy. Some evidence indicates that subclinical magnesium depletion might predispose patients with acute myocardial infarction to develop arrhythmias.

Other anti-arrhythmic effects of Mg^{2+} have been reported, although the underlying mechanisms have not been defined: restoration of sinus rhythm in critically ill medical and surgical patients with supraventricular tachycardias; suppression of intractable ventricular tachyarrhythmias; control of ventricular rate in new-onset atrial fibrillation (AF); prophylaxis of AF after coronary artery bypass grafting; slowing of digoxin-facilitated ventricular rate during AF in Wolff–Parkinson–White syndrome; abolition of pre-excitation (δ wave) in patients with Wolff–Parkinson–White syndrome during normal sinus rhythm; suppression of multifocal atrial tachycardia[28]; suppression of digoxin-induced ectopic tachyarrhythmias; prevention of bupivacaine-induced arrhythmias; and treatment of amitriptyline-induced ventricular fibrillation.

References

1. Rosanoff A. Magnesium and hypertension. *Clin Calcium* 1981;15:255–260.
2. Speich M, Bousquet B, Nicolas G. Reference values for ionized, complexed, and protein-bound plasma magnesium in men and women. *Clin Chem* 1981;27: 246–248.
3. Cox RH, Lozinskaya I, Dictz NJ. Calcium exerts a larger regulatory effect on potassium channels in small mesenteric artery mycoytes from spontaneously hypertensive rats compared to Wistar-Kyoto rats. *Am J Hypertens* 2003;16:21–27.
4. Sela S, Shurtz-Swirski R, Farah R, et al. A link between polymorphonuclear leukocyte intracellular calcium, plasma insulin, and essential hypertension. *Am J Hypertens* 2002;15:291–295.
5. Altura BM, Zhang A, Altura BT. Magnesium, hypertensive vascular diseases, atherogenesis, subcellular compartmentation of Ca^{2+} and Mg^{2+} and vascular contractility. *Miner Electrolyte Metab* 1993;19:323–336.
6. Shimosawa T, Takano K, Ando K, et al. Magnesium inhibits norepinephrine release by blocking N-type calcium channels at peripheral sympathetic, nerve endings. *Hypertension* 2004;44:897–902.
7. Resnick L. The cellular ionic basis of hypertension and allied clinical conditions. *Prog Cardiovasc Dis* 1999;42:1–22.
8. Kisters K, Wessels F, Kuper H, et al. Increased calcium and decreased magnesium concentrations and an increased calcium/magnesium ratio in spontaneously hypertensive rats versus Wistar-Kyoto rats: relation to arteriosclerosis. *Am J Hypertens* 2004;17:59–62.
9. Griendling KK, Sorescu D, Lassegue B, Ushio-Fukai M. Modulation of protein kinase activity and gene expression by reactive oxygen species and their role in vascular physiology and pathophysiology. *Arterioscler Thromb Vasc Biol* 2000;20: 2175–2183.
10. Woods KL, Fletcher S. Long-term outcome after intravenous magnesium sulphate in suspected acute myocardial infarction: the second Leicester Intravenous Magnesium Intervention Trial (LIMIT-2). *Lancet* 1994;343:816–819.
11. ISIS-4 (Fourth International Study of Infarct Survival) Collaborative Group. ISIS-4: a randomised factorial trial assessing early oral captopril, oral mononitrate, and intravenous magnesium sulphate in 58,050 patients with suspected acute myocardial infarction. *Lancet* 1995;345:669–685.
12. Cooper HA, Domanski MJ, Rosenberg Y, et al. Acute ST-segment elevation myocardial infarction and prior stroke: an analysis from the Magnesium in Coronaries (MAGIC) trial. *Am Heart J* 2004;148:1012–1019.
13. Matsusaka T, Hasebe N, Jin YT, et al. Magnesium reduces myocardial infarct size via enhancement of adenosine mechanism in rabbits. *Cardiovasc Res* 2002;54: 568–575.
14. Kitakaze M, Hori M, Takeshima S, Sato H, Inoue M, Kamada T. Ischemic preconditioning increases adenosine release and 5′-nucleotidase activity during

myocardial ischemia and reperfusion in dogs. Implications for myocardial salvage. *Circulation* 1993;87:208–215.
15. Oe H, Kuzuya T, Hoshida S, et al. Calcium overload and cardiac myocyte cell damage induced by arachidonate lipoxygenation. *Am J Physiol* 1994;267:H1396–H1402.
16. Chang C, Varghese PJ, Downey J, et al. Magnesium deficiency and myocardial infarct size in the dog. *J Am Coll Cardiol* 1985;5:280–289.
17. Smetana R, Brichta A, Glogar D, et al. Stress and magnesium metabolism in coronary artery disease. *Magnes Bull* 1991;13:125–127.
18. Ito H, Maruyama A, Iwakura K, et al. Clinical implications of the "no reflow" phenomenon. *Circulation* 1996;93:223–228.
19. Bolli R, Triana J, Jeroudi M. Prolonged impairment of coronary vasodilation after reversible ischemia: Evidence for microvascular stunning. *Circ Res* 1990;67:332–343.
20. Gomez MN. Magnesium and cardiovascular disease. *Anesthesiology* 1998;89:222–240.
21. Cohn JN, Levine TB, Olivari MT, et al. Plasma norepinephrine as a guide to prognosis in patients with chronic congestive heart failure. *N Engl J Med* 1984;311:819–823.
22. Hori M, Sato H, Kitakaze M, et al. ß-adrenergic stimulation disassembles microtubules in neonatal rat culcured cardiomyocytes through intracellular Ca^{2+} overload. *Circ Res* 1994;75:324–334.
23. Singal PK, Kapur N, Dhillon KS, et al. Role of free radicals in catecholamine-induced cardiomyopathy. *Can J Pharmacol* 1982;60:1390–1397.
24. Jin YT, Hasebe N, Matsusaka T, et al. Magnesium prevents left ventricular dysfunction induced by isoproterenol in dogs. *J Cardiovasc Pharmacol* 1999;34:S35–S39.
25. Hussain M, Orchard CH. Sarcoplasmic reticulum Ca^{2+} content, L-type Ca^{2+} current and the Ca2+ transient in rat myocytes during beta-adrenergic stimulation. *J Physiol* 1997;505:385–402.
26. Feldman RD. Beta-adrenergic desensitization reduces the sensitivity of adenylate cyclase for magnesium in permeabilized lymphocytes. *Mol Pharmacol* 1989;35:304–310.
27. Viskin S, Belhassen B, Sheps D, et al. Clinical and electrophysiologic effects of magnesium sulfate on paroxysmal supraventricular tachycardia and comparison with adenosine. *Am J Cardiol* 1992;70:879–885.
28. Kastor J. Multifocal atrial tachycardia. *N Engl J Med* 1990;322:1713–1717.

19
Magnesium: Forgotten Mineral in Cardiovascular Biology and Atherogenesis

Burton M. Altura and Bella T. Altura

In this review, a rationale is presented for how hypercholesterolemia, hypertension, diabetes mellitus (DM), end-stage renal disease (ESRD), prolonged stress, and exposure to magnesium (Mg)-wasting drugs can lead to atherosclerosis, ischemic heart disease, and stroke. The data, accumulated so far, indicate that Mg deficiency caused either by a poor diet or errors in Mg metabolism may be a missing link between diverse cardiovascular risk factors and atherogenesis. Early data from our laboratory and others indicate that reduction in extracellular free Mg ions can induce an entire array of pathophysiological phenomena known to be important in atherogenesis, that is, vasospasm, hypoxia, increased vascular reactivity, and elevation in intracellular calcium (Ca). More recent data has demonstrated molecular events, pointing the way to atherogenesis: that is, formation of pro-inflammatory agents, generation of free radicals, platelet aggregation, modulation of macrophage and leukocyte mobility, and emigration across the endothelial wall. Finally, oxidation of lipoproteins, changes in membrane fatty acid saturation, changes in membrane plasmalogens, and N-phospholipids suggest alterations in intracellular lipid signals. It has been shown that Mg deficiency can modulate membrane sphingomyelinase, generate vasoactive and pathogenic sphingolipids, which could alter multiple intracellular signaling pathways, modulate transcription factors, and thus cause intimal plaque formation.

Hypercholesterolemia, hypertension, DM, immune injury, ESRD, renal dialysis, prolonged stress, obesity, and smoking are widely accepted as risk factors for atherosclerosis. No common link has been identified that forms a rational basis to these disorders and atherogenesis. Moreover, it is not clear how lipoproteins, Ca^{2+}, and macrophages gain access to the normally impermeable arterial walls or what allows plyable, physiological vascular smooth muscle (VSM) cells to change their state (phenotype) from a contractile to a noncontractile, synthetic cell,[1,2] and how these diverse risk factors lead to ischemic heart disease (IHD) and stroke.

Until relatively recently, it was not believed that Mg, a common essential dietary element and electrolyte, played any role in maintaining normal cardiovascular dynamics. This lack of understanding is most likely a reflection of several factors: (1) scanty, uncontrolled experimental and clinical data regarding its precise actions on the heart and blood vessels; (2) poor appreciation of progressive shortfalls in Mg dietary intake since the turn of the century; (3) failure to recognize that patients with diets deficient in Mg are not necessarily hypomagnesemic; (4) an absence of the methodology needed to measure the free, ionized (IMg^{2+}) rather than total Mg (TMg) in tissues, cells, and body fluids.

Over the past decade, experimental, epidemiological and clinical studies point to the active role of magnesium ions (Mg^{2+}) in the maintenance of normal cardiovascular functions as well as in the etiology of cardiovascular pathology when problems arise in Mg intake and balance. Dietary Mg deficiency, as well as abnormalities in Mg metabolism, appear to play important roles in IHD, congestive heart failure, sudden death ischemic heart disease (SDIHD), atherosclerosis, a number of cardiac arrhythmias, hypertension, vascular complications in DM, immune injury, and ESRD. In this article, we attempt to review the cogent experimental, molecular, epidemiological, and clinical evidence that supports a role for Mg in atherogenesis and cardiovascular health and disease.

Progressive Decline in Dietary Intake of Magnesium: Relationship to Cardiovascular Disease (CVD)

At the turn of the century in the United States, ingestion of Mg was about 450 to 485 mg/day. This has been steadily decreasing each 5 to 10 years (Table 19.1). The most recent figures indicate values for men of ~185 to 260 mg/day and for women, ~172 to 235 mg/day. If we accept that the recommended daily allowance (RDA) = 350 mg/day, these values represent dietary shortfalls of 90 to 178 mg/day. Surveys in Europe and Canada reveal similar shortfalls in dietary intake of Mg (e.g., men ~189–262 mg/day; women ~143–283 mg/day); these values represent deficits of at least 67 to 161 mg/day. Many investigators and nutritionists believe that the current RDA recommended by the U.S. National

TABLE 19.1. Progressive decline in dietary intake of magnesium over the past 100 years.

Years	Mg intake/day
1900–1908	475–500
1909–1913	415–435
1915–1929	385–398
1935–1939	360–375
1947–1949	358–370
1957–1959	340–360
1965–1976	300–340
1978–1985	225–318
1987–1992	175–248

TABLE 19.2. Physiological functions of magnesium.

Enzyme functions	Structural functions
7 glycolytic enzymes	Proteins
T TCA cycle enzymes	Polyribosomes
Membrane-bound ATPases	Nuclei Acids
Kinases: creatinine kinase	Mitochondria
Alkaline P'Tase	Multienzyme Complexes, e.g., G-Proteins,
12 photosynthetic enzymes	N-MDA receptor complex
	Membrane channels
Membrane functions	Calcium antagonist
Hormone-receptor binding	Muscle contraction/relaxation
Gating of Ca^{2+} channels	Neurotransmitter release
Transmembrane flux of ions	Action potential conduction in nodal tissue
Adenylate/GMP cyclase system	
Ca^{2+}-Ca^{2+} release	

Academy of Sciences is too low and it should be 450 to 500 mg/day; thus, the deficits become very pronounced, thereby raising the possibility for development of disease processes, particularly when numerous biochemical (Sidebar) and physiological functions (Table 19.2) could be compromised.

According to recent epidemiological studies, in which dietary variables were assessed in human volunteers by the 24-h recall method, Mg had the strongest association with blood pressure.[3,4] It is known that strict vegetarians exhibit significantly decreased incidences of atherosclerosis and IHD, sudden-death ischemic heart disease, and high blood pressure. Is this a consequence of the increased dietary intake of Mg? Legumes, beans, nuts, soybeans, green leafy vegetables, and unprocessed cereals, which form the basis of vegetarian diets, are rich in Mg. Greenlanders who have a low incidence of IHD, SDIHD, and high blood pressure consume diets rich in Mg (both foods and drinking water) whereas Danes, who consume diets rich in processed food and low in Mg, have high incidences of these cardiovascular disorders. However, when Greenlanders move to Denmark and reside there for some years their incidences of these CVD are approximately equal to the indigenous Danes.[5] Similar patterns have been noted for South African Bantu natives, Bedouins in the Arabian desert, and Aborigines in Australia upon moving into Western societies.[6]

Pivotal to magnesium's wide-sweeping range of physiological effects is its role as a prime gatekeeper of cellular activities, membrane functions, and its function in regulating a great number of critical cellular and subcellular processes (see Table 19.2 and Sidebar).

Although it was recognized by two French physicians, Hazard and Wurmser, in 1932, that systemic administration of magnesium ions can produce rapid reduction in arterial blood pressure and vasodilatation, very little else was known about the cardiovascular actions of Mg^{2+} until the late 1960s. In 1961, one of us (BMA), while investigating the physiological responses of isolated blood vessels to vasoactive pressor and dilator substances, found that removal

Roles of Magnesium in Cell Metabolism

Two major roles for Mg exist in biological systems: (1) It can form chelates with various intracellular anionic ligands, for example, ADP, ATP. It is this ability to form pyrophosphates which is necessary to drive numerous biological reactions, but also is key for most cellular uptake systems of ions and foodstuffs, as well as for control of muscle tension and the shape of cells. (2) It can compete with calcium ions (Ca^{2+}) for binding sites on membranes (external and internal) and proteins.

Mg^{2+} catalyzes or activates more than 500 enzymes in the body and is pivotal in the transfer, storage, and utilization of energy. It activates phosphate groups and reactions that involve ATP as well as other nucleotide complexes. The intracellular level of free Mg^{2+} ($[Mg^{2+}]_i$) serves to regulate intermediary metabolism through activation of such rate-limiting enzymes as hexokinase, pyruvate dehydrogenase, enolase, and creatine phosphokinase. It regulates seven enzymes in glycolysis and four key enzymes in the tricarboxylic acid cycle. Mg^{2+} can cause a conformational change during catalytic processes (e.g., Na^{2+}-K^+ATPase), by promoting aggregation of multi-enzyme complexes (e.g., aldehyde dehydrogenase) or by a mixture of mechanisms (e.g., F_1-ATPase).

Within the cell nucleus, Mg^{2+} regulates DNA synthesis. Large numbers of Mg^{2+} ions are bound to the pentose-phosphate backbone of DNA. By regulating DNA and RNA (i.e., RNA synthetase, mRNA attachment to ribosomes, etc.) synthesis and structure, Mg^{2+} plays a vital role in regulating cell growth, reproduction, and membrane structure. Magnesium's role in regulating cell membrane permeability, transmembrane electrolyte flux, hormone and agonist-receptor binding, and cell adhesion are becoming widely accepted. Dietary deficiency of Mg (in different mammals) usually results in loss of intracellular K^+ from organs and cells; intracellular Ca^{2+} and Na^{2+} rise under such conditions (e.g., Table 19.2). Magnesium's role in regulating membrane permeability has been long recognized.

of the Mg^{2+} from an artificial physiological salt solution resulted in production of vasospasm and enhanced responses to vasoactive pressors and decreased responses to dilator agents.[3,7] Since no one, up to that time, had reported such dramatic results of a Mg^{2+}-free medium on either VSM tone or reactivity, we did not at first believe the findings. Repetition of our findings thus was undertaken over the next several years, with several different types of blood vessels, until verification of our data remained unquestionable. Such early and exciting findings, as well as others, became compelling to us.

Magnesium Depletion from the Body Is Common

A variety of commonly used drugs, such as alcohol, diuretics, chemotherapeutic agents (e.g., cisplatin, bleomycin), cyclosporine, certain antibiotics (amphotericin B), and fungicides, cardiac glycosides, and drugs of abuse (e.g., morphine, heroine, cocaine), among others,[3,6–10] produce a loss of body Mg; alcohol being the most well-known Mg waster. Diarrhea, malnutrition, vomiting, dehydration, high-salt diets, high intake of soft drinks, malabsorption syndromes, and certain renal disorders also result in considerable body loss of Mg.

Six to sixty-five percent of all hospitalized patients exhibit low serum Mg; patients with numerous cardiovascular disorders associated with atherogenesis demonstrate the most profound deficits in body Mg.[3,10–21] Did the disease(s) produce the Mg deficiency or was an existing Mg deficiency responsible, in part, for the etiology of the disease state?

Magnesium Exists in Several Forms in Blood and Body Fluids: Importance of Free Magnesium Ions

Blood normally contains magnesium ions in three states: bound to plasma proteins; complexed to small anion ligands, such as bicarbonate, citrate, sulfate, phosphate, lactate, or to peptides; and free. Most clinical laboratories measure total magnesium levels by colorimetry or atomic absorption spectrophotometry. However, it is the free ionized form of magnesium that is physiologically active. Usual estimates of free Mg^{2+} have relied upon total Mg measurements in protein-free ultrafiltrates, which exclude the protein-bound Mg.[22,23]

Because the levels of anions can vary significantly in pathological states, and in view of the role apparently played by Mg in cellular homeostasis (Table 19.2), it is desirable to directly measure free Mg^{2+} in blood[24] and other body fluids. With this aim in mind, Mg^{2+}-sensitive electrodes have been designed to obtain these measurements in the presence of potential cationic interferences.[25]

Twenty-five years ago (1980), we demonstrated that diverse large and small mammalian coronary arteries subjected to reductions in extracellular Mg^{2+} concentrations underwent rapid spasm and potentiation of circulating vasoconstrictor hormones.[26] Similar observations have been made in human coronary arteries in vitro and during anginal attacks.[12]

We therefore initiated ion-sensitive electrode studies to determine whether patients with acute myocardial infarction (AMI) or coronary heart disease exhibited significantly lowered levels of free Mg^{2+}.[27] We found that free, but not total, Mg was usually lowered in such patients and that free Mg^{2+} as a percentage of total Mg was also significantly lowered (Table 19.3). These studies have provoked a number of cardiologists to examine the potential use of Mg^{2+} in the therapy of acute myocardial infarction or congestive heart failure. The

TABLE 19.3. Plasma ionized magnesium versus total magnesium in ischemic heart disease patients scheduled for coronary bypass surgery.

Group	N	Mg (mM/L)	
		IMg^{2+}	TMg
Controls	42	0.60 ± 0.005	0.84 ± 0.008
IHD	35	0.51 ± 0.008^a	0.82 ± 0.006

Abbreviations: IHD, ischemic heart disease.
IMg^{2+} obtained with NOVA ISE; TMg obtained with AAS.
^{a}Significantly different from controls ($p < 0.001$).

results so far support the notion that Mg compounds might be of significant therapeutic value in these cardiac syndromes.

Mg^{2+} Controls Vascular Contractility, Basal Vascular Tone, and Produces Vasospasm

As early as 1971, Mg^{2+} ions have been shown to directly alter baseline tension or tone of blood vessels.[28-32] Decrements in [Mg^{2+}]$_o$ result, in a concentration-dependent manner, in rapid elevations in contractile tension development in numerous mammalian and human arteries and arterioles. Elevation in [Mg^{2+}]$_o$, on the other hand, above the physiological level inhibits spontaneous mechanical activity and lowers baseline tension. The elevations in mechanical activity, produced, in response to lowering [Mg^{2+}]$_o$, are inhibited when external Ca^{2+} ions are lowered or chelated. This demonstrates that influx of extracellular Ca^{2+} ions are necessary for these contractile responses. In addition, a variety of neurohumoral agents that induce a contraction in VSM exhibit heightened contractile responses as external Mg^{2+} is lowered, and attenuated contractile activity is observed as [Mg^{2+}]$_o$ is elevated. How are these actions of [Mg^{2+}]$_o$ brought about?

We first demonstrated in 1971, that Mg^{2+} regulates Ca^{2+} flux across VSM cell membranes,[28] as well as its release from intracellular storage sites.[33,34] Because a change in cytosolic free Ca^{2+} concentration is necessary for contraction or relaxation, it may be physiologically relevant that Mg^{2+} regulates the activity of VSM cells by competing with and modulating the levels of free intracellular Ca^{2+}. Precise signaling pathways appear to be activated in these Ca^{2+}-related changes (see below).

The vasodilator effect has been attributed, at least in part, to its Ca^{2+} channel-blocking property as well as to modulation of intracellular free Ca^{2+} in VSM cells. Recently, we have found that the precise concentration of [Mg^{2+}]$_o$ also clearly controls the subcellular distribution of both [Ca^{2+}]$_i$ and [Mg^{2+}]$_i$.[33,35]

The end result of lowered serum Mg^{2+} and intracellular VSM cell Mg^{2+} ([Mg^{2+}]$_i$) would be vasospasm, increased vascular tone, increased reactivity, reduction in peripheral blood flow (hypoxia), and, thus, local vascular injury, a stimulus for hypertension and atherogenesis.[1] Using a combination of radio-labeled ^{45}Ca, digital image analysis with fluorescent probes, and confocal laser-scanning microscopy with fluorescent probes, we found that Mg^{2+} gates special membrane Ca^{2+} channels in both VSM and endothelial cells and controls the release of intracellular free cytosolic Ca^{2+} ($[Ca^{2+}]_i$) from these cells.[3,7,33–35]

The entry and release of $[Ca^{2+}]_i$ and loss of membrane-Mg^{2+} most likely initiates a number of intracellular signaling pathways (i.e., growth factors, cell adhesion factors, cytokines, nuclear-factor κB, c-jun, c-fos and phospholipids, among others; see below) that could result in VSM cell proliferation (cell transformation), alteration in endothelial cell wall permeability, macrophage attraction, leukocyte adhesion, and Ca^{2+} cell entry. Low $[Mg^{2+}]_o$-induced vasospasm could accelerate uptake of oxidized lipids and/or stimulate generation of oxygen- and nitrosoradicals (see below).

Low Magnesium Levels Can Lead to Formation of Pro-Inflammatory Agents: Possible Importance in Atherogenesis

Inflammation is such an everyday occurrence that most people do not even notice it. It does not just occur in the skin, but internally as well, particularly when viruses or bacteria invade the body. Inflammation is actually one of the body's major defense mechanisms.

Over time, however, inflammation can cause great harm to bodily tissues it is supposed to heal. This destructive side of inflammation has long been known to be pivotal in diseases like rheumatoid arthritis and multiple sclerosis. Recently, evidence has accumulated to implicate inflammation in development of atherosclerosis, as well as in many other diseases. Entry of various types of inflammatory cells into the arterial wall characterizes formation of the fatty streak.[1,36] Droplets of fat in the blood stream are initially thought to be absorbed by arterial walls, inducing release of hormone-like mediators called cytokines. Various cytokines are then thought to cause the VSM cells of the arterial walls to become sticky, which then attracts inflammatory white cells called monocytes. These squeeze between the endothelial cells, and, once in the arterial wall, undergo transformation to scavenger cells called macrophages, which ingest the fat droplets. These cells, over time, become overloaded with fat and give rise to plaques. Eventually, the inflammatory process thins the fibrous cap of the plaques and rupture occurs, spilling the contents along with some cytokines into the blood, which induces clotting and blocking of an artery, causing a heart attack or a stroke, depending on its location.

Much speculation centers around what the initiation and sustaining factors are for the above process.

Evidence implicates matrix metalloproteases (MMP) in plaque rupture. These enzymes can act in the shoulder region of the plaques to degrade the extracellular matrix, thus weakening the fibrous cap.[36–38] Matrix metalloproteases are produced by VSM cells and macrophages. Inflammatory cytokines released in atherogenesis, such as tumor necrosis factor α (TNFα), interleukin-1 (IL-1), and T-cell–derived interferon γ can induce MMP expression. Thus, a balance of several forces predicts the state of the fibrous cap; matrix production and its degradation determining plaque stability. MMP is also regulated by production and release of platelet-derived growth factor (PDGF) secreted by both the platelets and VSM cells.[1,2,36]

Feeding rats and hamsters diets deficient in Mg, for periods of only 14 to 21 days, results in cardiomyopathic lesions, cytokine formation and release (IL-1, IL-2, IL-6; TNFα), and endothelin formation.[40–42] Using human umbilical vein endothelial cells (HUVE), Maier and colleagues[43] have shown that low $[Mg^{2+}]_o$ produces an upregulation of IL-1, as well as upregulation of vascular cell adhesion molecule-1 (VCAM-1) and plasminogen activator inhibitor-1 (PAI-1). Other workers have demonstrated that low $[Mg^{2+}]_o$ environments resulted in an upregulation of MMPs in cultured rat aortic VSM cells under basal and PDGF-stimulated conditions. Rayssiguier and Mazur[44] have reported that Mg deficiency in rats leads to a clinical type of inflammatory syndrome characterized by increased production of leukocytes, release of several cytokines, macrophage activation, and production of acute phase inflammatory proteins. Although other growth factors (PDGF, transforming growth factor B, fibroblast growth factors) have not yet been identified in Mg-deficient states, it is likely that Mg deficiency will be found to cause production of these well-known growth factors implicated in atherogenesis.[1,2,36]

Importance of Leukocytes, Adhesion Molecules, and Chemokines in Atherogenesis: Possible Roles of Magnesium

Although the major risk factors for atherogenesis have been identified above, the precise cellular and molecular mechanisms of an atherothrombotic lesion initiation and progression remain to be understood. It is now clear that recruitment of blood monocytes and lymphocytes to the arterial intimal surface is a key characteristic of atherothrombotic lesions.[1,36] Attached monocytes migrate into the endothelial intima, transform themselves into macrophages, and become foam cells. Subsequently, growth and expansion of early lesions occur, and then plaque destabilization, resulting in thrombotic complications.

A sine qua non of the atherosclerotic inflammatory disease is the recruitement of leukocytes to the site of inflammation. Leukocytes migrate along chemotactic gradients and induce tissue damage by releasing enzymes, chemical mediators, and toxic oxygen-free radicals. Diverse cytokines and chemokines produced locally, appear to regulate the type of leukocyte recruited. In this regard, IL-1 or TNFα induce endothelial cells to synthesize and express surface adhesion molecules, such as selectins, immunoglobulin gene superfamily proteins, integrins , and present chemokines (small heparin-binding proteins), which are bound to cell surface proteoglycans. The cell adhesion molecules and chemokines induce recruitment of leukocytes from the blood to the arterial walls. Different adhesion molecules mediate each stage of leukocyte emigration. The chemokines direct the migration to sites of inflammation or injury.

In the settings of hypercholesterolemia, VCAM-1 and intercellular adhesion molecule 1 (ICAM-1) are upregulated and expressed in the endothelial cells.[45,46]

Dietary Mg deficiency in rats has been shown, in a few days, to be associated with a marked increase in total circulating leukocytes, mostly neutrophils.[47,48] This Mg deficiency–induced leukocytosis is also associated with infiltration of the spleen with polymorphonuclear leukocytes (PMN) and macrophages concomitant with a rise in IL-6 levels. Using rabbits fed diets of low Mg, we have reported an activation of splenic and Kupfer cell macrophages.[49] Weglicki and his colleagues have shown that Mg deficiency in rats results in increased levels of IL-1, IL-6, and TNFα in T-lymphocytes.[50] To our knowledge, there is no data as yet to indicate whether a Mg-deficient state will give rise to increased levels of selectins, and integrin adhesion molecules or chemokines. Some studies have indicated that a Mg-deficient state in vitro, can result in upregulation of one key member of the immunoglobulin gene superfamily—VCAM-1.[43] It will be important to determine whether Mg deficiency in intact animals will result in early upregulation and expression of selectins, integrin molecules, and chemokines in intimal endothelium and VSM cells.

Low Levels of Magnesium Enhance Platelet Aggregation and Thrombotic Tendencies

A number of studies dating back to the late 1950s indicate that low $[Mg^{2+}]_o$ promotes blood coagulation.[6,51] More recently, several investigators have demonstrated that platelet aggregation is enhanced in low $[Mg^{2+}]_o$ environments[52,53] and that megakaryocyte numbers and white cell counts are increased concomitantly in Mg-deficient hamsters,[54] thus providing, potentially, an important link between early atherogenesis and subsequent thrombotic tendencies.

Low Magnesium Levels Result in Lipid Peroxidation and Free Radicals

Until recently, it was not known how Mg deficiency could promote cardiovascular damage. During the past 10 years, reports from several laboratories, including ours, have suggested that Mg deficiency can lead to formation of several different types of oxygen-free radicals, lipid peroxidation, and formation of ferrylmyoglobin.

In the course of normal metabolism, oxygen is responsible for formation of some highly toxic and reactive byproducts, such as superoxide (O_2^-), hydrogen peroxide (H_2O_2), hydroxyl radicals (OH·), and singlet oxygen ($^1O_2$). Oxidative stress causes extensive tissue damage, including lipid peroxidation of membranes, oxidation of proteins, and damage to DNA and RNA. H_2O_2 and OH· radicals are particularly reactive, H_2O_2 penetrating cell membranes easily. Other radicals that can be formed and are destructive are ·NO, peroxynitrite ($OONO^-$), and hypochlorite ($HOCl^-$). All of these molecules have been shown to promote vasoconstriction, spasm, and damage of small blood vessels.[62–66]

Because macrophages are the most potent producers of these reactive oxygen metabolites (ROMs), their invasion of blood vessel walls in the atherogenic process raises the possibility of ROMs in initiation of atherosclerosis. The major targets of ROMs in biological systems are lipids, proteins, and DNA. In the case of atherogenesis, one of the primary events is an attack on the low-density lipoproteins (LDL) in the blood.[67,68] Before LDL can be taken up by macrophages, it has to be oxidatively modified and subsequently gets endocytosed by macrophages to form foam cells. Although the two primary metabolites $O_2^{·-}$ and H_2O_2 each are not very reactive in this regard, the combination are thought to form the highly reactive hydroxyl radical (the Haber–Weiss reaction). Endothelial cells (EC) are known to generate the superoxide radical anion[69] and thus become a prime candidate for in situ modification of LDL. Activated macrophages and neutrophils, which appear in large numbers in atherogenesis, contain an enzyme myeloperoxidase (MPO) that converts H_2O_2 and chloride ions to $HOCl^-$, $^1O_2$ and OH·. All these can initiate lipid peroxidation and it is, thus, not surprising to find that MPO is involved in LDL oxidation.[70,71] Myeloperoxidase has been found in atherosclerotic plaques,[71] and when it combines with $HOCl^-$ and $O_2^{·-}$, it will produce OH· (the Long–Bielski reaction), thus causing potentially significant damage in the vascular wall. Most of these reactions are catalyzed by metal ions (Fe^{3+}, Cu^{2+}); surprisingly many of these ROMs, once formed, in turn release free Fe ions from ferritin. Such free Fe could combine in muscle tissues (heart and skeletal muscle) with myoglobin to form the reactive ferrylmyoglobin (see below).

With respect to Mg deficiency in rats, several reports have appeared that demonstrate that macrophages and PMNs exhibit release of superoxide anions and some free radical species in very early Mg deficiency[72–75] concomitant with increased numbers of PMNs and increased lipid peroxidation. Other studies performed by Rayssiguier's group in intact rats indicate that these phenomena are accompanied by changes in the physical status of cell membranes. Using rats, rabbits, and monkeys, others have demonstrated experimentally that Mg deficiency–induced hyperlipidemias are accompanied by vascular wall lipid infiltration and thick atheromatous plaques.[49,72] In some studies, correlations have been seen between the degree of Mg deficiency–induced lipid peroxidation (TBARS levels) and lipoprotein oxidizability.[72] Other investigators employing rats and hamsters also reported lipid peroxidation after 7 to 21 days.[42] Using bovine EC in culture in low $[Mg^{2+}]_o$ (0.4 mM), these workers demonstrated induced formation of oxygen radicals.[76] Using primary rat aortic VSM cells and canine VSM cells, our group has demonstrated that exposure of these cells to low $[Mg^{2+}]_o$ (0.15–0.48 mM) for 3 h resulted in concentration dependent rises in malondialdehyde (MDA) levels, with higher levels of MDA after 18 to 24h of low $[Mg^{2+}]_o$ exposure.[77] Using stroke-prone rats, Touyz and coworkers[78] have shown that low dietary Mg intake resulted in formation of ROMs concomitant with an exacerbation of elevated systolic blood pressure along with structural changes in the vascular wall. In this context, it should be pointed out that Mg deficient diet–induced thickening in arterial walls has been seen to occur prior to rises in arterial blood pressure,[79,80] suggesting that cell hyperplasia and proliferation of endothelial and VSM cells occurs earlier than thought in Mg-deficient states. To this, should be added the early findings of Gunther and coworkers, who reported, using human and rat erythrocytes, that oxygen-free radicals, like H_2O_2, by destroying the Na^+/Mg^{2+} anti-port in cells can result in efflux (and loss) of free Mg^{2+}.[81] Such a situation, if found in vivo, would exacerbate Mg deficiency–induced pro-oxidant states.

Lastly, we should mention another potential avenue for Mg deficiency–induced cardiovascular damage and atherogenesis via formation of the highly reactive ferrylmyoglobin.[82] The generation of free Fe mentioned above would be expected to combine with myoglobin found in cardiac and skeletal muscle to form the latter radical (see below). Experimentally, it has been shown that Mg deficiency–induced formation of ferrylmyoglobin in intact, perfused rat hearts will lead to cardiac failure.[82] This work also showed the link between Ca^{2+} overload induced by Mg deficiency and cardiac myocyte membrane damage.

All the above seem to point to a true link between Mg deficiency, lipid peroxidation, generation of ROMs, and atherogenesis, but definitive studies remain to be done.

Elevated Magnesium Levels Ameliorate Atherogenesis and Homocysteine-Induced Magnesium Depletion

Hypercholesterolemia has been widely accepted as a high risk factor for development of atherosclerosis. Evidence from both animals and humans suggests that the dietary and blood level of Mg^{2+} may modulate the serum levels of lipids and lipoproteins. Ever since the early experimental and clinical studies by Bejal and Czech (1956) and Gruneis (1953), respectively, there have been reports that either dietary or systemic administration of Mg salts might ameliorate atherogenesis.[49,79,90] Magnesium is necessary for the activity of lecithin cholesterol acyltransferase (LCAT) and lipoprotein lipase (LPL), which lower triglyceride levels and elevate high-density lipoprotein (HDL) cholesterol levels.[67,68] Our laboratory, using rabbits, has reported that the dietary level of Mg modulates the serum level of cholesterol and triglycerides in normal animals; the lower the intake of Mg, the higher the serum lipid levels.[49]

Dietary deficiency of Mg (compatible with reduced dietary intake of Mg seen in the adult population of the western world) in rabbits fed a high-cholesterol diet exacerbates atherogenesis and lipid deposition in arterial muscle, and stimulates (or activates) macrophages of the reticuloendothelial system. We have demonstrated, in this study, that pretreatment of animals with orally administered Mg aspartate HCL: (1) attenuated the atherosclerotic process markedly, and (2) lowered serum cholesterol and triglycerides in the atherosclerotic animals. These studies indicate that the extent of the atherogenic lesions were poorly correlated with the level of serum cholesterol and highly dependent on the level of dietary Mg and the Ca/Mg ratio. We noted that due to the high blood levels of lipids and lipoproteins, a true Mg-deficient state can often be masked or hidden.

A similar masked or hidden Mg deficiency has been observed, by our group, in renal transplant patients as well as in ESRD patients[17,18]; patients who are characterized by extensive, unexplained atherogenesis. Patient studies in our medical center with subjects having undergone renal transplantation or that have end-stage kidney failure, who are at high risk to develop atherosclerosis, demonstrate a concomitant marked, lowering of ionized Mg (but not TMg), decreased urinary output of Mg, hyperinsulinemia, elevated total serum cholesterol and/or triglycerides. ICa^{2+}/IMg^{2+} ratios, signs of elevated vascular tone, increased vascular reactivity, and risk of atherogenesis, were also significantly elevated in our stable renal transplant recipients and subjects with ESRD. The TMg in these patients is usually normal or elevated, respectively, suggesting that in advanced atherogenesis and states of hyperlipidemia, a normal TMg would often be seen, despite the fact that the body and vascular walls would be deficient in IMg^{2+}.

An increased plasma homocysteine (HC) concentration is common in patients with stroke, coronary heart disease, and peripheral vascular disease, and confers an independent risk for atherogenesis. Several studies have shown that HC causes vascular endothelial injury, proliferation of VSM cells, and altered blood coagulation.[55,56] Recent investigations have reported that elevated plasma HC result in production of phospholipids, activation of protein kinase C, and induction of c-fos and c-myb in VSM cells.[57] These findings suggest that hyperhomocysteinemia may contribute to progression of peripheral and cerebral atherosclerosis and infarction. However, the mechanism(s) whereby hyperhomocysteinemia induces heart attacks and stroke remain to be determined.

Existence of an interrelationship between vitamin status and plasma HC was noted by an inverse relationship between HC and plasma folate or vitamin B_{12} concentrations.[58] These studies were a direct result of early suggestions of McCully.[59] Because the major enzymes involved in HC metabolism are Mg^{2+} dependent, it is distinctly possible that Mg^{2+} plays an important role in HC-related atherogenesis, $[Ca^{2+}]_i$ overload, and coronary and cerebral vasospasm and stroke, particularly as Mg^{2+} depletion has been shown to initiate atherogenesis and vasospasm (see above). Using primary cerebral VSM cells, our laboratory demonstrated that HC itself induces Mg depletion.[60] However, concomitant addition of vitamin B_6, vitamin B_{12}, or folic acid, either alone or together, failed to prevent the HC-induced cellular loss of $[Mg^{2+}]_i$. These recent findings are thus compatible with the hypothesis that an increased serum HC concentration causes abnormal metabolism of Mg^{2+} in cerebral VSM cells, thus priming these cells for HC-induced atherogenesis, cerebrovasospasm, and stroke.

Although hyperhomocysteinemia (HHC) is an independent factor for coronary artery disease, some studies have indicated that patients with hyperhomocysteinemia are not susceptible to atherosclerosis. There may be an important link between HHC, Mg, and extracellular matrix metalloproteinase-2. In this context, it has been shown recently by Guo and coworkers,[61] using cultured rat aortic VSM cells, that HHC (50–1000 µM) increased the production of MMP-2 in a concentration-dependent manner, and folic acid plus Mg was found to reduce this increased MMP-2 production. These preliminary observations are suggestive of the possibility that therapeutic administration of folic acid with Mg may exert ameliorative actions on atherogenesis.

Deficits in Magnesium Can Induce Formation of Ferrylmyoglobin Radicals in Cardiac Muscle: New Possible Link to Coronary Arterial Atherogenesis

Cardiac tissue contains a large amount of myoglobin, which is involved in the intracellular transport and storage of oxygen. These functions depend on its ability to undergo reversible oxygenation by the ferrous heme group. The heme iron of deoxymyoglobin can be oxidized by hydrogen peroxide, which is currently thought to yield an oxy ferryl complex ($Fe^{IV} = O$), similar to compound II of peroxidases, and a transient protein radical.[83] This high oxidation state of myoglobin is known as ferrylmyoglobin. Because there is a lack of significant amounts of catalase in cardiac tissue, the heart has a characteristic susceptibility to hydroperoxide toxicity. The subsequent formation of this ferrylmyoglobin radical can further degrade deoxyribose after releasing iron from the porphyrin ring of the myoglobin, and thus lipid peroxidation can occur.[84-86] It has been suggested that oxidation of myoglobin may be involved in the initiation and propagation of free radical damage in some tissues, and such reactive species may cause cellular injury in muscle.[87-89] We hypothesize that such reactions might underlie initiation of myocardial dysfunction in ischemia and Mg^{2+} deficiency–induced cardiac injury and may play a key role in coronary artery atherogenesis.

Using reflectance spectrophotometry in isolated perfused working rat hearts, subjected to an acutely Mg-deficient environment (0.3 mM) for 30 min, we found that 80% of the myoglobin converted to its deoxygenated form; reduced cytochrome oxidase aa_3 also increased about 80% in low $[Mg^{2+}]_o$.[82] This process, set into motion by acute Mg deficiency, resulted from a direct accessibility of the exogenous peroxide to the cytosolic protein. Ferrylmyoglobin formation could be prevented or dissipated by one-electron reduction of this hypervalent form of myoglobin using ascorbate. Interestingly, the ferrylmyoglobin radical was formed before appearance of either creatinine phosphokinase or LDH, suggesting a cause and effect for cardiac myocyte membrane damage and, potentially, CAD.

We suggest the link between Mg deficiency–induced Ca^{2+} overload and cardiac myocyte membrane damage is formation and action of the ferrylmyoglobin radical; the latter myocardial injury should be prevented by use of one-electron reductants like ascorbate.

Multiple Cellular Signaling Pathways Are Activated in VSM by Alterations in Extracellular Magnesium

Although low levels of $[Mg^{2+}]_o$ lead to vasoconstriction, vasospasm, and decreased blood flow, forerunners of atherogenesis, whilst elevated levels of $[Mg^{2+}]_o$ lead to relaxation, vasodilatation, and increased peripheral blood flows, up until very recently, the signaling pathways (except for Ca^{2+}) for these diverse responses in vascular smooth muscle cells remained largely unknown. Using small and large peripheral and cerebral blood vessels, we and other workers have now identified a number of signaling pathways whereby low $[Mg^{2+}]_o$ leads to vasoconstriction.[8,33,34,77,78,91-96] These include activation of several protein kinase C isozymes, nonreceptor tyrosine kinases, P-I-3 kinases, mitogen-activated protein kinases (MAPK), and MAPK kinases. With respect to the relaxant effects of elevated $[Mg^{2+}]_o$, our laboratory found that high concentrations of $[Mg^{2+}]_o$ appear to eventuate in fresh rat aortic endothelial cell production and release of NO followed by activation of guanylate cyclase, leading to cGMP and vessel relaxation[34]; the smaller the blood vessel, the greater the dependence on mediation by NO. This work should be contrasted with recent work of others, using a passaged cell line of murine microvascular EC, who exposed the EC for 3 days to low $[Mg^{2+}]_o$.[97] These workers found that 0.1 mM $[Mg^{2+}]_o$, a very low concentration (incompatible with life), resulted in an elevation of NO synthesis. Such experiments should be repeated with fresh (or primary) EC and extracellular levels of $[Mg^{2+}]_o$ ranging from 0.3 to 0.48 mM. Whether or not the activation/deactivation of NO synthase is dependent on concentration and type of EC, and species, and its role in atherogenesis will have to await further study.

Magnesium-Dependent Changes in Fatty Acid Saturation, Synthesis and Release of Sphingolipids and Plasmalogen Content of VSM: Relation to Intracellular Signaling Pathways and Atherogenesis

In atherogenesis and vascular disease, it has been demonstrated that membrane phospholipids are altered. The failure to focus on the latter may be one of the chief reasons that neither the frequency of restenosis after percutaneous transluminal coronary angioplasty has improved nor have varied recommended treatments met with more success. Approximately 20 years ago, our laboratory reported that dietary Mg deficiency in rats resulted in alterations of membrane phospholipids and fatty acid saturation in intact cardiac tissue.[98] A few years later, it was shown that phosphoinositide-derived second messengers are important in regulation of Ca^{2+} in VSM.[99] Over the past decade these initial findings have given rise to numerous studies on this theme.

We investigated whether Mg^{2+} modulates fatty acid saturation and plasmalogen (α,β-unsaturated ether) content in rat aortic and canine cerebral VSM membranes.[100] In our studies, using ion exchange chromatography to separate lipids and proton nuclear magnetic resonance (NMR) spectroscopy, we noted that exposure of primary cultured aortic and cerebral arterial VSM cells to low $[Mg^{2+}]_o$ (0.3–0.48 mM) for 18 h increased the fatty acid content of these cells about threefold. Fatty acid saturation was increased about twofold when these cells were exposed to a medium containing around 0.2 mM $[Mg^{2+}]_o$, compared to one containing 1.2 mM $[Mg^{2+}]_o$. These data show that the level of fatty acid saturation in VSM cells is sensitive to $[Mg^{2+}]_o$ concentrations over the human pathophysiological range (0.3–0.6 mM) found by our ion-selective electrode work in patients with hypertension, IHD, AMI, DM, ESRD, and stroke, all subjects known to have increased levels of atherosclerosis.

In our studies, we found that the decrease in fatty acid unsaturation observed in VSM cells exposed to low $[Mg^{2+}]_o$ was accompanied by a two to three carbon decrease in average chain length.[100] In contrast to the Mg^{2+}-dependent increase in double bonds, the vinyl ether content decreased with increasing $[Mg^{2+}]_o$. In addition, we found that the ratio of fatty acid double bonds to vinyl ether double bonds in VSM cells increased about fivefold as $[Mg^{2+}]_o$ was increased from the pathophysiological to the physiological range (0.3–0.8 mM). We believe such data suggest that different desaturases must be involved in the synthesis of plasmalogens and unsaturated fatty acids, although one cannot rule out the possibility that Mg^{2+} has different effects on the same enzyme acting on different substrates. Because plasmalogens are high in excitable tissues like contractile tissues,[101] the increased contractility of VSM in low $[Mg^{2+}]_o$ in vivo (reviewed above), may be facilitated by increased plasmaogen content in Mg-deficient states, thus potentially playing an important role in the early atherogenic process.

The observed low Mg^{2+}-induced decrease in fatty acid chain length and in the number of double bonds most likely reflects free radical oxidation of the double bonds and chain shortening. It is possible that this oxidative modification of the polyunsaturated fatty acid residues of membrane phospholipids could generate platelet-activating factor (PAF)-like lipids,[102] believed to be associated with leukocyte emigration and diapedesis.[102] About 10 years ago, it was found that exposure of smooth muscle cells to oxidized LDL, which produces PAF-like lipids, was mitogenic; this is significant because atherogenesis requires mitosis of intimal VSM cells. Interestingly, the reported mitogenic activity was blocked by specific PAF-receptor antagonists.[103] We believe that the Mg^{2+}-dependent changes in membrane lipids reported by our group may link various CVD risk factors (hypercholesterolemia, hypertension, DM, ESRD, IHD, smoking, immune injury, obesity, prolonged stress) and atherogenesis. Such attributes of Mg^{2+} could be pivotal in generating and controlling the diverse intracellular signaling pathways.

In view of the above potential relationships between Mg regulation of membrane phospholipids, fatty acid content, and makeup, it seems logical to believe that Mg^{2+} might modulate atherogenesis through additional lipid second messengers like sphingolipids. Sphingomyelin (SM) is one of the major lipids in cell membranes and lipoproteins. Although direct links between atherogenesis and SM have not been established, experimental in vitro data suggest that SM and/or its metabolites may have proatherogenic potential. For example, SM-rich lipoproteins can be transformed to foam cell substrates by sphingomyelinase (SMase).[104] Second, the SM content of lipoprotein particles could influence lipid metabolism by influencing binding or activity of lecithin-cholesterol acyltransferase[105] and lipoprotein lipase.[106] Third, ceramide and/or other metabolic products of SM breakdown and synthesis are known to be potent regulators of cell proliferation, activation, and apoptosis,[107] possess vasoconstrictor and spasmogenic properties,[108-111] and thus may be initiators of plaque growth and stability. We have found that incorporation of [3H]-palmitic acid into phosphatidyl-choline (PC) and SM, in primary rat aortic VSM cells and canine cerebral arterial VSM cells, was altered within 15 to 30 min after modifying the extracellular Mg level.[112] In these studies, decreased $[Mg^{2+}]_o$ produced a fall in [3H]-SM and [3H]-PC over the first 2h. After 18-h incubation, the [3H]-PC/[3H]-SM ratio changed from 20:1 to 50:1. There was a reciprocal relationship between ([3H]-ceramide formation and [3H]-1,2-diacylglycerol (1,2-DAG) levels. Collectively, these results indicate that a fall in $[Mg^{2+}]_o$ produces a rapid and sustained decrease in membrane SM and a rise in intracellular ceramide. Thus, a major effect of lowering $[Mg^{2+}]_o$ appears to be a downregulation of SM synthase. Ceramide and its downstream products are known to activate different serine/threonine protein kinase cascades and protein phosphatases that regulate the function of several transcription factors, such as nuclear factor-κB (NF-κB), shown, recently, to be regulated by $[Mg^{2+}]_o$,[77] which coordinately controls genes encoding inflammatory cytokines and cell adhesion molecules. If low $[Mg^{2+}]_o$ is a trigger for promotion of transcription of inflammatory proteins and cell adhesion molecules, via a breakdown of SM to form ceramides (and other products), it could be a major contributing factor in atherogenesis, which as, reviewed above, has both inflammatory and leukocyte components. This should be a fruitful avenue for further investigation.

Acknowlegments. Some of the work described herein was supported in part by grants from the U.S. National Institutes of Health.

References

1. Ross R. Atherosclerosis: a defense mechanism gone awry. *Am J Pathol* 1993;143: 987–1001.
2. Schwartz SM, deBois D, O'Brien ERM. The Intima. Soil for atherosclerosis and restenosis. *Circ Res* 1995;77:445–465.
3. Altura BM, Altura BT. Role of magnesium in the pathogenesis of hypertension updated. In: Laragh JH, Brenner BM, eds. *Hypertentension: Pathophysiology, Diagnosis and Management*. 2nd ed. New York: Raven Press; 1995:1213–1242.
4. Seelig MS. Epidemiologic data on magnesium-deficiency-associated cardiovascular disease and osteoporosis. In: Rayssiguier Y, Mazur A, Durlach J, eds. *Advances in Magnesium Research. Nutrition, and Health*. London: John Libbey; 2001:177–190.
5. Jeppesen BB. Greenland, a soft water area with a low incidence of ischemic heart death. *Magnesium* 1987;6:307.
6. Seelig MS. *Magnesium Deficiency in the Pathogenesis of Disease*. New York: Plenum Press; 1980.
7. Altura BM, Altura BT. Magnesium and cardiovascular diseases. In: Berthon G, ed. *Handbook of Metal-Ligand Interactions in Biological Fluids*. Vol. 2. New York: Marcel Dekker; 1995:822–842.
8. Altura BM, Altura BT. Magnesium in cardiovascular biology. *Sci Am Sci Med* 1995;2:28–37.
9. Aikawa JK. *Magnesium: Its Biological Significance*. Boca Raton, FL: CRC Press; 1981.
10. Altura BM, Altura BT. New perspectives on the role of magnesium in the pathophysiology of the cardiovascular system. I. *Clin Magnes* 1985;4:226–244.
11. Whang R. Magnesium deficiency. *Am J Med* 1987;82(suppl. 3A):24–29.
12. Miyage H, Yasue H, Okumura K, Ogawa H, Goto K, Oshimia S. Effect of magnesium on anginal attack. *Circulation* 1989;79:597–602.
13. Gottlieb SS, Baruch L, Kukin ML, Bernstein JL, Fisher ML, Packer M. Prognostic importance of serum magnesium concentration in patients with congestive heart failure. *J Am Coll Cardiol* 1990;16:827–831.
14. Keller PK, Aronson RS. The role of magnesium in cardiac arrhythmias. *Prog Cardiovasc Dis* 1990;32:433–488.
15. Millane TA, Jennison SH, Mann JM, Holt DW, McKenna WJ, Camm AJ. Myocardial magnesium depletion associated with prolonged hypomagnesemia. *J Am Coll Cardiol* 1992;20:806–812.
16. Sheehan JP. Magnesium deficiency and diabetes mellitus. *Magnes Trace Elem* 1992;10:215–219.
17. Markell MS, Altura BT, Barbour RL, Altura BM. Ionized and total magnesium levels in cyclosporine-treated renal transplant recipients. *Clin Sci* 1993;85:315–318.
18. Markell MS, Altura BT, Sarn Y, Delany BG, Ifudo O, Friedman EA, Altura BM. Deficiency of serum ionized magnesium in patients receiving hemodialysis or peritoneal dialysis. *ASAIO J* 1993;39:M801–M804.
19. Altura BT, Bertschat F, Jeremias A, Ising H, Altura BM. Comparative findings on serum IMg^{2+} of normal and diseased subjects with NOVA and KONE ISE's for Mg^{2+}. *Scand J Clin Lab Invest* 1994;54(suppl. 217):77–82.
20. Amighi J, Sabett S, Schlager O, et al. Low serum magnesium predicts neurological events in patients with advanced atherosclerosis. *Stroke* 2004;35:22–27.

21. Tzankis I, Vividakis K, Tromi A, et al. Intra- and extracellular magnesium levels and atheromatosis in haemodialysis patients. *Magnes Res* 2004;17:102–108.
22. Shirey TL. Monitoring magnesium to guide magnesium therapy for heart surgery. *J Anesth* 2004;18:118–128.
23. Altura BT, Altura BM. A method for distinguishing ionized, complexed and protein-bound Mg in normal and diseased subjects. *Scand J Clin Lab Invest* 1969;454(suppl. 217):83–87.
24. Altura BM. Importance of Mg in physiology and medicine and the need for ion selective electrodes. *Scand J Clin Lab Invest* 1994;54(suppl. 217):5–10.
25. Altura BM, Lewenstein A, eds. Unique magnesium-sensitive electrodes [special issue]. *Scand J Clin Lab Invest* 1994;54(suppl. 217):1–100.
26. Turlapaty PD, Altura BM. Magnesium deficiency produces spasms of coronary arteries. *Science* 1980;208:198–200.
27. Altura BT, Altura BM. Measurement of ionized magnesium in whole blood, plasma and serum with a new ion-selective electrode in healthy and diseased human subjects. *Magnes Trace Elem* 1991;10:90–98.
28. Altura BM, Altura BT. Influence of magnesium on drug-induced contraction and ion content in rabbit aorta. *Am J Physiol* 1971;220:938–944.
29. Altura BM, Altura BT. Magnesium and contraction of arterial smooth muscle. *Microvasc Res* 1974;7:145–155.
30. Altura BM, Altura BT. Magnesium and vascular tone and reactivity. *Blood Vessels* 1978;25:5–16.
31. Altura BM, Altura BT. Magnesium ions and contraction of vascular smooth muscle. *Fed Proc* 1981;40:2672–2679.
32. Altura BM, Altura BT. Magnesium, electrolyte transport and coronary vascular tone. *Drugs* 1984;28(suppl. 1):120–142.
33. Altura BM, Zhang A, Altura BT. Magnesium, hypertensive vascular diseases, atherogenesis, subcellular compartment of Ca2+ and Mg2+ and contractility. *Miner Electrolyte Metab* 1993;19:323–336.
34. Yang ZW, Gebrewold A, Novakowski M, et al. Mg2+-induced endothelial-dependent relaxation of blood vessels and blood pressure lowering: role of NO. *Am J Physiol Regul Integr Comp Physiol* 2000;278:R628–R639.
35. Altura BM, Zhang A, Cheng TPO, et al. Extracellular magnesium regulates nuclear and perinuclear free ionized calcium in cerebral vascular smooth muscle. *Alcohol* 2001;23:83–90.
36. Tonkin A. *Atherosclerosis and Heart Disease*. London: Martin Dunitz; 2003.
37. Herman MP, Sukhova GK, Libby P, et al. Expression of neutrophil collagenase (matrix metalloproteinase-8) in human atheroma. *Circulation* 2001;104:1899–1904.
38. Gallis ZS, Sukhova GK, Lark MN, et al. Increased expression of matrix metalloproteinases and matrix degrading activity in vulnerable regions of human atherosclerotic plaques. *J Clin Invest* 1994;94:2493–2503.
39. Schonbeck D, Libby P. Cytokines and growth regulatory factors. In: Fuster V, Topol EJ, Nabel EG, eds. *Atherothrombosis and Coronary Artery Disease*. 2nd ed. Philadelphia: Lippincott Williams & Willcins; 2005:547–559.
40. Freedman AM, Atrachi AH, Cassidy MM, Weglicki WB. Magnesium deficiency-induced cardiomyopathy: protection by vitamin E. *Biochem Biophys Res Commun* 1990;170:1102–1106.

41. Freedman AM, Cassidy MM, Weglicki WB. Captopril protects against myocardial injury by magnesium deficiency. *Hypertension* 1991;18:142–147.
42. Weglicki WB, Phillips TM, Freedman AM, et al. Magnesium deficiency elevates circulating levels of inflammatory cytokines and endothelin. *Mol Cell Biochem* 1992;118:105–111.
43. Maier JAM, Malpuech-Brugere C, Zimowska W, Rayssiguier Y, Mazur A. Low magnesium promotes endothelial cell dysfunction: implications for atherosclerosis, inflammation and thrombosis. *Biochim Biophys Acta* 2004;689: 13–21.
44. Rayssiguier Y, Mazur A. Magnesium and inflammation: lessons from animal models. *Clin Calcium* 2005;15:245–248.
45. Cybulsky MI, Gimbrone MA Jr. Endothelial expression of mononuclear leukocyte adhesion molecule during atherogenesis. *Science* 1991;251:788–791.
46. Li H, Cybulsky MI, Gimbrone MA Jr., et al. An atherogenic diet rapidly induces VCAM-1, a cytokine-regulatable mononuclear leukocyte adhesion molecule, in rabbit aortic endothelium. *Arterioscler Thromb* 1993;13:197–204.
47. Kuranstin-Mills J, Cassidy MM, Stafford RE, Weglicki WB. Marked alterations in circulating inflammatory cells during cardiomyopathy development in a magnesium-deficient rat model. *Br J Nutr* 1997;78:845–855.
48. Malpuech-Brugere C, Novacki W, Daveau M, et al. Inflammatory response following acute magnesium deficiency in the rat. *Biochim Biophys Acta* 2000;1501: 91–98.
49. Altura BT, Bust M, Barbour RL, et al. Magnesium dietary intake modulates blood lipid levels and atherogenesis. *Proc Natl Acad Sci U S A* 1990;87:1840–1844.
50. Weglicki WB, Kramer JH, Mak IT, et al. Pro-oxidant and pro-inflammatory neuropeptides in magnesium deficiency. In: Rayssiguier Y, Mazur A, Durlach J, eds. *Advances in Magnesium Research: Nutrition and Health*. London: John Libbey; 2001:285–289.
51. Weaver K. A possible anticoagulant effect of magnesium in preeclampsia. In: Cantin M, Seelig MS, eds. *Magnesium in Health and Disease*. Holliswood: Spectrum; 1980:833–838.
52. Nadler JL, Buchanan T, Natarajan R, et al. Magnesium deficiency produces insulin resistance and increased thromboxane synthesis. *Hypertension* 1993;21: 1024–1029.
53. Gawaz M, Reininger A, Neumann FJ. Platelet function and platelet-leukocyte adhesion in symptomatic coronary heart disease. Effects of intravenous magnesium. *Thromb Res* 1996;83:341–349.
54. Rishi M, Ahmad A, Makheja A, et al. Effects of reduced dietary magnesium on platelet production in hamsters. *Lab Invest* 1990;63:717–721.
55. Stamler JS, Osborne JA, Jaraki O, et al. Adverse vascular effects of homocysteine are modulated by endothelium-derived relaxing factor and related oxides of nitrogen. *J Clin Invest* 1993;91:308–318.
56. Wrenn RW, Raeber CV, Herman LE, et al. Transforming growth factor-beta. *In Vitro Cell Dev Biol Anim* 1993;29:73–78.
57. Dalton ML, Gadson RF, Robert JR, et al. Homocysteine signal cascade production of phospholipids, activation of protein kinase C and induction of c-fos and c-myb in smooth muscle cells. *FASEB J* 1997;11:703–711.

58. Kang SS, Zhou J, Wong PWK, et al. Intermediate homocystenemia: a thermolabile variant of methylene-tetrahydrofolate reductase deficiency. *Pediatr Res* 1988;43: 414–421.
59. McKully KS. Vascular pathology of homocystenemia: implications for the pathogenesis of arteriosclerosis. *Am J Pathol* 1969;56:111–128.
60. Li W, Zheng T, Wang J, et al. Extracellular magnesium regulates effects of vitamin B6, B12 and folate in homocystenemia-induced depletion of intracellular free magnesium ions. *Neurosci Lett* 1999;274:83–86.
61. Guo H, Lee JD, Uzui H, et al. Effects of folic acid and magnesium on the production of homocysteine-induced extracellular matrix metalloproteinase-2 in cultured rat vascular smooth muscle cells. *Circ J* 2006;70:141–146.
62. Yang ZW, Zheng T, Zhang A, et al. Mechanism of hydrogen peroxide-induced contraction of rat aorta. *Eur J Pharmacol* 1998;344:169–181.
63. Yang ZW, Zheng T, Wang J, et al. Hydrogen peroxide induces contratction and rises in [Ca2+]I in canine cerebral arterial smooth muscle. *N S Arch Pharmacol* 1999;360:646–653.
64. Bharadwaj LA, Prasad K. Mechanism of hydroxyl radical-induced modulation of vascular tone. *Free Radic Biol Med* 1997;22:381–390.
65. Shen JZ, Zheng XF, Kwan CY. Differential contractile actions of reactive oxygen species on rat aorta. *Life Sci* 2000;66:PL291–PL296.
66. Li J, Li W, Liu W, et al. Mechanisms of hydroxyl radical-induced contraction of rat aorta. *Eur J Pharmacol* 2004;499:171–178.
67. Brown MS, Goldstein JL. Receptor-mediated control of cholesterol metabolism. *Science* 1976;191:150–154.
68. Goldstein JL, Ho YK, Basu SK, et al. Binding site on macrophages that mediates uptake and degradation of acetylated low density lipoprotein producing massive cholesterol deposition. *Proc Natl Acad Sci U S A* 1979;76:333–337.
69. Rosen GM, Freeman BA. Detection of superoxide generated by endothelial cells. *Proc Natl Acad Sci U S A* 1984;81:7269–7273.
70. Daugherty A, Dunn JL, Ratery DL, Heinecke JW. Myeloperoxidase, a catalyst for lipoprotein oxidation is expressed in human arteriosclerotic lesions. *J Clin Invest* 1994;94:437–444.
71. Savenkova ML, Mueller DM, Heinecke JW. Tyrosyl radical generated by myeloperoxidase is a physiological catalyst for the initiation of lipid peroxidation in low density lipoprotein. *J Biol Chem* 1994;269:20394–20400.
72. Rayssiguier Y, Gueux E, Bussiere L, et al. Dietary magnesium affects susceptibility of lipoproteins and tissues to peroxidation in rats. *J Am Coll Nutr* 1993;12:133–137.
73. Rock E, Astier C, Laub C, et al. Dietary magnesium deficiency in rats enhances free radical production in skeletal muscle. *J Nutr* 1995;195:1205–1210.
74. Malpuech-Brugere C, Kurysko J, Nowacki W, et al. Early morphological and immunological alteration in the spleen during magnesium deficiency in the rat. *Magnes Res* 1998;11:161–169.
75. Bussiere F, Tridon A, Malpuech-Brugere C, et al. Effect of magnesium on the production of reactive oxygen species by neutrophils. *Magnes Res* 2000;13:81.
76. Dickens BF, Weglicki WB, Li S, Mak IT. Magnesium deficiency in-vitro enhances free-radical intracellular oxidation and cytotoxicity in endothelial cells. *FEBS Lett* 1992;311:187–191.
77. Altura BM, Kostellow AB, Zheng A, et al. Expression of the nuclear factor-KB and proto-oncogenes c-fos and c-jun are induced by low extracellular Mg2+ in

aortic and cerebral vascular smooth muscle cells. *Am J Hypertens* 2003;16: 701–707.
78. Touyz RM, Pu Q, ChenY, et al. Effects of low dietary magnesium intake on development of hypertension in stroke-prone spontaneously hypertensive rats. *J Hypertens* 2002;11:2141–2143.
79. Sherer Y, Bitzur R, Cohen H, et al. Mechanisms of action of the antiatherogenic effect of magnesium: lessons from a mouse model. *Magnes Res* 2001;14:173–179.
80. Laurant P, Hayoz D, Brunner HR, Berthelot A. Effect of magnesium deficiency on blood pressure and mechanical properties of rat carotid artery. *Hypertension* 1999;29:1199–1203.
81. Gunther T, Vormann J, Foster RM. Effect of low oxygen free radicals on Mg2+ efflux from erythrocytes. *Eur J Clin Chem Clin Biochem* 1994;32:273–277.
82. Wu F, Altura BT, Gao J, et al. Ferrylmyoglobin formation induced by magnesium deficiency in perfused rat heart causes cardiac failure. *Biochim Biophys Acta Mol Dis* 1994;1225:158–164.
83. Giulivi C, Romero FJ, Cardenas E. The interaction of trolox c, a water-soluble vitamin E analog, with ferrylmyoglobin: reduction of the oxyferryl moiety. *Arch Biochem* 1992;299:302–312.
84. Grisham MB. Myoglobin-catalyzed hydrogen peroxide-dependent arachidonic acid peroxidation. *J Free Radic Biol Med* 1985;1:227–232.
85. Whitburn KD. The interaction of oxymyoglobin with hydrogen peroxide. *Arch Biochem Biophys* 1987;277:314–321.
86. Rice-Evans C, Okunada G, Kahn R. The suppression of iron release from activated myoglobin by physiological electron donors and by desferrioxamine. *Free Radic Res Commun* 1989;7:45–54.
87. Davies MJ, Garlick PB, Slater TF, Hearse DT. In: Rice-Evans C, Dormandy T, eds. *Free Radicals, Chemistry, Pathology and Medicine*. New York: Richelieu Press;1988:303–319.
88. Puppo A, Halliwell B. Formation of hydroxyl radicals in biological systems. *Free Radic Res Commun* 1988;4:415–422.
89. Galaris D, Savarian A, Cadenas E, Hochstein P. Ferrylmyoglobin-catalyzed linoleic acid peroxidation. *Arch Biochem Biophys* 1990;281:163–169.
90. Cohen H, Sherer Y, Shaish A, et al. Atherogenesis inhibition induced by magnesium-chloride fortification of drinking water. *Biol Trace Elem Res* 2002;90: 251–259.
91. Yang ZW, Altura BT, Altura BM. Low extracellular magnesium contraction of arterial smooth muscle: role of protein kinase C and protein tyrosine phosphorylation. *Eur J Pharmacol* 1999;378:273–281.
92. Yang ZW, Wang J, Altura BT, et al. Extracellular magnesium deficiency induces contraction of arterial smooth muscle: role of P-I-3 kinases and MAPK signaling pathways. *Pflugers Arch* 2000;439:240–247.
93. Yang ZW, Wang J, Zheng T, et al. Low extracellular magnesium induces contraction and [Ca2+]i rises in cerebral arteries. *Am J Physiol Heart Circ Physiol* 2000;279:H2898–H2407.
94. Yang ZW, Wang J, Zheng T, et al. Low extracellular magnesium induces contraction of cerebral arteries: role of tyrosine and mitogen-activated protein kinases. *Am J Physiol Heart Circ Physiol* 2000;279:H185–H194.

95. Laurant P, Touyz RM. Physiological and pathophysiological role of magnesium in the cardiovascular system: implications in hypertension. *J Hypertens* 2000;18:1177–1191.
96. He Y, Yao G, Savoia C, Touyz RM. Transient receptor potential melastatin 7 ion channels regulate magnesium homeostasis in vascular smooth muscle cells. *Circ Res* 2005;96:207–215.
97. Bernardini D, Nasulewicz A, Mazur A, Maier JAM. Magnesium and microvascular endothelial cells: a role in inflammation and angiogenesis. *Front Biosci* 2005;10:1177–1182.
98. Brautbar N, Altura BM. Hypophosphatemia and hypomagnesemia result in cardiovascular dysfunction. *Alcohol Clin Exp Res* 1987;11:118–126.
99. Nahorski SR, Wilcox RA, Mackrill JJ, et al. Phosphoinositide-derived second messengers and the regulation of Ca2+ in vascular smooth muscle. *J Hypertens* 1994;12:1024–1029.
100. Morrill GA, Gupta RK, Kostellow AB, et al. Mg2+ modulates membrane lipids in vascular smooth muscle: a link to atherogenesis. *FEBS Lett* 1997;408:191–194.
101. Horrocks LA, Sharma M. In: *New Comprehensive Biochemistry*. Vol. 4. Amsterdam: Elsevier; 1982:51–93.
102. Zimmerman GA, Prescott SM, McIntyre TM. Oxidatively fragmented phospholipids as inflammatory mediators: the dark side of polyunsaturated lipids. *J Nutr* 1995;125(suppl. 6):1661s–1665s.
103. Heery JM, Kozak M, Stafforini DM, et al. Oxidatively modified LDL contains phospholipids with platelet-activating factor-like activity. *J Clin Invest* 1995;96:2322–2330.
104. Schissel SL, Tweedie-Hardman J, Rapp JH, et al. Rabbit aorta and human atherosclerotic lesions hydrolyze the sphingolyelin of retained low-density lipoprotein. *J Clin Invest* 1996;98:1455–1464.
105. Bolin DJ, Jonas A. Sphingomyelin inhibits the lecithin-cholesterol acyltransferase reaction with reconstituted high-density lipoproteins by decreasing enzyme binding. *J Biol Chem* 1996;271:19152–19158.
106. Arimoto I, Saito H, Kawashima Y, et al. Effects of sphingomyelin and cholesterol on lipoprotein lipase-mediated lipolysis in lipid emulsions. *J Lipid Res* 1998;39:143–151.
107. Maceyka M, Payne SC, Milstein S, et al. Sphingosine kinase,sphingosine-1-phosphate,and apoptosis. *Biochim Biophys Acta* 2002;1585:193–201.
108. Zheng T, Li W, Wang J, et al. Sphyngomyelinase and ceramide analogs induce contraction and rises of [Ca2+]I in canine cerebral vascular smooth muscle. *Am J Physiol Heart Circ Physiol* 2002;278:H1421–H1428.
109. Bischoff A, Czyborra P, Fetscher C, et al. Sphingosine-1-phosphate and sphingosylphosphoryl-choline constrict renal and mesenteric microvessels in vitro. *Br J Pharmacol* 2000;130:1871–1877.
110. Todoroki-Ikeda N, Mizukami Y, Mogami K, et al. Sphingosylphosphorylcholine induces Ca2+-sensitization of vascular smooth muscle contraction: possible involvement of rho-kinase. *FEBS Lett* 2000;482:85–90.
111. Altura BM, Gebrewold A, Zheng T, et al. Sphingo-myelinase and ceramide analogs induce vasoconstriction and leukocyte-endothelial interactions in cerebral venules in intact rat brain. *Brain Res Bull* 2002;58:271–278.
112. Morrill GA, Gupta RK, Kostellow AB, et al. Mg2+ modulates membrane sphingolipid and lipid second messenger in vascular smooth muscle cells. *FEBS Lett* 1998;440:167–171.

Skeletal Diseases and Calcium Metabolism

20
Overview of Skeletal Diseases and Calcium Metabolism in Relation to Magnesium

Hirotoshi Morii

Whereas 69.3% of magnesium (Mg) exists in bone tissues and is incorporated in bone crystal as a constituent, Mg also regulates functions of osteoblasts and osteoclasts, as well as osteocytes. Calcium-sensing receptor (CaR) is a key regulator of parathyroid hormone (PTH) secretion and CaR requires magnesium for the action. Another important aspect of Mg is the role of this element as a nutrient. Mg is supplied as a nutrient from food.

Magnesium and Osteoporosis

Rude and colleagues dedicated a study to bone metabolism and magnesium in the present monograph. After publishing a review article entitled "Magnesium Deficiency: A Possible Risk Factor for Osteoporosis,"[1] the group has concentrated on this issue from many standpoints. After that one of the problems was the correlation between magnesium deficiency and cytokines related to bone turnover.[2] This study group demonstrated OPG decrease and RANKL increase in magnesium deficiency by the immunocytochemistry, thus reflecting the increased osteoclastogenesis and decrease in bone mass in magnesium deficiency. Ryder studied the effect of dietary magnesium intake and bone mineral density (BMD).[3] This group showed that in white, but not black, men and women, magnesium intake was positively associated with BMD of the whole body after adjustment of age, self-report of osteoporosis or fracture in adulthood, caloric intake, calcium and vitamin D intake, body mass index (BMI), smoking status, alcohol intake, physical activity, thiazide diuretic use, and estrogen use in women ($p = 0.05$ for men and $p = 0.005$ for women). Bone mineral density was $0.04\,g/cm^2$ higher in white women and $0.02\,g/cm^2$ higher in white men in the highest than in the lowest percentile of magnesium intake.[3]

However, there have been many discussions of the possibility of magnesium deficiency as a risk factor of osteoporosis. Morii cited a study from WHI-OS (Women's Health Initiative Observational Study), reported in 2002, in which the study group demonstrated that high magnesium intake does not confer a protective effect against fracture and in fact may increase the risk of wrist/lower arm fracture in the cohort.

Morii showed in the present review that Mg restored bone toughness partially compared to calcium in Ca–Mg deficient animals but plays a role in maintaining bone strength.

Magnesium and Bone Cells

Intracellular magnesium is important in the function of cells, including osteoblasts, osteoclasts, and osteocytes. Schwartz observed the effect of magnesium deficiency on bone-cell differentiation and bone formation using in vivo matrix-induced endochondral ossification. Demineralized bone matrix was implanted subcutaneously in young male rats fed a semisynthetic Mg-deficient diet (50 ppm) for 7 days. Control diet was supplemented to contain 1000 ppm Mg. The implants were harvested 7, 9, 11, 15, and 20 days after implantation. Implants removed from Mg-deficient animals showed retardation in cartilage and bone differentiation and matrix calcification. Magnesium content was markedly reduced compared to the controls. Bone marrow development was also retarded with Mg deficiency.[4]

Gruber induced dietary Mg deficiency in the mouse, which showed hypercalcemia. Animals received osteoprotegerin (OPG) injections for 12 days. Serum calcium was similar in Mg-deficient mice treated with OPG and control mice receiving OPG. Both groups had significantly higher serum calcium than controls or Mg-deficient animals receiving vehicle alone. Mg-depleted mice that received OPG in doses that inhibit osteoclastic bone resorption remained hyperclacemic. Thus OPG–osteoclast relation may have differed in Mg deficiency.

Calcium Sensing Receptor (CaR) and Magnesium

This issue was discussed in detail by Schlingman in this monograph (Chapter 22). CaR plays important roles in the regulaton of PTH and re-absorption mechanisms of divalent cations from tubules. Magnesium plays inpotant roles in these processes. Morii also reviewed some aspects of CaR in relation to magnesium. Kawata showed that CaR is activated by Ca^{++}, Mg^{++}, $Ga3^+$, and neomycin dose in adependent manner. It was also demonstrated calcimimetic administration induced decrease in serum Mg levels.[5] It was postulated that the inhibition of PTH secretion or some mechanism related to the actions on renal tubules may have participated.[5]

Magnesium and Vitamin K in Relation to Bone Metabolism

This issue is fully discussed by Amizuka in the present monograph. Kobayshi and colleagues also concentrated on this issue in recent years. It is very interesting that bone abnormalities are reversed by vitamin K. Further studies are needed from this standpoint.

References

1. Burckhardt P, Dawson-Hughes B, Heaney RP, eds. *Nutritional Aspects of Osteoporosis.* San Diego: Academic Press; 2001.
2. Rude RK, Gruber HE, Wei LY, Frausto A. Immunoloxalization of RANKL is increased and OPG decreased during dietary magnesium deficiency in rat. *Nutr Metab* 2005;14:24.
3. Ryder KM, Shorr RI, Bush AJ, et al. Magnesium intake from food and supplements is associated with bone mineral density in healthy older white subjects *J Am Geriat Soc* 2005;53:1875–1880.
4. Schwartz R, Reddi AH. Influence of magnesium depletion on matrix-induced endochondral bone formation. *Calcif Tissue Int* 1979;29:15–20.
5. Kawata T, Nagano N. The calcium receptor and magnesium metabolism. *Clin Calcium* 15:1805–1812.

21
Magnesium and Osteoporosis

Hirotoshi Morii, Takehisa Kawata, Nobuo Nagano, Takashi Shimada, Chie Motonaga, Mariko Okamori, Takao Nohmi, Takami Miki, Masatoshi Kobayashi, K. Hara, and Y. Akiyama

Magnesium is the second most abundant cation in bone and has been recognized as one of essential elements of bone. It is estimated that 1300 g of calcium, 14 g of magnesium, and 600 g of phosphorus exist in bones of a 70-kg human adult.[1] Basic structure of bone is composed of hydroxyapatite [$Ca_5(OH)(PO_4)_3$], but Newman proposed that magnesium is incorporated in bone crystal in the following formula: [$Ca_9^{++}(H_2O^+)_2(PO_4^=)_6(OH^-)_2$][$Ca^{++} \cdot Mg_{0.3}^{++} \cdot Na_{0.3}^+ \cdot CO_3^- \cdot Cit_{0.3}$].[2]

Thus, magnesium may be involved in bone metabolism as a component of bone mineral, as well as by regulating functions of osteoblast and osteoclast. Magnesium has been known to regulate parathyroid hormone (PTH) secretion and function directly of indirectly via the regulation of calcium-sensing receptor (CaR).

Calcium-Sensing Receptor and Magnesium

Calcium-sensing receptor was cloned from bovine parathyroid cells by Brown and colleagues in 1993[3] and was shown to respond to changes in extracellular concentration of calcium and influence on PTH secretion. Magnesium has an action of suppressing PTH secretion as a ligand of CaR, just as calcium. However, the affinity of magnesium to CaR is much less compared with calcium and Murakami showed that EC_{50} in raising intracellular level was 0.9 mM for calcium and 7.9 mM for magnesium (Figure 21.1).[4]

Calcimimetics were developed to increase the sensitivity of CaR to calcium so that hyperparathyroidism may be controlled, especially in secondary hyperparathyroidism in chronic renal failure (Figure 21.2).[5]

Thus, magnesium plays a role in the mechanism of suppressive effect of calcium in controlling the secretion of PTH and in the action of calcimimetics.

...dogene Östrogenexposition

...edrige Knochendichte korreliert mit ...siko für Demenz vom Alzheimer-Typ

Mg!

...eine Hormonersatztherapie (HRT) das Risiko für eine Demenz vom Alzheimertyp ...T) vermindern kann, wird seit dem Abbruch der WHI-Studie und anderer pro...ktiver Studien zur HRT wohl vorerst nicht direkt zu klären sein. Nun wurde der ...luss der Knochendichte als Surrogatmarker für die kumulative endogene Östro...exposition auf die Entwicklung einer DAT prospektiv untersucht.

...der bevölkerungsbasierten Kohor...tudie wurde bei 987 zum Rekrutierungszeitpunkt kognitiv gesunden Personen die Knochendichte an Oberschenkelhals, Trochanter ...Radiusschaft bestimmt. Die Werte ...en zur Demenzinzidenz im anschl...nden achtjährigen Beobachtungs...aum ins Verhältnis gesetzt.

...justiert auf Einflussfaktoren wie Al...Apolipoprotein E4-Status, Homocy...spiegel, Ausbildungsgrad, Östrogen...ahme, Rauchen und Schlaganfalla...ese wiesen Frauen mit einer Schen...ls-Knochendichte in der untersten ...tile eine mehr als zweimal so hohe ...scheinlichkeit für eine DAT und eine ...enz jeglicher Genese auf als Teilneh...nnen mit höherer Knochendichte ...rd ratio 2,04 bzw. 2,01). Ein Trend er...ch für die Trochanter-, nicht aber für ...adius-Werte. Diese Zusammenhän...nnen als eine präventive Wirkung ...dogenen Östrogenexposition inter...rt werden. Bei den Männern errech...ch für alle drei Lokalisationen eine ...e Korrelation zwischen Knochen...und DAT-Risiko, die aber keine Sig...nz erreichte.

Surrogatmarker / Demenzinzidenz / Hormonersatztherapie

FAZIT: Frauen mit niedriger Knochendichte wiesen ein verdoppeltes Demenzrisiko auf. Es kann spekuliert werden, dass diese Frauen von einer – ansonsten risikobehafteten – Hormonersatztherapie profitieren könnten. (bk)

🄺 *Tan ZS et al.: Bone mineral density and the risk of Alzheimer disease.* **Arch Neurol** 62 (2005) 107-111

✗ *Bestellnummer der Originalarbeit 051223*

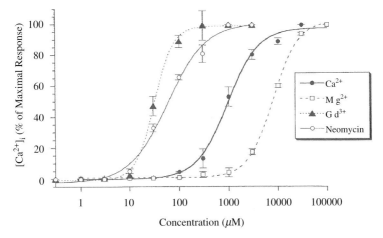

FIGURE 21.1. Effect of Ca^{2+}, Mg^{2+}, Gd^{2+}, and neomycin on $[Ca^{2+}]i$ in HuCaR-HEK293. Hu CaR-HEK 293 cells loaded wuith fura2 were suspended in buffer containing 1 mM CaVl2. The cells were treated with indicated concentrations of Ca^{2+}, Mg^{2+}, Gd^{2+}, and neomycin. The percentage of maximal peak increase in $[Ca^{2+}]i$ was plotted. Each point is the mean ± standard error (SE) of three experiments. Ca^{2+}, Mg^{2+}, Gd^{2+}, and neomycin increased $[Ca^{2+}]i$. The EC_{50} values for Ca^{2+}, Mg^{2+}, Gd^{2+}, and neomycin were 0.9, 7.9, 0.032, and 0.06 mM, respectively.

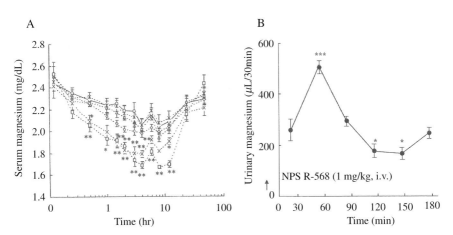

FIGURE 21.2. Effect of NRS-568 on serum total magnesium and urinary total excretion. Single oral administration of NRS-568 (1, 3, 10, 30, or 100 mg/kg) decreased serum total magnesium (A) (○, vehicle; *, 1 mg/kg; △, 3 mg/kg; ◇, 10 mg/kg; X, 30 mg/kg; □, 100 mg/kg) Single injection of NPS-568 (1 mg/kg) increased urinary magnesium excretion (B).

Magnesium Intake and Osteoporosis

Effect of Magnesium Deficiency on Bone in Ovariectomized Rats

Sprague-Dawley rats 9 weeks of age were ovariectomized and fed a calcium- and magnesium-free diet (AIM-93M) for 10 weeks, which included 3 weeks before ovariectomy. For the calcium group, whey calcium (at 1.09%, as this compound contains 28% as elemental calcium in this solution) was mixed, and for the magnesium group, magnesium sulfate at 0.64% (0.1% as magnesium) mixed in addition to calcium carbonate group ($CaCO_3$) to which 0.79% calcium carbonate was administered. Each group consisted of 6 animals. Body weight, food consumption, and ingested water were estimated. Strength and toughness of femur were tested after sacrifice at the end of 10 weeks. Strength was tested by braking femur by pressing at three points of the femur. Toughness was tested by twisting femur.

Body weight showed steady increase until 10 weeks in the four groups of animals. There were no significant differences among the groups (Figure 21.3a). Significant increase was demonstrated in the strength between the control ovariectomized group (OVX) and two kinds of calcium groups (Whey and $CaCO^3$) or the magnesium group (Figure 21.3b), although the increase was not so pronounced in magnesium group compared with the calcium groups. There were significant differences in the toughness of femur between OVX and two calcium groups, respectively. The difference in toughness between OVX and Mg, was not significant, although there was a slight increase in the value of power of toughness in the Mg group.

It may be concluded from this small study that calcium is effective in promoting bone strength and toughness in any form of the compound and that magnesium has similar effect as calcium, but the effect is rather permissive compared with the effect of calcium.

Magnesium Nutrition and Osteoporosis

Rude published an interesting review article on magnesium deficiency as a risk factor for osteoporosis, collecting papers until 2000.[6] In his review article, Rude cited a paper in which magnesium intake was positively related to quantitative ultrasound properties of bone. Regarding the effect of magnesium on bone turnover markers, Rude stated that magnesium supplementation did not have effect on seum osteocalcin, bone-alkaline phosphatase, or urinary pyridinoline and deocypyridinolone excretion. As to magnesium therapy in osteoporosis, Rude reported a significant increase in bone density of the proximal femur and lumbar spine in celiac sprue patients who received approximately 575 mg of magnesium per day.[7]

In experimental study, Rude showed in 2003 that in all magnesium-depleted mice hypomagnesemia developed and skeletal magnesium content fell significantly. Serum calcium increased but PTH was normal and osteoprotegerin did

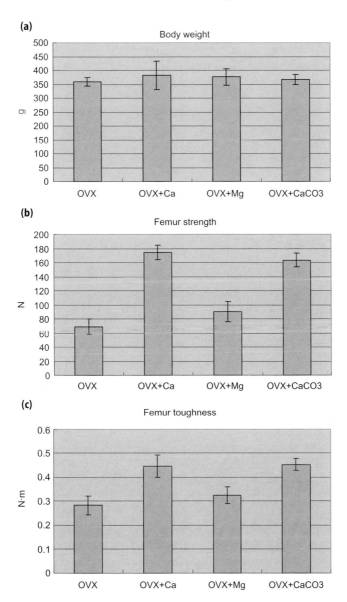

FIGURE 21.3. (a) Body weight in ovariectomized rats fed calcium- and magnesium-free diet for 10 weeks (OVX) after ovariectomy, added whey calcium (OVX+Ca), added magnesium sulfate (OVX+Mg), and added calcium carbonate (OVX+CaCO$_3$). (b) Femur strength in OVX, OVX+Ca, OVX+Mg, and OVX+CaCO$_3$ OVX versus OVX+Ca, $p < 0.001$; OVX versus OVX+Mg, $p < 0.05$; OVX versus OVX+CaCO$_3$, $p < 0.001$. (c) Femur toughness in OVX, OVX+Ca, OVX+Mg, and OVX+CaCO$_3$. OVX versus OVX+Ca, $p < 0.001$; OVX versus OVX+CaCO$_3$, $p < 0.001$.

not improve hypercalcemia. Serum calcium did not change after dietary calcium was reduced to 50% and the cause of hypercalcemia was supposed to be compensatory increase of intestinal absorption of calcium in magnesium deficiency. Growth plate was decreased, trabecular bone volume decreased, and osteoclast number increased. Substance P increased in megakaryocyte and lymphocyte, interleukin 1 (IL-1), and tumor necrosis factor α (TNFα) increased in osteoclasts.[8]

Jackson investigated the relationship between magnesium intake and hip, wrist/lower arm, and other clinical fractures in 89,717 postmenopausal women aged 50 to 79 years enrolled in WHI-OS (Women's Health Initiative Observational Study). It was shown that high magnesium intake does not confer a protective effect against fracture and, in fact, may increase the risk of wrist/lower arm fracture in the cohort.[9]

Magnesium Deficiency and Vitamin K

There appeared interesting studies on the relationship between magnesium deficiency and vitamin K. It was demonstrated that the maximum load decreased and elasticity decreased in magnesium-deficient rats, in spite that cortical thickness and mineral contents of femur were preserved.[10] Kobayashi also showed mineral/matrix was higher in magnesium deficiency compared with controls using Fourier transform infrared microscopy (FTIRM), but that the ratio was reversed to normal by administration of menatetrenone (Figure 21.4).[11] In a further study Kobayashi measured amino acid composition in the

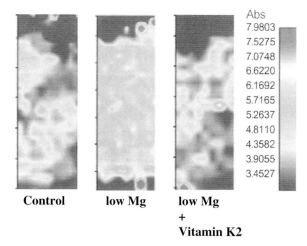

FIGURE 21.4. Mineral/matrix analyzed by FTIRM. Magnesium deficiency causes increased mineral/matrix but is reversed to the normal level by administration of menatetrenone.

femoral diaphysis. The total amino acid level in the magnesium-deficient rats was significantly lower than that in the normal rats. It was speculated that an increase in the mineral/matrix in the magnesium-deficient rat was associated with a decrease in the collagen area, but not with an increase in the mineral area.

Summary and Conclusion

Magnesium plays an important role in the action of CaR, which regulates PTH secretion. Epidemiological studies showed inconclusive effect of magnesium deficiency in the pathogenesis and supplementation in the treatment of osteoporosis. Experimental studies demonstrated that magnesium deficiency causes osteoporosis and magnesium supplementation results in the amelioration of the disease. However, the effect of magnesium seems permissive compared with calcium in restoring bone fragility in calcium- and magnesium-deficient states. It was discovered that bone abnormalities due to magnesium deficiency would be compensated by menatetrenone in experimental animals.

References

1. Aurbach GD, Marx SJ, Stephen AM. Parathyroid hormone, calcitonin, and calciferols. In: Williams RH, ed. *Textbook of Endocrinology*. 6th ed. Philadelphia: Saunders; 1981.
2. Newman WF, Neuman MW. *The Chemical Dynamics of Bone Mineral*. Chicago: The University of Chicago Press; 1958.
3. BROWN EM, Gamba G, Riccardi D. Cloning and characterization of an extracellular Ca-sensing receptor from bovine parathyroid. *Nature* 1993;366:575–580.
4. Murakami Y, Furuya Y, Wada M. Pharmacological properties of the calcimimetic compound NPS R-568 in vitro and in vivo. *Clin Exp Nephrol* 2000;4:293–299.
5. Nagano N. Pharmacological and clinical properties of calcimimetics: calcium receptor activators that afford an innovative approach to controlling hyperparathyroidism. *Pharmacol Ther* 2005.
6. Rude RK. Magnesium deficiency: a possible risk factor for osteoporosis. In: Burckhardt P, Dawson-Hughes B, Heaney RP, eds. *Nutritional Aspects of Osteoporosis*. San Diego: Academic Press; 2001.
7. Rude RK, Olreich M. Magnesium deficiency: possible role in osteoorosis associated with gluten-sensitive enteropathy. *Osteoporos Int* 1996;6:453–461.
8. Rude RK, Gruber HE, Wei LY, Frausto A, Mills BG. Magnesium deficiency: effect on bone and mineral metabolism in the mouse. *Calcif Tissue Int* 2003;72:32–41.
9. Jackson RD, Bassford T, Cauley J, et al. The impact of magnesium intake on fractures: results from Womens's Health Initiative Observational Study (WHI-OS) [abstract 1089]. *J Bone Miner Res* 2002;17(suppl. 19):S146.
10. Kobayashi M, Hara K, Akiyama Y. Effect of vitamin K2 (menatetrenone) and alendronate on bone mineral density and bopne strength in rats fed a low-magnesium diet. *Bone* 2004;35:1136–1143.
11. Kobayashi M, Hara K, Akiyama Y. Vitamin K2 and bone quality. *Clin Calcium* 2005;15:1147–1153.

22
Calcium-Sensing Receptor and Magnesium

Karl Peter Schlingmann

Divalent cation metabolism is a complex process involving the coordinated function of several organ systems and endocrine glands. The calcium-sensing receptor (CaSR) plays a central role for the homeostasis of both calcium and magnesium as it mediates not only the level of parathyroid hormone (PTH) secretion from the parathyroid, but also directly regulates the rate of divalent cation re-absorption in the kidney. Moreover, its expression in intestinal and bone cells suggests a role in all of the organs involved in maintaining systemic calcium and magnesium homeostasis. Several deactivating as well as activating CaSR mutations have been identified leading to hereditary hyperparathyroidism and autosomal-dominant hypocalcemia (ADH), respectively. Furthermore, the Bartter-like phenotype observed in patients with activating CaSR mutations demonstrates a functional link between renal calcium and magnesium handling and urine concentrating mechanism. This review focuses on the role of the CaSR for body magnesium homeostasis with special respect to renal magnesium handling and summarizes the diverse biological functions of the CaSR in regulating divalent mineral metabolism.

Magnesium is the dominant divalent intracellular cation and is essential for a variety of cellular processes such as enzyme function, DNA and protein synthesis, or the regulation of ion channels. Under physiological conditions, serum magnesium levels are maintained at almost constant values. Magnesium homeostasis primarily depends on the balance between intestinal absorption and renal excretion. Magnesium deficiency can result from reduced dietary intake, intestinal malabsorption, or renal loss. The control of body magnesium homeostasis primarily resides in the kidney. Of the magnesium filtered in the glomerulus, the vast majority is re-absorbed along the tubule. Whereas in the thick ascending limb of the loop of Henle (TALH), magnesium is reclaimed passively via the paracellular route, magnesium re-absorption in the distal convolute (DCT) involves active transcellular transport. Magnesium transport processes in both the TALH and the DCT sensitively respond to changes in body magnesium in order to maintain magnesium balance.

In contrast to calcium, there are no hormones specifically regulating magnesium homeostasis. Rather, magnesium metabolism is influenced by a number of hormones primarily regulating calcium metabolism, renal salt and water handling, and glucose homeostasis, such as PTH, vitamin D_3 metabolites, and calcitonin, as well as ADH and aldosterone, and insulin and glucagone.[1] In addition to these hormonal stimuli, epithelial magnesium transport in gastrointestinal tract and kidney is directly influenced by extracellular magnesium and calcium levels. Furthermore, PTH secretion is regulated by extracellular calcium and magnesium in such a way that levels exceeding a certain threshold inhibit PTH secretion.

The molecular identity of the cellular sensor transducing the information of extracellular divalent cation levels to the interior of target cells was elucidated by Brown and coworkers, who cloned and characterized the CaSR.[2] The CaSR belongs to the family of G-protein–coupled receptors and bears several low-affinity cation binding sites in the extracellular domain, allowing the cooperative interaction with multiple cations at physiological, millimolar concentrations.[3] Mutations affecting the CaSR making it either less or more sensitive to calcium and magnesium cause various clinical disorders: Individuals heterozygous for inactivating mutations, that is, rendering the CaSR to be less sensitive to extracellular divalent cations, show familial hypocalciuric hypercalcemia (FHH), while carriers of inactivating mutations in the homozygous state exhibit severe neonatal hyperparathyroidism (NSHPT). Mutations causing increased sensitivity of the CaSR to extracellular divalent cations produce a hereditary form of hypoparathyroidism called ADH. In addition to the pathognomonic changes in calcium metabolism, patients also exhibit concomitant changes in magnesium homeostasis.

The aim of this review is to summarize the data from molecular genetic, biochemical, and electrophysiological studies on the diverse biological roles of the CaSR in regulating divalent mineral metabolism.

Ca^{2+}/Mg^{2+}-Sensing Receptor

Modulation of PTH secretion from the parathyroid gland by extracellular calcium and magnesium involves the interaction with a specific cell-surface receptor. Experimental data had indicated the existence of such an divalent cation-sensing mechanism since raising extracellular calcium lead to the activation of phospholipase C, which in turn resulted in the accumulation of inositol trisphosphate and the release of calcium from intracellular stores.[4] Finally, Brown and coworkers used a functional cloning approach in *Xenopus* oocytes to clone the CaSR-cDNA from bovine parathyroid gland.[2] Subsequently, CaSR expression was demonstrated in different human tissues, including parathyroid, kidney, gastrointestinal tract, thyroidal C-cells, brain, and bone cells.[5,6]

The CaSR belongs to family C of the G-protein–coupled superfamily, which further includes eight metabotropic glutamate receptors, two GABA receptors, three taste receptors, and six orphan receptors.[7] Structural homologies and conservation of specific domains in some family members suggest an evolutionary link between inorganic cation and amino acid sensing.[8] The deduced amino acid sequence of the CaSR protein shows the characteristic signature of all G-protein–coupled receptors with seven membrane-spanning domains, a very large extracellular N-terminal domain, and an intracellular carboxy terminus. Analysis of the amino acid sequence reveals potential sites for N-linked glycosylation, as well as phosphorylation by protein kinase C.[5,9] Biochemical analyses indicate that the CaSR mainly exists as a dimer on the cell surface with the two molecules covalently linked by disulfide bonds.[10] The extracellular domain of the CaSR does not contain any of the known high-affinity calcium binding motifs but instead bears several regions rich in negatively charged amino acids that allow the cooperative interaction with multiple cations in the physiological, millimolar concentration range.[3]

After ligand binding, CaSR signal transduction involves coupling through G proteins to at least three phospholipases, providing for coordinate, receptor-mediated regulation of multiple signal transduction pathways. Activation of phospholipase C (PLC) and consecutive mobilization of intracellular calcium appear to be the key mechanism by which the CaSR exerts its biological actions. Activations of phospholipases A_2 and D seem to be rather secondary events largely depending on prior PLC activation.[11]

Functional characteristics of wild-type CaSR, as well as various naturally occurring and genetically engineered mutants, have been studied extensively in heterologous expression systems including HEK293 cells and *Xenopus* oocytes.[12] CaSR activation is usually quantified via monitoring ocillations of intracellular calcium or via measurement of inositol trisphosphate production. Using this approach, increasing extracellular calcium yields a sigmoidal dose-response curve consistent with functional studies of native parathyroid CaSR in vivo and in vitro.[5,9,13]

Although calcium is regarded as the primary physiological ligand, other divalent and trivalent cations can also activate the receptor. In addition, CaSR activity is allosterically modulated by amino acids with a preference for aromatic and small aliphatic L-amino acids, as well as by structurally related phenylalkylamines called calcimimetics and calcilytics.[14,15]

Does magnesium serve as a physiologically relevant CaSR agonist in vivo? Studies on CaSR activation by extracellular magnesium reveal several interesting CaSR characteristics. Generally, magnesium is considered a partial agonist exerting an intrinsic activity of half to two thirds compared to that of calcium at equimolar concentrations.[16,17] However, it is still unclear whether this lesser potency is due to the cations capacity to interact with the receptor binding site(s) or due to other contributing factors. Due to the lack of direct binding assays, many studies used rather indirect techniques to study CaSR activation and measured downstream signals like inositol trisphoshate production, intracellular calcium release, or suppression of cyclic andenosine monophosphate (AMP) levels. It is well imaginable that addition of extracellular magnesium possibly exerts a number of additional effects on CaSR-expressing cells that interfere with CaSR-mediated intracellular events like blocking calcium influx pathways.[6] Furthermore, cationic CaSR agonists ability to potentiate each other's action on the CaSR and CaSR sensitivity to magnesium greatly depends on extracellar calcium levels. In general, the activation of PLC and A$_2$ by the CaSR requires a certain extracellular calcium level with a threshold of around 1.5 mM. Accordingly, magnesium is substantially more potent in activating the CaSR in the presence of 1.5 mM than 0.5 mM calcium.[16] On the other hand, magnesium could potentially activate the CaSR simply by rendering it more sensitive to extracellular calcium.

Three serine residues (Ser147, Ser169, and Ser170) as well as one proline (P823) in the seven transmembrane region have been identified as pivotal for full responsiveness to extracellular calcium.[17,18] It remains to be clarified if CaSR activation by various cations (and especially magnesium) results in a transduction of distinctive intracellular signals, that is, by recognition via distinctive binding sites and/or activation of differing second messenger profiles. Yet, no binding sites specific for magnesium have been reported. However, Ruat and colleagues demonstrated that the relative degree of PLC and PLA$_2$ activation significantly differed depending on the cation activating the CaSR.[16]

A strong argument for the CaSR being a physiologically relevant magnesium sensor in vivo are the marked changes in magnesium metabolism observed in individuals with inactivating or activating CaSR mutations. These will be discussed below.

Physiology of Renal Magnesium Handling

The kidney plays a central role in maintaining body magnesium homeostasis. Renal magnesium handling is a filtration–re-absorption process as there is little evidence for tubular magnesium secretion. Approximately 80% of total serum magnesium (the ultrafiltrable fraction) is filtered at the glomeruli, of which more than 95% is re-absorbed along the nephron. Magnesium re-absorption differs in quantity and kinetics depending on the different nephron segments. Five to 15% are re-absorbed by the convoluted and straight portions of the proximal tubule. The majority of filtered magnesium (70%–80%) is re-absorbed in the loop of Henle, especially in the cortical thick ascending limb (cTALH). Of the 10% to 15% of the filtered magnesium delivered to the DCT from the loop of Henle, 70% to 80% is re-absorbed so that approximately 3% of the filtered load normally appears in the urine. The re-absorption rate in the DCT defines the final urinary magnesium excretion rate as there is no significant re-absorption of magnesium beyond this nephron segment. However, magnesium transport processes in both the TALH and the DCT respond to changes in magnesium to effect sensitive control of magnesium balance.

Magnesium transport processes differ depending on the different nephron segments.[19,20] Magnesium re-absorption in the TALH is passive, occurring through the paracellular pathway. Magnesium movement is influenced by electrostatic charges of tight junction proteins that comprise this route and specifically confer ion selectivity of the pathway to divalent cations. A paracellular protein, paracellin-1 or claudin-16, has been identified in the TAL that is involved in controlling magnesium and calcium permeability of the paracellular pathway. Mutations in *CLDN16* encoding paracellin-1 lead to a combined form of renal magnesium and calcium wasting, nephrocalcinosis, and progressive renal failure.[21] Paracellular divalent cation transport is driven by the lumen-positive transepithelial voltage, which is generated by transcellular salt re-absorption and potassium recycling across the apical membrane (Figure 22.1). Therefore, any influence that alters transepithelial voltage or the permeability of the paracellular pathway will alter magnesium re-absorption in the TALH.

In contrast, magnesium transport in the DCT is active and transcellular in nature. Physiological studies indicate that apical entry into the DCT cell is mediated by a specific and regulated magnesium channel driven by favorable transmembrane voltage.[1] Magnesium entry into the DCT cell is the rate-limiting step of transcellular magnesium re-absorption. The mechanism of basolateral transport into the interstitium is unknown. Magnesium has to be extruded against an unfavorable electrochemical gradient. Most physiological studies favor a sodium-dependent exchange mechanism. Driving force for magnesium entry into DCT cells and transcellular magnesium transport is the lumen-negative transepithelial potential in the DCT generated by transcellular sodium chloride re-absorption. A large number of hormones influence magnesium re-absorption within the DCT among others, including the

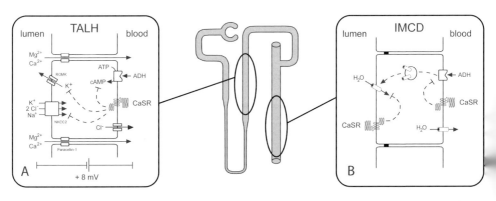

Figure 22.1. Effects of CaSR on electrolyte transport in TALH and IMCD. (A) TALH. Basolaterally expressed CaSR inhibits transcellular salt re-absorption upon activation by extracellular calcium and magnesium. Diminished active salt transport leads to a decrease in transepithelial lumen-positive potential, which represents the driving force for passive paracellular re-absorption of calcium and magnesium. (B) IMCD. Activaton of basolateral CaSR by peritubular calcium and magnesium inhibits ADH-induced incorporation of AQP-2 water channels into the apical IMCD membrane. In addition, apically expressed CaSR specifically reduces ADH-elicited osmotic water permeability.

calciotropic hormones PTH, vitamin D_3, and calcitonin, as well as ADH and aldosterone.[1] Interestingly, all of the hormones regulating transcellular magnesium transport in the DCT also influence paracellular magnesium transport in the TALH. Hormonal actions on magnesium transport in both nephron segments are mediated by changes in transepithelial voltage and paracellular permeability.[19]

CaSR in the Kidney

Besides hormonal controls, the distal nephron is able to regulate magnesium uptake in direct response to diminished extracellular magnesium. The CaSR plays a crucial role in this regulatory process. The CaSR is expressed nearly along the entire nephron. Its highest expression level is at the basolateral membrane of the cTALH.[22] The CaSR's basolateral expression in this nephron segment is consistent with a model that it senses interstitial or serum calcium and magnesium levels and mediates a negative feedback control of divalent cation re-absorption via the paracellular pathway (Figure 22.1). The cellular events following an activation of the CaSR by extracellular magnesium and/or calcium are complex. Two effects have been attributed to account for the inhibitory effect of CaSR activation on paracellular calcium and magnesium transport in the TALH: (1) a decrease in divalent selective paracellular permeability is suspected; (2) a decrease in transepithelial lumen-positive voltage secondary to inhibition of transcellular salt transport.[6] Inhibition of salt transport is thought to occur indirectly via inhibition of the sodium-potassium-2chloride contransporter WKCC2 at the apical membrane (Figure 22.2). Intracellular signaling events following CaSR activation probably include a reduction in cAMP formation, activation of PLC, and/or stimulation of arachidonic acid production via PLA_2.[6] Metabolites of arachidonic acid probably inhibit the apical potassium channel ROMK.[23] This functional coupling of inhibited salt transport and impaired divalent cation re-absorption in the TALH is also seen with furosemide administration and, furthermore, explains renal calcium and magnesium losses in patients with antenatal Bartter syndrome.[20]

Less is known about the functional consequences of CaSR activation in the DCT. CaSR activation is thought to interfere with hormone-stimulated calcium and magnesium uptake into DCT cells via inhibition of cAMP formation.[24] If there are also functional consequences of CaSR activation on transcellular salt re-absorption in the DCT remains to be clarified.

A further important role for renal salt and water handling is attributed to the CaSR in IMCD, where it is expressed both at the apical and basolateral membrane (Figure 22.1). Activation of CaSR results in a reduction of ADH-mediated aquaporin-2 (AQP-2) incorporation into the apical membrane and consecutively in reduced water permeability of the inner medullary collecting duct (IMCD).[25] The inhibitory

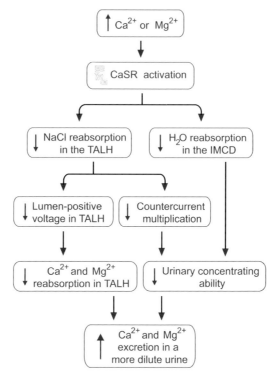

FIGURE 22.2. Intregrative role of the CaSR in renal divalent mineral and water handling. In the presence of increased serum calcium and/or magnesium, CaSR activation leads to a reduction of divalent cation re-absorption in the TALH and to decreased water re-absorption in the IMCD. The net effect is an increased excretion of divalent minerals in a more dilute urine, preventing renal stone formation.

effect of CaSR on ADH-stimulated AQP-2 membrane trafficking is probably mediated by PLC activation and/or reduction in cAMP levels.[3,25]

It is proposed that the coordinated action of the CaSR in TALH and IMCD allows for a coordinate control of divalent mineral and body water metabolism and minimizes the risk of urinary tract stone formation in the face of an increase in calcium and magnesium excretion (Figure 22.2).[26]

Divalent cation-sensing mechanisms have been demonstrated in a number of other cell types in the body of which not all are directly involved in divalent cation metabolism. Control of osteoblast as well as osteoclast function by extracellular divalent cations should especially be mentioned. Elevations of extracellular divalents are able to directly inhibit osteoclast function[27] and stimulate osteoblast proliferation in cell cultures,[28] pointing to a role not only as substrate but also regulator of bone resorption and formation. However, the pharmacological profile of some of the cation-sensing mechanisms demonstrated in vivo and in vitro substantially differs from CaSR characteristics. In addition, CaSR knockout mice display an essentially normal skeletal phenotype when preventing the deleterious effects of PTH.[29] These findings together point to the existence of additional molecules involved in the sensing of extracellular divalent cations, at least in bone. Indeed, additional molecules have been identified, including a ryanodine receptor, RyR-2, in osteoclasts as well as a novel G-protein–coupled receptor, GPRC6a, in osteoblasts as potential divalent cation sensors besides the CaSR.[7,30]

Ca^{2+}/Mg^{2+}-Sensing Receptor–Associated Disorders

Several diseases associated with both activating and inactivating mutations in the CaSR gene have been described (Table 22.1). These genetic alterations of CaSR activity are presented here with a special focus on disturbances in magnesium metabolism.

Activating mutations of the CaSR result in ADH.[31] Patients typically manifest during childhood with seizures or carpopedal spasms. Laboratory evaluation reveals the typical combination of hypocalcemia and low PTH levels, but the majority patients also exhibit moderate hypomagnesemia with serum levels around 0.5 to 0.6 mmol/L.[32] Affected individuals are often given the diagnosis of primary hypoparathyroidism on the basis of inadequately low PTH levels despite their hypocalcemia. Serum calcium levels are typically in a range of 6 to 7 mg/dL. The differentiation from primary hypoparathyroidism is of particular importance because treatment with vitamin D can result in a dramatic increase in hypercalciuria and the occurrence of nephrocalcinosis and impairment of renal function in ADH patients. Therefore, therapy with vitamin D or calcium supplementation should be reserved for symptomatic patients with the aim of maintaining serum calcium levels just sufficient for the relief of symptoms.[32]

Activating CaSR mutations lead to a lower setpoint of the receptor or an increased affinity for extracellular divalent cations. This inadequate activation by physiological extracellular calcium and magnesium levels then results in

TABLE 22.1. CaSR-associated disorders.

CaSR-associated disorders	OMIM no.	Inheritance	Age at onset	Serum Mg^{2+}	Serum Ca^{2+}	Urine Mg^{2+}	Urine Ca^{2+}	Nephro-calcinosis	Renal stones
Autosomal-dominant Hypoparathyroidism (ADH)	601198	AD	Infancy	↓	↓	↑	↑– ↑↑	Yes[a]	Yes[a]
Familial hypocalciuric hypercalcemia (FHH)	145980	AD	Often asymptomatic	N to ↑	↑	↓	↓	No	?
Neonatal severe hyperparathyroidism (NSHPT)	239200	AR	Infancy	N to ↑	↑↑↑	↓	↓	No	?

[a] Frequent complication under therapy with calcium and vitamin D.

diminished PTH secretion and decreased re-absorption of both divalent minerals, mainly in the cTAL. For magnesium, the inhibition of PTH-stimulated re-absorption in the DCT may significantly contribute to an increased renal loss in addition to the effects observed in the TAL.[33]

Recently, a Bartter-like phenotype in patients with activating CaSR mutations has been described.[33,34] In addition to hypocalcemia and deficient PTH secretion, these patients developed renal salt and water loss associated with hypokalemic alkalosis. Furthermore, all patients were found to be profoundly hypomagnesemic. Functional expression of the underlying mutations revealed a complete CaSR activation under physiologic serum calcium concentrations (threshold for CaSR activation at <0.5 mmol/L calcium).[33,34] As described above, this probably leads to an inhibition of salt re-absorption in the cTAL and a decrease in water permeability of the IMCD, resulting in renal salt and water losses with secondary hyperaldosteronism and hypokalemia.

Recently, mice with an activating CaSR mutation have been reported that display features of the human disease phenotype, including hypocalcemia, inappropriately low PTH levels, and ectopic calcifications.[35] However, urine calcium and magnesium levels were not significantly reduced compared to control animals and serum magnesium levels were found to be normal. Consistently, functional expression of the mutant CaSR in HEK293 cells displayed only a mild activation with a slight reduction of the EC_{50} for calcium (1.95 mmol/L compared to 2.25 mmol/L in wild-type mice).

Familial hypocalciuric hypercalcemia (FHH) and neonatal severe hyperparathyroidism (NSHPT) result from inactivating mutations present in either the heterozygous or homozygous (or compound heterozygous) state, respectively.[36,37]

Familial hypocalciuric hypercalcemia patients typically present with mild-to-moderate hypercalcemia, accompanied by few if any symptoms and often do not require treatment. Urinary excretion rates for calcium and magnesium are markedly reduced and serum PTH levels are inappropriately high. In addition, affected individuals also show mild hypermagnesemia, which has been attributed to an increase in tubular magnesium re-absorption in the distal nephron.[38]

In contrast, NSHPT patients with homozygous CaSR mutations usually present in early infancy with polyuria and dehydration due to severe symptomatic hypercalcemia. Unrecognized and untreated, hyperparathyroidism and hypercalcemia result in skeletal deformities, extraosseous calcifications, muscle wasting, and a devastating neurodevelopmental deficit. Early treatment with partial to total parathyroidectomy seems to be essential for outcome.[39] Data on magnesium metabolism in NSHPT patients is sparse. However, significant elevations of serum magnesium have been reported.[38,40] A Japanese patient has been reported with an inactivating CaSR mutation (P39A) in homozygous state that exhibits a mild phenotype without typical symptoms of primary hyperparathyroidsm.[41] Next to moderate hypercalcemia, the patient shows pronounced hypermagnesemia (S_{Mg} 1.36 mmol/L).

These findings concerning magnesium handling in patients with CaSR inactivation are also supported by observations in knockout mice, which carry a heterozygous or homozygous deletion of the *CaSR* gene. These mice also show significant hypermagnesemia therefore providing additional biological evidence for the CaSRs role in magnesium homeostasis.[42]

CaSR Polymorphisms

In addition to numerous disease causing mutations in the *CaSR* gene, the NCBI database lists six non-synonymous single nucleotide polymorphisms. Three clustered polymorphisms in the C terminus of the CaSR, A986S, G990R, and Q1011E, are of special interest as they are common in the Caucasian population and have been linked to serum calcium levels in healthy probands.[43,44] The CaSR A986S polymorphism with an allele frequency of around 15% was especially considered a likely candidate locus for a genetic predisposition to various bone and mineral disorders.[45] However, subsequent studies in populations at risk for osteoporosis could not confirm this suspicion.[46,47] Unfortunately, there is no data on serum magnesium in these genotyped cohorts.

Pharmacological CaSR Modulation

The discovery of the CaSR also lead to the development of drugs that specifically alter its sensitivity to extracellular divalent cations that have now entered routine practice. Calcimimetics that amplify the sensitivity of the CaSR to extracellular calcium can suppress PTH levels with a resultant fall in serum calcium levels. Their therapeutic potential has been evaluated in patients with primary hyperparathyroidism as well as secondary hyperparathyroidism linked to renal disease. In contrast, calcilytics inhibit the effects of extracellular cations on CaSR, thereby increasing PTH secretion. Intermittent administration mimicking the cyclical pattern of endogenous PTH potentially promotes anabolic effects on bone that could be used for prevention and treatment of osteoporosis. Unfortunately, data on the effects of a pharmacological manipulation of the CaSR on magnesium metabolism is poor. However, changes in body magnesium paralleling those of calcium are expected. Although surely outweighed by the positive effects of a reduction of calcium phosphorus product on clinical outcome,[48] especially magnesium deficiency following calcimimetic administration, might be of clinical relevance in patients on hemodialysis. If calcilytics might exert beneficial effects on calcium and magnesium metabolism besides their anabolic effects on bone remains to be studied. Besides targeted pharmacological intervention, other polyvalent cationic drugs are able to activate the CaSR. For example, the renal wasting of calcium and magnesium, which is a common side effect of aminoglycosides like gentamicin, is attributed to an activation of the CaSR.[49]

Conclusions

The discovery of the CaSR as a cellular sensor for extracllular divalent cations has greatly improved our understanding of divalent mineral metabolism. By CaSR activation, extracellular calcium as well as magnesium can directly serve as first messengers in and between a variety of tissues which, in turn, can directly respond to changes in ambient divalent cations. Extracellular calcium and magnesium are not only able to sensitively adapt the level of PTH secretion in the paratyroid but directly adjust the level of their excretion in the kidney by influencing tubular re-absorption processes. Moreover, the role of the CaSR along the kidney tubule provides insight into the close relationship between divalent mineral handling and renal concentrating mechanism. The phenotype of patients with inactivating and activating CaSR mutations that show profound changes in magnesium metabolism clearly indicates the importance of the CaSR as a magnesium sensor in vivo.

The role of the CaSR in tissues besides parathyroid and kidney is much less understood so far. Yet, the discovery of additional molecules sensing extracellular divalents demonstrates the complexity and differentiation of signaling pathways in response to extracellular divalents. Finally, drugs specifically altering CaSR sensitivity offer new therapeutic options for the treatment of primary and secondary forms of hyperparathyroidism, osteoporosis, and potentially further disorders involving disturbances in divalent cation metabolism.

References

1. Dai LJ, Ritchie G, Kerstan D, et al. Magnesium transport in the renal distal convoluted tubule. *Physiol Rev* 2001;81:51–84.
2. Brown EM, Gamba G, Riccardi D, et al. Cloning and characterization of an extracellular Ca(2+)-sensing receptor from bovine parathyroid. *Nature* 1993;366:575–580.
3. Hebert SC. Extracellular calcium-sensing receptor: implications for calcium and magnesium handling in the kidney. *Kidney Int* 1996;50:2129–2139.
4. Brown E, Enyedi P, LeBoff M, et al. High extracellular Ca2+ and Mg2+ stimulate accumulation of inositol phosphates in bovine parathyroid cells. *FEBS Lett* 1987;218:113–118.
5. Riccardi D, Park J, Lee WS, et al. Cloning and functional expression of a rat kidney extracellular calcium/polyvalent cation-sensing receptor. *Proc Natl Acad Sci U S A* 1995;92:131–135.
6. Brown EM, MacLeod RJ. Extracellular calcium sensing and extracellular calcium signaling. *Physiol Rev* 2001;81:239–297.
7. Pi M, Faber P, Ekema G, et al. Identification of a novel extracellular cation-sensing G-protein-coupled receptor. *J Biol Chem* 2005;280:40201–40209.
8. Mun HC, Franks AH, Culverston EL, et al. The Venus Fly Trap domain of the extracellular Ca2+-sensing receptor is required for L-amino acid sensing. *J Biol Chem* 2004;279:51739–51744.

9. Garrett JE, Capuano IV, Hammerland LG, et al. Molecular cloning and functional expression of human parathyroid calcium receptor cDNAs. *J Biol Chem* 1995;270:12919–12925.
10. Bai M, Trivedi S, Brown EM. Dimerization of the extracellular calcium-sensing receptor (CaR) on the cell surface of CaR-transfected HEK293 cells. *J Biol Chem* 1998;273:23605–23610.
11. Kifor O, Diaz R, Butters R, et al. The Ca2+-sensing receptor (CaR) activates phospholipases C, A2, and D in bovine parathyroid and CaR-transfected, human embryonic kidney (HEK293) cells. *J Bone Miner Res* 1997;12:715–725.
12. Bai M, Quinn S, Trivedi S, et al. Expression and characterization of inactivating and activating mutations in the human Ca2+o-sensing receptor. *J Biol Chem* 1996;271:19537–19545.
13. Pearce SH, Bai M, Quinn SJ, et al. Functional characterization of calcium-sensing receptor mutations expressed in human embryonic kidney cells. *J Clin Invest* 1996;98:1860–1866.
14. Conigrave AD, Quinn SJ, Brown EM. L-amino acid sensing by the extracellular Ca2+-sensing receptor. *Proc Natl Acad Sci U S A* 2000;97:4814–4819.
15. Nemeth EF, Steffey ME, Hammerland LG, et al. Calcimimetics with potent and selective activity on the parathyroid calcium receptor. *Proc Natl Acad Sci U S A* 1998;95:4040–4045.
16. Ruat M, Snowman AM, Hester LD, et al. Cloned and expressed rat Ca2+-sensing receptor. *J Biol Chem* 1996;271:5972–5975.
17. Brauner-Osborne H, Jensen AA, Sheppard PO, et al. The agonist-binding domain of the calcium-sensing receptor is located at the amino-terminal domain. *J Biol Chem* 1999;274:18382–18386.
18. Zhang Z, Qiu W, Quinn SJ, et al. Three adjacent serines in the extracellular domains of the CaR are required for L-amino acid-mediated potentiation of receptor function. *J Biol Chem* 2002;277:33727–33735.
19. Cole DE, Quamme GA. Inherited disorders of renal magnesium handling. *J Am Soc Nephrol* 2000;11:1937–1947.
20. Schlingmann KP, Konrad M, Seyberth HW. Genetics of hereditary disorders of magnesium homeostasis. *Pediatr Nephrol* 2004;19:13–25.
21. Simon DB, Lu Y, Choate KA, et al. Paracellin-1, a renal tight junction protein required for paracellular Mg2+ resorption. *Science* 1999;285:103–106.
22. Riccardi D, Hall AE, Chattopadhyay N, et al. Localization of the extracellular Ca2+/polyvalent cation-sensing protein in rat kidney. *Am J Physiol* 1998;274:F611–F622.
23. Wang W, Lu M, Balazy M, et al. Phospholipase A2 is involved in mediating the effect of extracellular Ca2+ on apical K+ channels in rat TAL. *Am J Physiol* 1997;273:F421–F429.
24. Bapty BW, Dai LJ, Ritchie G, et al. Activation of Mg2+/Ca2+ sensing inhibits hormone-stimulated Mg2+ uptake in mouse distal convoluted tubule cells. *Am J Physiol* 1998;275:F353–F360.
25. Sands JM, Naruse M, Baum M, et al. Apical extracellular calcium/polyvalent cation-sensing receptor regulates vasopressin-elicited water permeability in rat kidney inner medullary collecting duct. *J Clin Invest* 1997;99:1399–1405.
26. Hebert SC, Brown EM, Harris HW. Role of the Ca(2+)-sensing receptor in divalent mineral ion homeostasis. *J Exp Biol* 1997;200:295–302.

27. Zaidi M, Kerby J, Huang CL, et al. Divalent cations mimic the inhibitory effect of extracellular ionised calcium on bone resorption by isolated rat osteoclasts: further evidence for a "calcium receptor." *J Cell Physiol* 1991;149:422–427.
28. Dvorak MM, Siddiqua A, Ward DT, et al. Physiological changes in extracellular calcium concentration directly control osteoblast function in the absence of calciotropic hormones. *Proc Natl Acad Sci U S A* 2004;101:5140–5145.
29. Kos CH, Karaplis AC, Peng JB, et al. The calcium-sensing receptor is required for normal calcium homeostasis independent of parathyroid hormone. *J Clin Invest* 2003;111:1021–1028.
30. Zaidi M, Shankar VS, Tunwell R, et al. A ryanodine receptor-like molecule expressed in the osteoclast plasma membrane functions in extracellular Ca2+ sensing. *J Clin Invest* 1995;96:1582–1590.
31. Pollak MR, Brown EM, Estep HL, et al. Autosomal dominant hypocalcaemia caused by a Ca(2+)-sensing receptor gene mutation. *Nat Genet* 1994;8:303–307.
32. Pearce SH, Williamson C, Kifor O, et al. A familial syndrome of hypocalcemia with hypercalciuria due to mutations in the calcium-sensing receptor [see comments]. *N Engl J Med* 1996;335:1115–1122.
33. Vargas-Poussou R, Huang C, Hulin P, et al. Functional characterization of a calcium-sensing receptor mutation in severe autosomal dominant hypocalcemia with a Bartter-like syndrome. *J Am Soc Nephrol* 2002;13:2259–2266.
34. Watanabe S, Fukumoto S, Chang H, et al. Association between activating mutations of calcium-sensing receptor and Bartter's syndrome. *Lancet* 2002;360:692–694.
35. Hough TA, Bogani D, Cheeseman MT, et al. Activating calcium-sensing receptor mutation in the mouse is associated with cataracts and ectopic calcification. *Proc Natl Acad Sci U S A* 2004;101:13566–13571.
36. Pollak MR, Brown EM, Chou YH, et al. Mutations in the human Ca(2+)-sensing receptor gene cause familial hypocalciuric hypercalcemia and neonatal severe hyperparathyroidism. *Cell* 1993;75:1297–1303.
37. Pollak MR, Chou YH, Marx SJ, et al. Familial hypocalciuric hypercalcemia and neonatal severe hyperparathyroidism. Effects of mutant gene dosage on phenotype. *J Clin Invest* 1994;93:1108–1112.
38. Marx SJ, Attie MF, Levine MA, et al. The hypocalciuric or benign variant of familial hypercalcemia: clinical and biochemical features in fifteen kindreds. *Medicine (Baltimore)* 1981;60:397–412.
39. Cole DE, Janicic N, Salisbury SR, et al. Neonatal severe hyperparathyroidism, secondary hyperparathyroidism, and familial hypocalciuric hypercalcemia: multiple different phenotypes associated with an inactivating Alu insertion mutation of the calcium-sensing receptor gene. *Am J Med Genet* 1997;71:202–210.
40. Bai M, Pearce SH, Kifor O, et al. In vivo and in vitro characterization of neonatal hyperparathyroidism resulting from a de novo, heterozygous mutation in the Ca2+-sensing receptor gene: normal maternal calcium homeostasis as a cause of secondary hyperparathyroidism in familial benign hypocalciuric hypercalcemia. *J Clin Invest* 1997;99:88–96.
41. Aida K, Koishi S, Inoue M, et al. Familial hypocalciuric hypercalcemia associated with mutation in the human Ca(2+)-sensing receptor gene. *J Clin Endocrinol Metab* 1995;80:2594–2598.

42. Ho C, Conner DA, Pollak MR, et al. A mouse model of human familial hypocalciuric hypercalcemia and neonatal severe hyperparathyroidism. *Nat Genet* 1995;11:389–394.
43. Cole DE, Vieth R, Trang HM, et al. Association between total serum calcium and the A986S polymorphism of the calcium-sensing receptor gene. *Mol Genet Metab* 2001;72:168–174.
44. Scillitani A, Guarnieri V, De Geronimo S, et al. Blood ionized calcium is associated with clustered polymorphisms in the carboxyl-terminal tail of the calcium-sensing receptor. *J Clin Endocrinol Metab* 2004;89:5634–5638.
45. Cole DE, Peltekova VD, Rubin LA, et al. A986S polymorphism of the calcium-sensing receptor and circulating calcium concentrations. *Lancet* 1999;353:112–115.
46. Takacs I, Speer G, Bajnok E, et al. Lack of association between calcium-sensing receptor gene "A986S" polymorphism and bone mineral density in Hungarian postmenopausal women. *Bone* 2002;30:849–852.
47. Bollerslev J, Wilson SG, Dick IM, et al. Calcium-sensing receptor gene polymorphism A986S does not predict serum calcium level, bone mineral density, calcaneal ultrasound indices, or fracture rate in a large cohort of elderly women. *Calcif Tissue Int* 2004;74:12–17.
48. Young EW, Albert JM, Satayathum S, et al. Predictors and consequences of altered mineral metabolism: the Dialysis Outcomes and Practice Patterns Study. *Kidney Int* 2005;67:1179–1187.
49. Kang HS, Kerstan D, Dai L, et al. Aminoglycosides inhibit hormone-stimulated Mg2+ uptake in mouse distal convoluted tubule cells. *Can J Physiol Pharmacol* 2000;78:595–602.

Kidney

23
Magnesium and the Kidney: Overview

Linda K. Massey

Normal human serum magnesium concentration is about 0.7 to 0.85 mmol/L. About 20% to 30% of serum magnesium is protein bound; this leaves 70% to 80% that is ultrafiltrable, which in turn is made up of 55% ionized magnesium and 15% magnesium complexed with anions such as phosphate and citrate. About 100 to 120 mmol per day is filtered, and about 4 to 6 mmol is excreted, so that about 95% of the glomerular filtrate is re-absorbed.[1]

The renal handling of magnesium in humans is a filtration–resorption process; there is no tubular secretion of magnesium. About 15% to 20% of the filtered magnesium is re-absorbed in the proximal convoluted tubule. Then approximately 65% to 75% of filtered magnesium is re-absorbed in the thick cortical ascending loop of Henle. Another 5% to 10% is re-absorbed in the distal convoluted tubule. Finally the collecting duct is only a minor site for magnesium re-absorption.[2,3]

Magnesium re-absorption in the proximal tubule appears to be passive. It follows changes in salt and water re-absorption and is associated with the rate of fluid flow. In the loop of Henle, there appears to be an additional active transport system; a decrease in magnesium re-absorption in this segment is independent of sodium chloride transport in either hypermagnesemia or hypercalcemia.[2] Recently, a protein, claudin 16, has been identified in the thick ascending loop of Henle (TALH) that forms a tight junction of the paracellular pathway.[3,4] Magnesium is re-absorbed in the distal convoluted tubule (DCT) through a transcellular, active transport process. In the DCT, magnesium transport into the cytosol is through selective channels and extrusion into the interstitium and is apparently mediated by a sodium/magnesium cotransporter. In vivo studies in animals and humans have demonstrated a tubular maximum for magnesium that reflects a composite of these tubular re-absorptive processes.

The role of the kidney in conserving magnesium during chronic mild magnesium deficiency is controversial. During acute experimental magnesium depletion in humans, urinary magnesium decreases to very low levels, less than 1 mmol/day within 3 to 4 days. However, the effects of long-term low dietary magnesium intake on renal conservation have been little studied. The few studies available are reviewed in the Dietary Reference Intakes.[5]

Despite the tight regulation of magnesium by the kidney, no one has described a dominant hormone or factor that is responsible for renal magnesium homeostasis. Because patients with either primary hyper- or hypoparathyroidism usually have normal serum magnesium concentrations and a normal tubular maximum for magnesium, it is unlikely that parathyroid hormone (PTH) is an important regulator of magnesium homeostasis. Although PTH increases magnesium re-absorption, the PTH-induced hypercalcemia opposes this increase in re-absorption. Glucagon, calcitonin, vasopressin, aldosterone, prostaglandins, insulin, and vitamin D also affect magnesium transport in the loop of Henle, but their physiological relevance is uncertain.[2]

The calcium-sensing receptor is also sensitive to magnesium, so an elevated serum magnesium concentration will decrease potassium movement into the lumen of the TALH, leading to a decreased lumen-positive voltage and a decrease in magnesium re-absorption.[2]

Diet and Drugs Affect Urinary Magnesium Excretion

Any changes in the cotransport of sodium, chloride, and potassium, as well as active sodium re-absorption, results in changes in the transepithelial voltages in the loop of Henle, which will affect magnesium re-absorption. Although acutely increased dietary sodium chloride increases magnesium excretion, the chronic effects are largely unstudied.

Caffeine acutely increases urinary magnesium excretion relative to creatinine excretion for 9h after its consumption.[6] Night time compensatory conservation was insufficient to offset these losses, resulting in a net 24-h urinary increase of 0.16 mm of magnesium Adaptation of magnesium homeostasis to chronic consumption of caffeine is unstudied.

Diuretic therapy affects magnesium renal handling. Loop diuretics, such as furosemide, decrease magnesium absorption in the TALH by inhibiting sodium–potassium–chlorine movement into the cell, resulting in a diminished transepithelial voltage that leads to a decrease in both calcium and magnesium re-absorption. Whereas amiloride and tramterene increase magnesium transport in the DCT, chronic chlorothiazide use may result in renal magnesium wasting. These diuretics, which are used commonly in the treatment of hypertension, heart failure, and other edematous states, may cause hypermagnesuria, leading to possible hypomagnesemia and tissue magnesium deficiency.[7]

Diseases Impacted by Abnormal Urinary Magnesium

Magnesium depletion is a common feature of diabetes mellitus, apparently related to glycemic control. Djurhuss and colleagues[8] directly showed the hypermagnesuric effect of elevated blood glucose by infusing 200% glucose into 10 patients with type 1 diabetes. Blood glucose increased from 5.3 to

12.3 mmol/L, but urinary glucose was only slightly increased. Hyperglycemia increased renal magnesium excretion and clearance 2.4-fold over a euglycemic day. Plasma magnesium decreased 3% during hyperglycemia. This study clearly separated the influence of plasma insulin from that of blood glucose, which proved to be the major effecter of renal magnesium loss. This finding does not negate the additional role of insulin in magnesium metabolism, including the distribution of magnesium in cells versus plasma.

Low urinary magnesium is a risk factor for calcium oxalate nephrolithiasis. Magnesium acts as a calcium antagonist in oxalate binding. Magnesium oxalate is far more soluble than calcium oxalate, so reduces the saturation limit of calcium oxalate in the blood. Daily urines with a magnesium output of less than 2.0 mmol/day are the most likely to result in a risk index above the saturation limit for calcium oxalate.[9] The cause of low urinary magnesium in otherwise healthy stoneformers is likely to be dietary magnesium deficiency.

The extensive use of non–potassium-sparing diuretics predisposes many patients with chronic heart failure to magnesium deficiency. With reduced renal function, re-absorption of filtered magnesium is reduced in chronic renal failure, which may lead to magnesium deficiency.[7]

There are a number of primary inherited disorders of magnesium re-absorption, which are generally associated with disorders in calcium transport.[4,10,11] Cole and Quamme[12] have described extensively the five most characterized. First there is hypomagnesemia with secondary hypocalcemia, an autosomal-recessive disease segregatin with chromosome 9q12–22.2. Second, an autosomal-dominant hypomagnesemia caused by isolated renal magncsium wasting that maps to chromosome 11q23. Third, a recessive hypomagnesmia with hypercalciuria and nephrocalcinosis caused by a mutation of the claudin 16 gene (3q27) that codes for a tight junction protein that regulates paracellular Mg^{2+} transport in the loop of Henle. Fourth, an autosomal-dominant hypoparathryroidism, a variably hypomagnesemic disorder caused by mutations in the extracellular Ca^{2+}/Mg^{2+}-sensing receptor, the Casr gene at 3q13.3–21. Finally, Gitelman's syndrome presents with renal magnesium wasting and is associated with mutations in the chlorothiazide sensitive NaCl cotransporter expressed in the distal convoluted tubule (SLC12A3 at 16q13).

References

1. Bushinsky DA. Calcium, magnesium and phosphorus: renal handling and urinary excretion. In: Favus M, ed. *Primer on the Metabolic Bone Diseases and Disorders of Mineral Metabolism.* 5th ed. Washington, DC: American Society for Bone and Mineral Research; 2003:97–105.
2. Quamme GA, de Rouffignac C. Epithelial magnesium transport and regulation by the kidney. *Front Biosci* 2000;5:694-711.
3. Satoh J-I, Romero MF. Mg^{2+} transport in the kidney. *BioMetals* 2002;15:285–295.
4. Hoendrerop JGJ, Bindels RJM. Epithelial Ca^{2+} and Mg^{2+} channels in health and disease. *J Am Soc Nephrol* 2005;16:15–26.

5. Institute of Medicine. Magnesium. In: *Dietary Reference Intakes for Calcium, Phosphorus, Magnesium, Vitamin D and Fluoride.* Washington DC. National Academic Press, 1997.
6. Kynast-Gales SA, Massey LK. Effect of caffeine on circadian excretion of urinary calcium and magnesium. *J Am Coll Nutr* 1994;13:467–472.
7. Weglicki W, Quamme G, Tucker K, Haigney M, Resnick L. Potassium, magnesium and electrolyte imbalance and complications in disease management. *Clin Exp Hypertens* 2005;1:95–112.
8. Djurhuss MS, Skott P, Vaag A, et al. Hyperglycaemia enhances renal magnesium excretion in Type I diabetic patients. *Scand J Clin Lab Invest* 2000;60:403–409.
9. Massey L. Magnesium therapy for nephrolithiasis. *Magnes Res* 2005;18:123–126.
10. Konrad M, Schlingmann KP, Gudermann T. Insights into the molecular nature of magnesium homeostasis. *Am J Physiol Renal Physiol* 2004; 286:F599–F605.
11. Cole DEC, Quamme GA. Inherited disorders of renal magnesium handling. *J Am Soc Nephrol* 2000;11:1937–1947.

24
Cellular Basis of Magnesium Transport

Pulat Tursun, Michiko Tashiro, Masaru Watanabe, and Masato Konishi

About 40% of magnesium that is contained in food and drinking water is absorbed from the gastrointestinal tract. On the other hand, the kidneys are the most important for control of body magnesium balance through its urinary excretion, which is primarily regulated by tubular re-absorption. The amount of magnesium excretion in the urine is regulated by hormones and other factors, and can vary widely.

Total serum magnesium levels lie in the range of 0.7 to 1.1 mM (i.e., 1.4–2.2 mEq/L), about 30% of which is bound to proteins (mainly albumin) and the rest (70%) is in the ionized form (Mg^{2+}). Because magnesium molecules bound to macromolecules (such as proteins) do not permeate glomerular membranes, about 80% of serum magnesium is ultrafilterable. While the glomerular filtrate flows through the renal tubules, more than 95% of filtered magnesium is re-absorbed, and the rest is excreted in the urine; 15% to 20% of magnesium in the glomerular filtrate is re-absorbed in the proximal tubules, 65% to 75% is re-absorbed in the loop of Henle, and 5% to 10% is re-absorbed in the distal tubules.[1] Little absorption occurs in the collecting ducts. Fractions of magnesium re-absorbed in various portions of renal tubules are different from those for other ions; for example, about two thirds of sodium is re-absorbed in proximal tubules.

Epithelial cells of renal tubular walls are connected to each other by tight junctions, through which some solutes and water can permeate (the paracellular pathway). Transport through the paracellular pathway is passive; that is, it is driven by the difference of electrochemical potential between the tubular lumen and interstitial fluid. The other pathway is transcellular and active: movement of substances across luminal and basolateral membranes through epithelial cells. A large fraction of filtered magnesium is re-absorbed in the loop of Henle and the distal tubules, and re-absorption in these portions is thought to play important roles in regulation of urinary magnesium excretion. For example, it is known that re-absorption in the loop of Henle and the distal tubules increases in magnesium deficiency, and decreases when the serum magnesium level is elevated.

In the loop of Henle, the thick ascending limb (TAL) plays a key role in magnesium re-absorption. In the TAL, about 65% to 75% of ultrafiltered magnesium is re-absorbed passively through paracellular pathway.[2] The Mg^{2+} flux is thought to be driven by the potential difference because potential of the luminal side is slightly more positive than that of the basolateral side. Various hormones, such as parathormone, calcitonin, glucagons, and vasopressin, stimulate magnesium re-absorption in the TAL by an increase in the potential difference across the tubular walls or an increase in Mg^{2+} permeability of tight junctions.[3] The regulatory mechanisms of paracellular permeability are not fully understood, but paracellin-1, a tight junction protein, seems to provide an important clue. Familial renal hypomagnesemia is a hereditary disease that causes low serum magnesium levels through an abnormally increased urinary excretion of magnesium. By analyzing genes of families with this disease, Simon and colleagues[4] found various mutations of a gene, CLDN16, which encodes a protein localized in renal tight junctions. This protein, paracellin-1 or claudin-16, consists of 305 amino acids and has four putative transmembrane domains. The function of paracellin-1 is not yet clear, but it is speculated that the protein somehow contributes to a selective paracellular conductance by building a pore permitting fluxes of magnesium (and calcium). Lowering the Mg^{2+} permeability of tight junctions by mutations of paracellin-1 would cause a massive magnesium loss into the urine.[4]

In the distal convoluted tubules (DCT), 5% to 10% of filtered magnesium is re-absorbed.[2] In contrast to the case of the TAL described above, re-absorption in the DCT is thought to be transcellular and active.[3] Magnesium ion passively enters the cell through the luminal membrane (driven by a negative intracellular potential) and actively extruded across the basolateral membrane (Figure 24.1). For passive entry in the luminal side, Mg^{2+}-permeable channels have

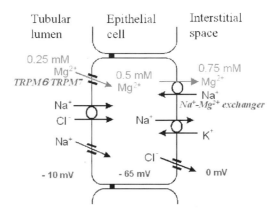

Figure 24.1. A scheme of transcellular magnesium re-absorption in the distal convoluted tubules. Magnesium passively enters into tubular cells through TRPM6 and/or TRPM7 channels localized in the luminal membrane, and is actively extruded from the cells by the Na^+/Mg^{2+} exchanger localized in the basolateral membrane.

been postulated, and recent studies highlight two members of the transient receptor potential (TRP) channel superfamily as candidates for a physiological Mg^{2+} entry pathway[5]: TRPM6 and TRPM7 channels that permeate divalent cations, including Mg^{2+}. Schlingmann and colleagues[6] and Walder and colleagues[7] showed that primary hypomagnesemia with secondary hypocalcemia, a rare familiar disease in which excess urinary excretion of magnesium leads to low serum magnesium levels, is caused by mutations of the *TRPM6* gene that encodes the TRPM6 channel. It was subsequently shown that the TRPM6 channel is localized in the DCT of kidneys, as well as the small intestine and colon, sites of transcellular re-absorption of magnesium.[8] TRPM7, the closest homologue of TRPM6, with 52% homology of the amino acid sequence, is ubiquitously distributed, and appears to have a central role in Mg^{2+} uptake in cells.[9,10] Both TRPM6 and TRPM7 channels permeate Mg^{2+} and are regulated by low intracellular Mg^{2+} concentration ($[Mg^{2+}]_i$).[8,11] It is not clear, however, if TRPM6 alone can form functional channels. Some investigators have suggested the possibility of the assembly of functional complexes of TRPM6 and TRPM7.[2,2]

On the basolateral side, on the other hand, extracellular Na^+-dependent Mg^{2+} efflux (the putative Na^+/Mg^{2+} exchange) has been postulated as an active Mg^{2+} extrusion mechanism.[13,14] Although the transport has been characterized functionally in renal tubular cells and other types of cells, the transporter molecules are not yet identified.

In this chapter, we will focus on the properties of the putative Na^+/Mg^{2+} exchange of the basolateral membrane, based on our recent works on renal tubular cells and cardiac myocytes.

Na^+/Mg^{2+} Exchange in Renal Tubular Cells

Although the Na^+/Mg^{2+} exchange has been postulated as an active Mg^{2+} transport in the basolateral membrane, the transporter molecules are not yet identified, hindering further studies on molecular mechanisms of the transport. As the first step towards molecular cloning of the Na^+/Mg^{2+} exchanger, we established a mutant cell line from mouse cortical renal tubular (MCT) cells that could grow in culture media with very high extracellular Mg^{2+} concentration ($[Mg^{2+}]_o$) through stepwise increases of Mg^{2+} concentration in the culture media and selection of high Mg^{2+}-tolerant cells.[15]

The standard culture medium containing $1\,\mathrm{m}M$ Mg^{2+} for MCT cell culture was replaced by media containing various concentrations of Mg^{2+} and incubated replacing the medium with fresh media twice a week. In a medium containing $41\,\mathrm{m}M$ or a lower concentration of Mg^{2+}, cells showed no apparent changes in their shape or growth rate. However, approximately 80% of the cells in the $61\,\mathrm{m}M$ Mg^{2+}-containing medium and all the cells in $81\,\mathrm{m}M$ or higher Mg^{2+}-containing medium died within 7 days. The surviving cells in $61\,\mathrm{m}M$ Mg^{2+}-containing medium started dividing thereafter, and the cells reached confluence 21 days after increasing the Mg^{2+} concentration. The cells were then dissociated with trypsin and diluted 10-fold into fresh medium. After passages in medium containing 61 mM Mg^{2+} for 6 weeks, the culture medium was changed to medium containing 71 mM Mg^{2+}. Thereafter, the Mg^{2+} concentration of the medium was further increased every 10 to 12 weeks in increments of 10 mM. Massive cell death was observed when Mg^{2+} concentration was elevated from 71 mM to 81 mM and from 81 mM to 91 mM. The cells that had adapted to 81 mM Mg^{2+} and 101 mM Mg^{2+} could be dislodged and stored in 10% dimethylsulfoxide and 90% fetal calf serum (FCS) in liquid nitrogen, with 20% to 50% of viability at retrieval. Mouse cortical renal tubular cells could be adapted in Delbecco's modified Eagle's medium (DMEM) containing as high as 121 mM Mg^{2+} without losing their growing capacity.

We measured [Mg^{2+}]$_i$ with a fluorescent indicator furaptra (Mag-fura-2) in clusters of wild-type cells and high Mg^{2+}-tolerant cells. In the presence of 150 mM extracellular Na$^+$, [Mg^{2+}]$_i$ in the highly Mg^{2+}-tolerant cells was significantly lower than that in the wild-type cells either at 1 mM or 51 mM [Mg^{2+}]$_o$. In the absence of extracellular Na$^+$, however, there was no significant difference in [Mg^{2+}]$_i$ between the high Mg^{2+}-tolerant cells and wild-type cells.

To evaluate the rate of Mg^{2+} efflux in the high Mg^{2+}-tolerant and the wild-type cells, the rates of change in [Mg^{2+}]$_i$ upon reduction of [Mg^{2+}]$_o$ from 51 mM to 1 µM were compared between the high Mg^{2+}-tolerant cells and wild-type cells. In the presence of 150 mM extracellular Na$^+$, reduction of [Mg^{2+}]$_o$ to 1 mM caused a decrease in [Mg^{2+}]$_i$ with the average rate much greater in the high Mg^{2+}-tolerant cells than in the wild-type cells. The decrease in [Mg^{2+}]$_i$ was largely diminished in the absence of extracellular Na$^+$ in both cell types, indicating the Na$^+$-dependent Mg^{2+} extrusion activity.

To analyze the Na$^+$ dependence of the Mg^{2+} efflux, we measured the rate of decrease in [Mg^{2+}]$_i$ in the high Mg^{2+}-tolerant cells at various extracellular Na$^+$ concentration ([Na$^+$]$_o$). Extracellular Na$^+$ accelerated, in a concentration-dependent manner, the rate of decrease in [Mg^{2+}]$_i$ induced by reduction of [Mg^{2+}]$_o$ from 51 mM to 1 mM (Figure 24.2). The data were explained by a Hill-type curve with a Hill coefficient of about 2 and half activation at 25 mM [Na$^+$]$_o$.

We also examined the effects of imipramine, a known inhibitor of the Na$^+$/Mg^{2+} exchange in erythrocytes[16,17] and cardiac myocytes,[18,19] on the rate of decrease in [Mg^{2+}]$_i$ at 150 mM [Na$^+$]$_o$. Imipramine slowed the decrease in [Mg^{2+}]$_i$ in a concentration-dependent manner with a half-inhibitory concentration between 50 to 200 µM, the range roughly comparable to reported IC$_{50}$ values of the agent for putative Na$^+$/Mg^{2+} exchange in red blood cells and cardiac myocytes.[16,17,19]

The Na$^+$/Mg^{2+} exchange activity is observed in the wild-type MCT cells, and the transport probably serves as a physiological Mg^{2+} extrusion pathway. In the high Mg^{2+}-tolerant cells, the greatly enhanced activity of the active Mg^{2+} transport prevents increase in [Mg^{2+}]$_i$ to high levels and is likely responsible for the Mg^{2+} tolerance. This mutant cell line may be useful to identify Mg^{2+}

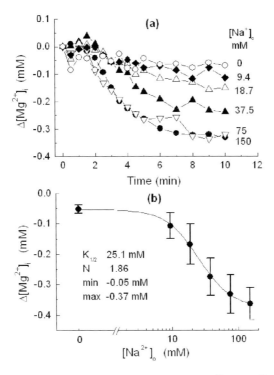

FIGURE 24.2. Effect of extracellular Na$^+$ on the rate of changes in $[Mg^{2+}]_i$ ($\Delta[Mg^{2+}]_i$) in the high Mg^{2+}-tolerant cells. (a) Results from a series of experiments carried out on the same subculture. Extracellular Mg^{2+} was reduced from 51 mM to 1 mM at time zero in the presence of various Na$^+$ concentrations: 150 mM (●), 75 mM (▽), 37.5 mM (▲), 18.8 mM (▼), 9.4 mM (◆), and 0 mM (○). Each symbol represents data obtained from a different cell cluster. (b) $\Delta[Mg^{2+}]_i$ values measured in six series of experiments were plotted as a function of $[Na^+]_o$. Each symbol represents a mean ± standard deviation (SD) of four cell clusters. A smooth line indicates the least squares fit by a Hill-type curve with parameters shown in the panel: $\Delta[Mg^{2+}]_i = \min + (\max - \min) \times [Na^+]_o^N/(K_{1/2}^N + [Na^+]_o^N)$, where min and max denote, respectively, $\Delta[Mg^{2+}]_i$ values at 0 and saturating $[Na^+]_o$, N is the Hill coefficient, and $K_{1/2}$ is $[Na^+]_o$ that gives a midpoint value of $\Delta[Mg^{2+}]_i$ between min and max. (Source: Reproduced from Watanabe et al.[15] by copyright permission of the American Physiological Society.)

transporter molecules and to understand molecular mechanisms of Na$^+$/Mg^{2+} exchange.[15]

Characteristics of Na$^+$/Mg^{2+} Exchange

Because significant leakage of the indicator from MCT cells hampered long-term measurements of $[Mg^{2+}]_i$ in clusters of the small number of cells, we further studied detailed characteristics of the Na$^+$/Mg^{2+} exchange transport using cardiac ventricular myocytes as a model system of the transcellular magnesium transport.[20,21] Cardiac myocytes are advantageous for accurate measurements of $[Mg^{2+}]_i$ with optical signals, because the indicator fluorescence signals with good signal-to-noise ratio can be stably measured from single cells for a period of many hours, and fluorescence intensity ratios of the indicator can be converted to $[Mg^{2+}]_i$ using the calibration curve constructed in the myocytes.[22] After the myocytes were loaded with Mg^{2+} by exposure to high $[Mg^{2+}]_o$, reduction of $[Mg^{2+}]_o$ to $1\,mM$ (in the presence of extracellular Na^+) induced a decrease in $[Mg^{2+}]_i$. The rate of decrease in $[Mg^{2+}]_i$ was higher at higher initial $[Mg^{2+}]_i$. We further calculated changes in total magnesium concentration to obtain quantitative information on Mg^{2+} flux across the cell membrane by utilizing concentration and dissociation constant values of known cytoplasmic Mg^{2+} buffers. The calculations indicated that, in the presence of extracellular Na^+, Mg^{2+} efflux is markedly activated by cytoplasmic Mg^{2+} with a half-maximal activation of about $1.9\,mM$.[20] Thus, the Na^+/Mg^{2+} exchange activity is negligible at basal $[Mg^{2+}]_i$, and is strongly activated by small elevations of $[Mg^{2+}]_i$. On the other hand, raising $[Mg^{2+}]_o$ slowed the decrease in $[Mg^{2+}]_i$ with 50% reduction of the rate at about $10\,mM$ $[Mg^{2+}]_o$.[20]

Dependence of the Na^+/Mg^{2+} exchange on intracellular and extracellular concentrations of Na^+ was also studied in cardiac myocytes.[21] The myocytes were loaded with Mg^{2+} by exposure to high $[Mg^{2+}]_o$ either in the presence of $106\,mM$ Na^+ plus $1\,mM$ ouabain (Na^+ loading) or in the presence of only $1.6\,mM$ Na^+ to deplete the cells of Na^+ (Na^+ depletion). The initial rate of decrease in $[Mg^{2+}]_i$ from the Mg^{2+}-loaded myocytes were compared between the cells loaded with Na^+ and those depleted with Na^+. Average $[Na^+]_i$, when estimated from fluorescence signals of the Na^+ indicator SBFI, increased from $12\,mM$ to $31\,mM$ and $47\,mM$ after Na^+ loading for 1 and 3h, respectively, and decreased to near $0\,mM$ after 3h of Na^+ depletion. The intracellular Na^+ loading significantly reduced the initial rate of decrease in $[Mg^{2+}]_i$ on average, by 40% at 1h and by 64% at 3h, suggesting that the Mg^{2+} transport was inhibited by intracellular Na^+ with 50% inhibition of the rate at about $40\,mM$. Extracellular Na^+ concentration dependence of the rate of Mg^{2+} efflux revealed that the Mg^{2+} transport was activated by extracellular Na^+ with half-maximal activation at $55\,mM$, a value somewhat higher than that obtained in the MCT cells.[15] Thus, the Mg^{2+} extrusion transport critically depends on extracellular Na^+ for its activity, and elevated intracellular Na^+ can strongly inhibit the transport.[21] These results are consistent with the putative Na^+/Mg^{2+} exchange as the primary transport mechanism for active extrusion of cellular Mg^{2+}.

Regarding the stoichiometry of the exchange, recent experiments estimated 1 Na^+/1 Mg^{2+} (37°C)[18,23] or 1 to 2 Na^+/1 Mg^{2+} (25°C).[24] However, more complex stoichiometries involving ions other than Na^+ and Mg^{2+} have also been proposed. First, based on the Ca^{2+} dependence of noradrenaline-induced Mg^{2+} efflux observed in rat ventricular myocytes, Romani and colleagues[25] postulated that the Na^+/Ca^{2+} exchanger might play a role in the Mg^{2+} transport. Tashiro and colleagues[26] experimentally showed that Na^+-dependent Mg^{2+} transport activity developed after overexpression of the Na^+/Ca^{2+} exchanger in CCL39 cells, suggesting that Mg^{2+} is transported by the Na^+/Ca^{2+} exchanger at least under some experimental conditions. Later quantitative analysis, however, raised the question of the physiological significance of the Na^+/Ca^{2+} exchanger for cellular Mg^{2+} transport.[27] Furthermore, Rasgado-Flores and coworkers[28] found that K^+ and Cl^- were involved in extracellular Mg^{2+}-dependent Na^+ efflux in squid giant axons, and proposed the putative Na^+-K^+-Cl^-/Mg^{2+} exchanger that carries 1 Mg^{2+} in exchange for 2 Na^+, 2 K^+, and 2 Cl^-.[29]

To explore possible roles of the Na^+/Ca^{2+} exchanger and the putative Na^+-K^+-Cl^-/Mg^{2+} exchanger, we tested if the rate of Mg^{2+} efflux was modified by intracellular and extracellular concentrations of Ca^{2+} (expected for the Na^+/Ca^{2+} exchanger)[30] and K^+ and Cl^- (expected for the Na^+-K^+-Cl^-/Mg^{2+} exchanger).[31] We measured the $[Mg^{2+}]_i$ of rat ventricular myocytes with the fluorescent indicator furaptra, and estimated the initial rate of Mg^{2+} efflux, as described above, with varied intracellular and extracellular concentrations of Ca^{2+}, K^+, and Cl^-. The initial rate of Mg^{2+} efflux was essentially unchanged by the addition of extracellular Ca^{2+} up to $2\,mM$.[30] Heavy intracellular loading of a Ca^{2+} chelator, either BAPTA or dimethyl BAPTA, by incubation with its acetoxy-methyl ester form ($5\,\mu M$ for 3.5h) did not significantly change the initial rate of Mg^{2+} efflux.[30]

Extracellular or intracellular, or both, concentrations of K^+ and Cl^- were modified, while membrane potential was set at $-13\,mV$ with amphotericin B-perforated patch clamp technique.[31] None of the following conditions significantly changed the efflux rate: (1) changes in $[K^+]_o$ between $0\,mM$ and $75\,mM$; (2) intracellular perfusion with K^+-free (Cs^+-substituted) solution from the patch pipette in combination with removal of extracellular K^+; or (3) extracellular and intracellular perfusion with K^+-free and Cl^--free solutions.[31] It is thus likely that Mg^{2+} is transported in exchange with Na^+, but not with Ca^{2+}, K^+ or Cl^-, in mammalian cardiac myocytes.

The effects of intracellular and extracellular concentrations of Na^+, Mg^{2+}, Ca^{2+}, K^+ and Cl^- have been studied (Table 24.1). Among these ion species, the

TABLE 24.1. Effects of extracellular and intracellular ions on Na^+/Mg^{2+} exchange (25°C).

Ion	Extracellular	Intracellular
Na^+	Required for activity[15,21]	Inhibition[21]
	($K_a = 25$–$55\,mM$)	($K_i = 40\,mM$)
K^+	No effect[31]	No effect[31]
	(0–$75\,mM$)	(Cs^+ replacement)
Mg^{2+}	Inhibition[20]	Required for activity[20]
	($K_i = 10\,mM$)	($K_a = 1.9\,mM$)
Ca^{2+}	No effect[30]	No effect[30]
	(0–$2\,mM$)	(Inhibition by overload?)
Cl^-	No effect[31]	No effect[31]
	(0–$142\,mM$)	(Ms^- replacement)

Abbreviations: K_a, concentration for half-maximal activation; K_i, concentration for 50% inhibition; Ms, methanesulfonate.

Mg^{2+} extrusion appears to require only intracellular Mg^{2+} and extracellular Na^+ for its activity, while clear inhibition of the Mg^{2+} transport is associated only with extracellular Mg^{2+} and intracellular Na^+. These results do not support complex exchange stoichiometries that involve ion species other than Na^+ and Mg^{2+} in mammalian cardiac myocytes. The simple Na^+/Mg^{2+} exchanger appears to be functionally distinct from any other transporters so far identified in mammalian cells. Further studies are needed to determine the exchange stoichiometry of Na^+ and Mg^{2+}.

Conclusions

- It has been shown that TRPM6 and TRPM7 channels open when $[Mg^{2+}]_i$ decreases to low levels,[8,11] and that Na^+/Mg^{2+} exchange is activated when $[Mg^{2+}]_i$ increases above its basal level.[20] It is thus tempting to speculate that $[Mg^{2+}]_i$ plays a pivotal role as a regulator of Mg^{2+} re-absorption in renal tubules; Mg^{2+} influx through apical TRPM channels raises $[Mg^{2+}]_i$, which then activates the Na^+/Mg^{2+} exchanger to extrude Mg^{2+} through the basolateral membrane. However, more quantitative information needs to be accumulated to describe how TRPM channels and the Na^+/Mg^{2+} exchanger constitute a physiological Mg^{2+} re-absorption pathway.

Acknowledgments. The authors are indebted to Prof. J. Patrick Barron, International Medical Communications Center of Tokyo Medical University, for his review of this manuscript. This study was supported by the High-Tech Research Center Project for Private Universities; matching fund subsidy from Ministry of Education, Culture, Sports, Science and Technology, 2003–2007.

References

1. Schlingmann KP, Konrad M, Seyberth HW. Genetics of hereditary disorders of magnesium homeostasis. *Pediatr Nephrol* 2004;19:13–25.
2. Schlingmann KP, Gudermann T. A critical role of TRPM channel-kinase for human magnesium transport. *J Physiol* 2005;566:301–308.
3. Quamme GA, de Rouffignac C. Epithelial magnesium transport and regulation by the kidney. *Front Biosci* 2000;5:d694–d711.
4. Simon DB, Lu Y, Choate KA, et al. Paracellin-1, a renal tight junction protein required for paracellular Mg^{2+} reabsorption. *Science* 1999;285:103–106.
5. Fleig A, Penner R. The TRPM ion channel superfamily: molecular, biophysical and functional features. *Trends Pharmacol Sci* 2004;25:633–639.
6. Schlingmann KP, Weber S, Peters M, et al. Hypomagnesemia with secondary hypocalcemia is caused by mutations in *TRPM6*, a new members of the *TRPM* gene family. *Nat Genet* 2002;31:166–170.
7. Walder RY, Landau D, Meyer P, et al. Mutations of TRPM6 causes familial hypomagnesemia with secondary hypocalcemia. *Nat Genet* 2002;31:171–174.

8. Voets T, Nilius B, Hoefs S, et al. TRPM6 forms the Mg^{2+} influx channel involved in intestinal and renal Mg^{2+} reabsorption. *J Biol Chem* 2004;279:19–25.
9. Runnels LW, Yue L, Clapham DE. TRP-PLIK, a bifunctional protein with kinase and ion channel activities. *Science* 2001;291:1043–1047.
10. Schmitz C, Perraud A-L, Johnson CO, et al. Regulation of vertebrate cellular Mg^{2+} homeostasis by TRPM7. *Cell* 2003;114:191–200.
11. Nadler MJS, Hermosura MC, Inabe K, et al. LTRPCR is a Mg-ATP-regulated divalent cation channel required for cell viability. *Nature* 2001;411:590–595.
12. Chubanov V, Waldegger S, Mederos y S, et al. Disruption of TRPM6/TRPM7 complex formation by a mutation in the *TRPM6* gene causes hypomagnesemia with secondary hypocalcemia. *Proc Natl Acad Sci U S A* 2004;101:2894–2899.
13. Flatman PW. Mechanisms of magnesium transport. *Annu Rev Physiol* 1991;53: 259–271.
14. Romani A, Scarpa A. Regulation of cellular magnesium. *Front Biosci* 2000;5: D720–D734.
15. Watanabe M, Konishi M, Ohkido I, Matsufuji S. Enhanced sodium-dependent extrusion of magnesium in mutant cells established from a mouse renal tubular cell line. *Am J Physiol* 2005;289:F742–F748.
16. Feray JC, Garay R. Demonstration of a $Na^+:Mg^{2+}$ exchange in human red cells by its sensitivity to tricyclic antidepressant drugs. *Arch Pharmacol* 1988;338:332–337.
17. Flatman PW, Smith LM. Magnesium transport in ferret red cells. *J Physiol* 1990;431:11–25.
18. Handy RD, Gow IF, Ellis D, Flatman PW. Na-dependent regulation of intracellular free magnesium concentration in isolated rat ventricular myocytes. *J Mol Cell Cardiol* 1996;28:1641–1651.
19. Tashiro M, Konishi M. Sodium gradient-dependent transport of magnesium in rat ventricular myocytes. *Am J Physiol* 2000;279:C1955–C1962.
20. Tursun P, Tashiro M, Konishi M. Modulation of Mg^{2+} efflux from rat ventricular myocytes studied with the fluorescent indicator furaptra. *Biophys J* 2005;88: 1911–1924.
21. Tashiro M, Tursun P, Konishi M. Intracellular and extracellular concentrations of Na^+ modulate Mg^{2+} transport in rat ventricular myocytes. *Biophys J* 2005;88:3235–3247.
22. Watanabe M, Konishi M. Intracellular calibration of the fluorescent Mg^{2+} indicator furaptra in rat ventricular myocytes. *Pflugers Arch* 2001;442:35–40.
23. Almulla HA, Bush PG, Steele MG, Flatman PW, Ellis D. Sodium-dependent recovery of ionised magnesium concentration following magnesium load in rat heart myocytes. *Pflugers Arch* 2006;451:657–667.
24. Tshiro M, Tursun P, Miyazaki T, Watanabe M, Konishi M. Effects of membrane potential on Na^+-dependent Mg^{2+} extrusion from rat ventricular myocytes. *Jpn J Physiol* 2002;52:541–551.
25. Romani A, Marfella C, Scarpa A. Regulation of magnesium uptake and release in the heart and in isolated ventricular myocytes. *Circ Res* 1993;72:1139–1148.
26. Tashiro M, Konishi M, Iwamoto T, Shigekawa M, Kurihara S. Transport of magnesium by two isoforms of the Na^+-Ca^{2+} exchanger expressed in CCL39 fibroblasts. *Pflugers Arch* 2000;440:819–827.
27. Konishi M, Tashiro M, Watanabe M, Iwamoto T, Shigekawa M, Kurihara S. Cell membrane transport of magnesium in cardiac myocytes and CCL39 cells

expressing the sodium-calcium exchanger. In: Rayssiguier Y, Mazur A, Durlach J, eds. *Advances in Magnesium Research: Nutrition and Health*. London: John Libbey; 2001:53–57.

28. Rasgado-Flores H, Gonzalez-Serratos H, DeSantiago J. Extracellular Mg^{2+}-dependent Na^+, K^+, and Cl^- efflux in squid giant axons. *Am J Physiol* 1994;266: C1112–C1117.
29. Rasgado-Flores H, Gonzalez-Serratos H. Plasmalemmal transport of magnesium in excitable cells. *Front Biosci* 2000;5:d866–d879.
30. Tashiro M, Tursun P, Konishi M. Effects of intracellular and extracellular Ca^{2+} on Mg^{2+} efflux from rat ventricular myocytes. *Jpn J Physiol* 2005;55:S76.
31. Tashiro M, Tursun P, Miyazaki T, Watanabe M, Konishi M. Effects of intracellular and extracellular concentrations of Ca^{2+}, K^+, and Cl^- on the Na^+-dependant Mg^{2+} efflux in rat ventricular myocytes. *Biophys J* 2006;91:244–254.

25
Magnesium in Chronic Renal Failure

Juan F. Navarro and Carmen Mora-Fernández

Magnesium (Mg) is the fourth most abundant cation in the body. An average adult contains approximately 25 g (2000 mEq) of Mg.[1,2] Of the body's Mg, the vast majory of this ion is in the intracellular compartment (99%), and the remaining 1% is in the extracellular fluid.[3] The principal site of intracellular Mg is bone (60%–65%) where two distinct pools, cortical and trabecular, have been described. It is thought that Mg forms a surface constituent of the hydroxyapatite mineral component. Initially, much of this magnesium is readily exchangeable with serum and therefore represents a moderately accessible magnesium store, which can be drawn on in times of deficiency. Approximately 25% to 30% is localized within the skeletal muscle, and about 10% to 15% in other non-muscle soft tissues.[4,5]

The majority of the intracellular Mg is bound to several chelators, such as citrate, proteins, adenosine diphosphate (ADP) and triphosphate (ATP), RNA, and DNA. Only 5% to 10% is free, which is essential for the regulation of the intracellular Mg content.[6] The normal total plasma Mg concentration ranges from 0.62 to 1.02 mmol/L, where about 5% to 10% is complexed as salts (bicarbonate, citrate, phosphate, sulphate), 30% is protein bound, and 60% is present as free Mg ions, the biologically active form.[7,8] Magnesium functions as a cofactor of many enzymes involved in energy metabolism, protein synthesis, RNA and DNA synthesis, and maintenance of the electrical potential of nervous tissues and cell membranes.

Magnesium Homeostasis

Magnesium balance, like that of other ions, is a function of intake and excretion (Figure 25.1). Magnesium is widely distributed in plant and animal foods, and geochemical and other environmental variables rarely have a major influence on its content in foods. Most green vegetables, legume seeds, peas, beans, and nuts are rich in magnesium, as are some shellfish, spices, and soya flour, all of which usually contain more than 500 mg/kg fresh weight. Although most unrefined cereal grains are reasonable sources, many highly refined flours,

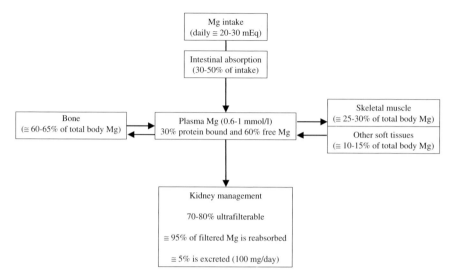

FIGURE 25.1. Schematic representation of magnesium homeostasis.

tubers, fruits, and fungi and most oils and fats contribute little dietary magnesium (<100 mg/kg fresh weight). Corn flour, cassava and sago flour, and polished rice flour have an extremely low magnesium content. The average daily intake of magnesium is approximately 20 to 30 mEq (240–365 mg), and in order to maintain an optimal balance, a daily intake of 0.5 to 0.7 mEq/kg is critical.

Negligible amounts of Mg are absorbed in the stomach. The major site of digestive absorption of Mg is the small intestine, particularly in the proximal portions, but also occurs in the illeum as well as in the colon.[9,10] Absorption may involve active transport, facilitated diffusion, and a passive process by the electrochemical gradient and the solvent drag.[11] Under normal conditions, 30% to 50% of the ingested Mg is absorbed,[12] but gastrointestinal absorption can adapt to altered intake. Thus, under low Mg intake approximately up to 80% of Mg can be absorbed,[9] while only 25% is absorbed when intake is high.[13]

Scarce investigations have analyzed the intestinal absorption of Mg in chronic renal failure (CRF), with great controversy in the results. Early reports indicated that there were not significant differences in net Mg absorption between normal subjects and CRF patients.[14,15] Randall[16] reported only slightly reduced absorption of Mg in chronic renal disease, while other authors found average net absorption rates ranging between 17% to 38%.[17,18] Strictly controlled balance studies showed that the average net absorption of Mg in patients with CRF was significantly lower than the absorption in subjects with normal renal function,[12] results that have been confirmed using radioisotope techniques, even when age was taken into consideration.[19] Factors determining the reduced intestinal Mg absorption are not completely known. However, the absorption of Mg in the jejunum has been demonstrated to be dependent on vitamin D, and that administration of 1-25-dihydroxyvitamin D_3 to patients with CRF resulted in an increased intestinal absorption of Mg.[20] Furthermore, successful renal transplantation is able to restore the impaired Mg absorption in these patients.[16] Therefore, deficient synthesis of the active metabolite of vitamin D by the kidneys may play a significant causative role in Mg malabsorption in CRF.

Another aspect of interest regarding Mg absorption in the context of renal disease is the effect of protein intake, because protein restriction is widely recommended in patients with renal insufficiency. Early works by McCance and colleagues[21] demonstrated that diets containing 0.7 to 0.9 g of protein/kg body weight were associated with a magnesium absorption of 32%, whereas when the diets contained 2.3 to 2.6 g of protein/kg body weight, magnesium absorption was 41%. These initial results were confirmed by later studies, indicating that when intake of protein is low, the intestinal absorption of Mg is reduced. All these studies were performed in subjects with normal renal function, and therefore, no data are available about the relationship between protein intake and intestinal Mg absorption in patients with CRF.

The kidney plays a critical role in Mg homeostasis and in determining the concentration of Mg in the extracellular fluid. Approximately 70% to 80% of plasma Mg is ultrafilterable, with about 2 g of Mg filtered daily. Under normal circumstances, 95% of the filtered load of Mg is re-absorbed by the kidney, and only 5% (100 mg) is excreted in the urine.[1,22] Magnesium transport differs from that of most other ions in that the proximal tubule is not the major site of re-absorption, because only 15% to 25% of the filtered Mg is re-absorbed in the proximal tubule. Micropuncture studies have showed that most of this re-absorption (60%–70%) occurs at the thick ascending limb of the loop of Henle, whereas the distal tubules and collecting ducts play a limited role in Mg handling.[23,24] Plasma Mg concentration is a major determinant of the renal handling. During significant hypermagnesemia, the kidney can excrete up to 70% of the filtered load. In contrast, during Mg deficiency the kidney can compensate by reducing the Mg excreted in the urine to less than 0.5% of the filtered load.[25] Furthermore, active re-absorption of magnesium has been reported to be influenced by acid–base balance.[26] Thus, diverse studies have showed that dietary changes which result in increased urinary pH and decreased titratable acidity are associated with reduced urinary magnesium output by 35%, despite marked increases in dietary magnesium input from vegetable protein diets.[27]

In CRF, although urinary Mg excretion may be normal or even increased in many patients with glomerular filtration rate (GFR) lower than 30 mL/min, it is usually low in advanced renal insufficiency.[28] As renal function deteriorates, the fraction of filtered Mg excreted increases as a result of impaired tubular re-absorption, which becomes more marked when GFR is below 10 mL/min. However, this compensatory rise of fractional Mg excretion is insufficient, and therefore the serum Mg concentration increases.

Another aspect of interest is that increased dietary calcium, particularly if associated with high sodium intakes, contributes to enhance urinary Mg output and to a shift toward negative Mg balance.[26,29] A close relationship between the urinary excretion of sodium and calcium and Mg has been also observed in patients with advanced CRF.[30] Thus, whereas in patients with polycystic kidney disease and a GFR higher than 20 mL/min there was no relationship between sodium and Mg or phosphate excretion, in patients with a GFR below 20 mL/min, the excretion of calcium, Mg, and phosphate was significantly related to that of sodium. These findings suggest that impairment of tubular re-absorption of these electrolytes is the result of a common mechanism.[31]

Serum Magnesium Levels and Tissue Magnesium Content in Chronic Renal Failure

Because renal capacity for Mg elimination is impaired due to the reduction of the net Mg filtration, intestinal absorption will always exceed renal excretion despite a compensatory increase in the fractional excretion of Mg as renal function deteriorates. Thus, dietary Mg intake is a major determinant of serum and total body Mg levels because no other alternative routes for Mg excretion are available in patients with CRF.

Until severe reductions of GFR (<30 mL/min, chronic kidney disease stage 4), serum Mg levels are usually normal.[32,33] With lower rates of renal function, serum Mg is increased, with a greater rise of total one than that of ionic Mg. The typical patient with end-stage renal disease has a plasma Mg concentration between 1 to 1.5 mmol/L, but severe, symptomatic, and even fatal hypermagnesemia can be induced by the exogenous administration of antacids or laxatives containing Mg in usual therapeutic doses.[34,35] Other situations that have been reported to be produce a rise in serum Mg are the administration of thiazide diuretics [an effect attributed to extrarenal mechanisms, such as the potentiation of parathyroid hormone (PTH) action on bone],[36] and treatment with 1-α-hydroxyvitamin D_3.[37]

Hypermagnesemia (serum Mg >1.02 mmol/L) is not infrequent in dialysis population. Saha and colleagues[38] found that both total and ionized serum Mg were significantly higher in hemodialysis patients compared to control subjects. In patients of our unit dialyzed with a standard dialysate Mg (0.75 mmol/L), the mean serum Mg concentration was 1.15 mmol/L, and moreover, 68% of subjects had hypermagnesemia.[39] In uremic patients on hemodialysis the serum Mg concentration parallels the dialysate Mg level because Mg readily crosses the dialysis membrane with its movement determined by the gradient between the concentration of diffusible Mg in blood and the level of Mg in dialysate. Several authors have reported that a reduction in the Mg dialysate produces a significant decrease in the serum Mg concentration as early as the first month.[40–42] In contrast, in another study, changing the dialysate Mg from

0.75 to 1.5 mmol/L the mean serum Mg concentration increased from 1.25 to 1.7 mmol/L.[43] These data demonstrate that dialysate Mg plays a critical role in maintaining magnesium homeostasis in end-stage renal disease patients treated with hemodialysis.[44]

In peritoneal dialysis (PD) patients, elevated serum Mg levels have been also reported,[45] which largely depends on the concentration of the ion in the dialysate. Ahsan Ejaz and colleagues[46] studied the evolution of serum Mg concentration in 33 patients dialyzed with a low Mg PD solution (0.5 mmol/L). All subjects had serum Mg levels in the normal or elevated range prior to institution of PD (0.65–1.15 mmol/L). However, after dialysis with a low Mg fluid, 64% of them developed hypomagnesemia. Similar findings have been recently reported by Hutchinson and coworkers.[47] These authors found that the serum Mg concentration in 12 patients on dialysis with a standard PD fluid (Mg content, 0.75 mmol/L) was 1.24 mmol/L. However, 1 month after changing to a reduced Mg formulation (0.25 mmol/L), the mean serum Mg decreased to 0.89 ± 0.04 mmol/L ($p < 0.001$). These data are in according with initial studies in continuous ambulatory PD patients showing that mild hypermagnesemia was common when Mg concentration in the dialysate was 0.75 mmol/L.[48]

Results about tissue Mg content in uremic patients have shown conflicting results. Evaluation of the concentration of Mg in different tissues, such as skeletal muscle, has been reported to be either low, normal, or elevated.[33] On the other hand, normal Mg levels have been noted in lymphocytes of hemodialysis patients, in spite of enhanced serum Mg concentrations.[49] On the contrary to these findings, increased bone Mg content is an universal finding, being this excess of Mg distributed in both the rapidly exchangeable and the nonexchangeable pools.[4,33] Contiguglia and coworkers[50] reported a mean increase of 66% in the Mg content of bone.

Consequences of Hypermagnesemia: Implications for Renal Patients

Mild hypermagnesemia (plasma Mg lower than 1.5 mmol/L) is usually asymptomatic. When plasma Mg is between 2 to 3 mmol/L, drowsiness, lethargy, and diminished deep tendon reflexes can be observed. The most consistent complication of hypermagnesemia is neuromuscular toxicity, which is due to a reduction in the impulse transmission across the neuromuscular junction, and decreased tendon reflexes is the initial clinical manifestation.[51] Hypermagnesemia with serum Mg level between 3 to 5 mmol/L can result in somnolence, loss of deep tendon reflexes, hypocalcemia, and cardiovascular effects including hypotension, bradycardia, and electrocardiographic changes such as prolongation of the P-R interval, increase in QRS duration, and Q-T interval.[52] Finally, extreme hypermagnesemia with plasma Mg concentration above 5 mmol/L is associated with muscle paralysis, potentially leading to flaccid quadriplegia and apnea, complete heart block, and cardiac arrest.

TABLE 25.1. Consequences of hypermagnesemia and potential effects in chronic renal failure.

Mild hypermagnesemia
    Drowsiness
    Lethargy
    Hyporeflexia
Moderate hypermagnesemia
    Somnolence
    Areflexia
    Hypocalcemia
    Hypotension
    Bradycardia
    Electrocardiographic changes (prolongation of P-R and Q-T interval, increase in QRS duration)
Severe hypermagnesemia
    Muscle paralysis
    Quadraplegia
    Apnea
    Complete heart block
    Cardiac arrest
Potential effects in chronic renal failure
    Retarding vascular calcifications
    Reduction of nerve conduction velocity
    Increased pruritus
    Mineralization defects
    Contribution to osteomalacic renal osteodystrophy
    Bone pain
    Impairment of parathyroid gland function
    Pathogenic factor for adynamic bone disease

In patients with CRF, hypermagnesemia has been implicated in both deletereous and beneficial effects (Table 25.1).[44] Purposed benefits of hypermagnesemia included supression of PTH[53] and retarding vascular calcifications.[54] In contrast, potential harmful effects include altered nerve conduction velocity[55] and increased pruritus.[56] Initial studies reported that dialysis against a low Mg dialysate had beneficial effects on uremic itch,[57] although later prospective double-blind controlled trials failed to show any improvement of pruritus after Mg-free dialysis.[58] However, the most significant consequences of elevated Mg levels in renal patients have been related to osseous metabolism.

Magnesium, Bone Disease, and Parathyroid Hormone in Chronic Renal Failure

Progressive CRF is accompanied almost universally by abnormalities in bone structure, with multiple factors involved in the pathogenesis of this disorder. Based on the histological characteristics, bone disease in patients with renal insufficiency can be divided in two broad groups: (1) disorders characterized by normal or increased bone turnover, such as osteitis fibrosa, mixed bone pathology and other forms of hyperparathyroid bone disease; and (2) disorders characterized by low rates of bone remodelling, such as osteomalacia and adynamic bone disease.[59,60]

Magnesium ions are not incorporated directly into the apatite mineral structure of bone, but are concentrated in the hydration shell around hydroxyapatite crystals forming part of the surface-bound ion complex.[44,61] This physicochemical property of Mg influences the release of Mg from bone and its availability when Mg absorption is restricted.[61] Magnesium concentration in the physiological level (Mg/Ca ratio between 0.004–0.04) do not interfere with the crystallization of amorphus calcium phosphate to hydroxyapatite.[62]

Since the early 1970s, a high bone Mg content in patients with renal disease has been consistently reported by different authors.[4,33,50] In 1986, Brautbar and Gruber outlined the hypothesis that increased bone Mg arises from elevated serum Mg, and that increased bone Mg plays a direct role in mineralization defects and the pathogenesis of renal osteodystrophy.[61] Thus, high bone Mg content has been implicated in the pathogenesis of osteomalacic renal osteodystrophy.[63,64] In vitro studies have demonstrated that bone Mg excess can inhibit the formation of hydroxyapatite crystals by two mechanisms: (1) by interference with calcium, which produces a decrease in the calcium concentration in the mediating solution and reduces the amorphus calcium phosphate solubility[61,62,65,66]; and (2) by combination of Mg with pyrophosphate, which results in an insoluble salt that is resistant to hydrolysis by pyrophosphatase.[44,67] Furthermore, supportive data on the potential noxius effects of an increased bone Mg content are also suggested by the beneficial effects of a reduction in the dialysate Mg concentration. Brunner and Thiel[63] observed the resolution of clinical symptoms, such as bone pain, as well as healing of osteomalacic fractures in dialysis patients after reduction of dialysate Mg level. Moreover, Gonella and coworkers[64] reported a significant improvement in bone histology by normalizing serum Mg after 1 year in hemodialysis with a low Mg dialysate.

Besides these direct effects of increased Mg on bone histology, another aspect of special interest is the influence of hypermagnesemia on parathyroid gland function. It is well established that PTH production is stimulated by hypocalcemia and decreased by elevated serum calcium levels, with an inverse correlation between serum calcium and PTH concentrations in CRF patients. However, conflicting data have been reported about the relationship between Mg and PTH. Investigations in animal models[53] and non-uremic patients[68,69] have showed evidence for supression of parathyroid gland function by hypermagnesemia. However, severe hypomagnesemia has been associated with profound hypocalcemia in patients with CRF, which has been commonly attributed to impaired PTH release and/or peripheral resistance to its action.[70-72] In this situation, Mg replacement was associated in most cases with increase in PTH levels, despite a concomitant increase in serum calcium.[70,72] Important information has been obtained from dialysis patients, as well as uremic patients treated with Mg salts.

Kenny and colleagues[73] suggested that a low Mg dialysate might be expected to have a suppressive effect on parathyroid gland function. However, most of studies failed to confirm this hypothesis. Several trials on the successful use of Mg salts to control hyperphosphatemia and hypeparathyroidism[74–77] support the hypothesis that hypermagnesemia may have a suppressive effect on PTH secretion in dialysis patients.[78,79] O'Donovan and coworkers[74] compared the evolution of patients treated with Mg carbonate respect to a control group on therapy with aluminum hydroxide. After 2 years, the authors observed a significant reduction in the mean serum PTH concentration in patients receiving Mg salts, but not in controls. Morinière and coworkers[75] evaluated the effects of the administration of Mg hydroxide in association with moderate doses of calcium carbonate. They found that plasma concentration of Mg progressively increased concomitantly with a progressive reduction of PTH. Delmez and colleagues[76] observed that the administration of Mg carbonate allowed a reduction in the doses of calcium carbonate without any increase in the serum PTH level. Furthermore, these findings are concordant with the variation of PTH after modifications of dialysate magnesium content. Parsons and colleagues[80] and Nilsson and colleagues[81] reported that the correction of raised serum Mg concentration in patients dialyzed against a low dialysate Mg fluid resulted in a significant increase in serum PTH. In contrast, MacGonicle and coworkers[82] observed that in hypermagnesemic patients a further rise in serum Mg caused a subsequent fall of PTH.

We have analyzed the serum Mg concentration in dialysis patients classified according their PTH level.[39,45,83] We observed that patients with inadequately low PTH (<120 pg/mL) had a serum Mg concentration significantly higher that patients with adequate (120–250 pg/mL) or high (>250 pg/mL) PTH. Moreover, we found a significant negative linear relationship between the serum concentrations of PTH and Mg. Similar findings have been reported by Saha and colleagues,[84] who also found a significant inverse correlation between intact PTH and serum Mg. Finally, this relationship between Mg and PTH is independent of other factors that can affect the PTH synthesis or secretion. Thus, plasma Mg has been demonstrated as a significant determinant of serum PTH concentration independently of calcium and phosphorus.[39,45]

In summary, serum Mg concentration is frequently elevated in CRF. In these patients, hypermagnesemia is usually mild, without relevant acute clinical symptoms, although further increased plasma Mg level may have relevant consequences, especially on bone metabolism. Hypermagnesemia can have direct effects on bone histology, playing a significant pathogenic role in osteomalacic renal osteodystrophy. On the other hand, diverses evidences have been reported for supression of parathyroid gland activity by hypermagnesemia. Further studies are necessary to clarify the mechanisms of action of hypermagnesemia on parathyroid gland function in uremic patients.

References

1. Slatopolsky E, Klahr F. Disorders of calcium, magnesium and phosphorus metabolism. In: Shrier RW, Gottschalk CW, eds. *Diseases of the Kidney.* Boston: Little Brown; 1988:2902–2920.
2. Sutton R, Dakhaee K. Magnesium balance and metabolism. In: Jacobson HR, Stirker GE, Klahr S, eds. *The Principles and Practice of Nephrology.* Philadelphia: B.C. Decker; 1991:136–139.
3. Lowenthal DT. Clinical pharmacology of magnesium chloride. In: Giles TD, Seelig MS, eds. *The Role of Magnesium Chloride Therapy in Clinical Practice.* Clifton, NJ: Oxford Health Care; 1998:9–10.
4. Alfrey AC, Miller NL. Bone magnesium pools in uremia. *J Clin Invest* 1973;52:3019–3023.
5. Webster PO. Magnesium. *Am J Clin Nutr* 1987;45:1305–1312.
6. Gunther T. Mechanisms of regulation of Mg2+ efflux and Mg2+ influx. *Miner Electrolyte Metab* 1993;19:259–265.
7. Speich M, Bousquet B, Nicolas G. Reference values for ionized, complexed, and protein-bound plasma magnesium in men and women. *Clin Chem* 1981;27:246–248.
8. Elin RJ. Magnesium metabolism in health and disease. *Dis Mon* 1988;34:165–218.
9. Graham LA, Caesar JJ, Burgen ASV. Gastrointestinal absorption and excretion of Mg^{28} in man. *Metabolism* 1960;9:646–659.
10. Brannan PG, Bergne-Marini P, Pak CYC, Hull AR, Fodtran JS. Magnesium absorption in the human small intestine. *J Clin Invest* 1976;57:1412–1418.
11. Kayne LH, Lee DBN. Intestinal magnesium absroption. *Miner Electrolyte Metab* 1993;19:210–217.
12. Spencer H, Lesniak M, Gatza CA, Osis D, Lender M. Magnesium absorption and metabolism in patients with chronic renal failure and in patients with normal renal function. *Gastroenterology* 1980;79:26–34.
13. Alfrey AC, Miller NL, Trow R. Effect of age and magnesium depletion on bone magnesium pools in rats. *J Clin Invest* 1974;54:1074–1081.
14. Clarkson EM, McDonald SJ, DeWardener HE, Warren R. Magnesium metabolism in chronic renal failure. *Clin Sci* 1965;28:107–115.
15. Kopple JD, Coburn JW. Metabolic studies of low protein diets in uremia. II. Calcium, phosphorus, and magnesium. *Medicine* 1973;52:597–607.
16. Randall RE Jr. Magnesium metabolism in chronic renal disease. *Ann N Y Acad Sci* 1969;162:831–846.
17. Mountokalakis TD, Virvidakis CE, Singhellakis PN, Alevizaki CC, Ikkos DC. Magnesium absorption in CRF. *Gastroenterology* 1981;80:632–635.
18. Spencer H, Osis D. Studies of magnesium metabolism in man. *Magnesium* 1988;7:271–280.
19. Mountokalakis TD, Virvidakis CF, Singhellakis PN, Alevizaki GG, Ikkos DG. Relationship between degree of renal failure and impairment of intestinal magnesium absorption. In: Cantin M, Seelig MS, eds. *Magnesium in Health and Disease.* New York: SP Medical and Scientific Books; 1980:453–458.
20. Schmulen AC, Lerman M, Pak CYC, et al. Effect of 1,25-$(OH)_2D_3$ on jejunal absorption of magnesium in patients with chronic renal failure. *Am J Physiol* 1980;238:G349–G352.

21. McCance RA, Widdowson EM, Lehmann H. The effect of protein intake on the absorption of calcium and magnesium. *Biochem J* 1942;36:686–691.
22. Quamme GA, Dirks JH. Magnesium metabolism. In: Maxwell MH, Kleeman CR, Narins RG, eds. *Disorders of Fluid and Electrolyte Metabolism*. New York: McGraw Hill; 1987:297–316.
23. Quamme GA, de Rouffignac C. Transport of magnesium in renal epithelial cells. In: Birch NJ, ed. *Magnesium and the Cell*. London: Academic; 1993:235–262.
24. Al-Ghamdi SM, Cameron EC, Sutton RA. Magnesium deficiency: pathophysiologic and clinical overview. *Am J Kidney Dis* 1994;24:737–752.
25. Lowenthal DT, Ruiz JG. Magnesium deficiency in the elderly. *Geriatric Urol Nephrol* 1995;5:105–111.
26. Quamme GA, Dirks JH. The physiology of renal magnesium handling. *Renal Physiol* 1986;9:257–269.
27. Hu JF, Zhao XH, Parpia B, Campbell TC. Dietary intakes and urinary excretion of calcium and acids: a cross-sectional study of women in China. *Am J Clin Nutr* 1993;58:398–406.
28. Popovtzer MM, Schainuck LI, Massry SG, Kleeman CR. Divalent ion excretion in chronic kidney disease. Relation to degree of renal insufficiency. *Clin Sci* 1970;38:297–307.
29. Kesteloot H, Joosens JV. The relationship between dietary intake and urinary excretion of sodium, potassium, calcium and magnesium. *J Hum Hypertens* 1990;4:527–533.
30. Popovtzer MM, Massry SG, Coburn JW, Kleeman CR. The interrelationship between sodium, calcium and magensium excretion in advanced renal failure. *J Lab Clin Med* 1969;73:763–771.
31. Martinez-Maldonado M, Yium JJ, Suki WN, Eknoyan G. Electrolyte excretion in polycystic kidney disease. Interrelationship between sodium, calcium, magnesium and phosphate. *J Lab Clin Med* 1977;90:1066–1075.
32. Kleeman CR, Better OS. Disordered divalent ion metabolism in kidney disease. Comments on pathogenesis and treatment. *Kidney Int* 1973;4:73–79.
33. Mountokalakis TD. Magnesium metabolism in chronic renal failure. *Magnes Res* 1990;3:121–127.
34. Zaman F, Abreo K. Severe hypermagnesemia as a result of laxative use in renal insufficiency. *South Med J* 2003;96:102–103.
35. Schelling JR. Fatal hypermagnesemia. *Clin Nephrol* 2000;53:61–65.
36. Kopple MH, Massry SG, Shinaberger JH, Hartenbower DL, Coburn JW. Thiazide-induced rise in serum calcium and magnesium in patients on maintenance hemodialysis. *Ann Intern Med* 1970;72:895–901.
37. Sörensen E, Tougaard L, Bröchner-Mortensen J. Iatrogenic magnesium intoxication during 1-α-hydroxycholecalciferol treatment. *Br Med J* 1976;2:215.
38. Saha H, Harmoinen A, Pietilä K, Mörsky P, Pasternack A. Measurement of serum ionized versus total levels of magnesium and calcium in hemodialysis patients. *Clin Nephrol* 1996;46:326–331.
39. Navarro J, Mora C, Macía M, García J. Serum magnesium concentration is an independent predictor of parathyroid hormone levels in peritoneal dialysis patients. *Perit Dial Int* 1999;19:455–461.
40. Gonella M, Moriconi L, Betti G, et al. Serum levels of PTH, Mg, Ca, inorganic phosphorus and alkaline phosphatase in uremic patients on different Mg dialysis. *Proc Eur Dial Transplant Assoc* 1980;17:362–366.

41. Nilsson P, Johansson SG, Danielson BG. Magnesium studies in hemodialysis patients before and after treatment with low dialysate magnesium. *Nephron* 1984;37:25–29.
42. Kancir CB, Wanscher M. Effect of magnesium gradient concentration between plasma and dialysate on magnesium variations induced by hemodialysis. *Magnesium* 1989;89:132–136.
43. Nair KS, Holdaway IM, Evans MC, Cameron AD. Influence of Mg on the secretion and action of parathyroid hormone. *J Endocrinol Invest* 1979;2:267–270.
44. Vaporean ML, Van Stone JC. Dialysate magnesium. *Semin Dial* 1993;6:46–51.
45. Navarro J, Mora C, Jiménez A, Torres A, Macía M, García J. Relationship between serum magnesium and parathyroid hormone levels in hemodialysis patients. *Am J Kidney Dis* 1999;34:43–48.
46. Ahsan Ejaz A, McShane AP, Gandhi VC, Leehey DJ, Ing TS. Hypomagnesemia in continous ambulatory peritoneal dialysis patients dialyzed with a low-magnesium peritoneal dialysis solution. *Perit Dial Int* 1995;15:61–64.
47. Hutchinson AJ, Were AJ, Boulton HF, Mawer EB, Laing I, Gokal R. Hypercalcaemia, hypermagnesaemia, hyperphosphataemia and hyperaluminaemia in CAPD: improvement in serum biochemistry by reduction in dialysate calcium and magnesium concentrations. *Nephron* 1996;75:52–58.
48. Blumenkrantz JM, Kopple JD, Moran JK, Coburn JW. Metabolic balance studies and dietary protein requeriments in patients undergoing continous ambulatory peritoneal dialysis. *Kidney Int* 1982;21:849–861.
49. Nilsson P, Johanson SG, Danielson BG. Magnesium studies in hemodialysis patients before and after treatment with low dialysate magenesium. *Nephron* 1984;37:25–29.
50. Contiguglia SR, Alfrey AC, Miller N, Butkus D. Total-body magnesium excess in chronic renal failure. *Lancet* 1972;ii:1300–1302.
51. Krender DA. Hypermagnesemia and neuromuscular transmission. *Semin Neurol* 1990;10:42–47.
52. Agus ZS, Morad M. Modulation of cardiac ion channels by magnesium. *Annu Rev Physiol* 1991;53:299–305.
53. Massry SG, Coburn JW, Kleeman CR. Evidence for suppression of parathyroid gland activity by hypermagnesemia. *J Clin Invest* 1970;49:1619–1629.
54. Meema HE, Oreopoulos DG, Rapoport A. Serum magnesium level and arterial calcification in end-stage renal disease. *Kidney Int* 1987;32:388–394.
55. Cisari C, Gasco P, Calabrese G, Pratesi G, Gonella M. Serum magnesium and nerve conduction velocity in uraemic patients on chronic haemodialysis. *Magnes Res* 1989;2:267–269.
56. Carmichael AJ, McHugh MM, Martin AM, Farrow M. Serlogical markers of renal itch in patients receiving lon term haemodialysis. *BMJ* 1988;296:1575.
57. Graf H, Kovarik J, Stummvell HK, Wolf A. Disappearance of uraemic pruritus after lowering dialysate Mg concentration. *BMJ* 1979;ii:1478–1479.
58. Carmichael AJ, Dickinson F, McHugh MI, Martin AM, Farrow M. Magnesium free dialysate for uraemic pruritus. *BMJ* 1988;297:1584–1585.
59. Felsenfeld AJ, Rodríguez M, Dunlay R, Llach F. A comparison of parathyroid-gland function in haemodialysis patients with different forms of renal osteodystrophy. *Nephrol Dial Transplant* 1991;6:244–251.
60. Sherrard D, Hercz G, Pei Y, et al. The spectrum of bone disease in end-stage renal failure—an evolving disorder. *Kidney Int* 1993;43:436–442.

61. Brautbar N, Gruber H. Magnesium and bone disease. *Nephron* 1986;44:1–7.
62. Blumenthal N, Posner A. Hydroxyapatite: mechanism of formation and properties. *Calcif Tissue Res* 1973;13:235–243.
63. Brunner FP, Thiel G. The use of magnesium-containing phosphate binders in patients with end-stage renal disease on maintenance haemodialysis. *Nephron* 1982;32:266.
64. Gonella M, Ballanti P, Della Roca C, et al. Improved bone morphology by normalizing serum magnesium in chronically hemodialyzed patients. *Miner Electrolyte Metab* 1988;14:240–245.
65. Termine J, Peckauskas R, Posner A. Calcium phosphate formation in vitro. II. Effects of environment on amorphous-crystalline transformation. *Arch Biochem Biophys* 1970;140:318–325.
66. Boskey A, Posner A. Effect of magnesium on lipid-induced calcification: an in vitro model for bone mineralization. *Calcif Tissue Int* 1980;32:139–143.
67. Drueke T. Does magnesium excess play a role in renal osteodystrophy? *Contrib Nephrol* 1984;38:195–204.
68. Ferment O, Garnier PE, Touitou Y. Comparison of the feedback effect of magnesium and calcium on parathyroid hormone secretion in man. *J Endocrinol* 1987;113:117–123.
69. Gough IR, Balderson GA, Lloyd HM, Galligan J, Willgoss D, Fryar BG. The effect of intravenous magnesium sulphate on parathyroid function in primary hyperparathyroidism. *World J Surg* 1988;12:463–469.
70. Mennes P, Rosenbaum R, Martin K, Slatopolski E. Hypomagnesemia and impaired parathyroid hormone secretion in chronic renal disease. *Ann Intern Med* 1978;88:206–209.
71. Lien JWK. Hypomagnesaemic hypocalcaemia in renal failure. *BMJ* 1978;iv:1400.
72. Miller PD, Krebs RA, Neal BJ, McIntyre DO. Hypomagnesemia. Supression of secondary hyperparathyroidism in chronic renal failure. *JAMA* 1979;241:953–955.
73. Kenny MA, Casillas E, Ahmad S. Magnesium, calcium and PTH relationship in dialysis patients after magnesium repletion. *Nephron* 1987;46:199
74. O'Donovan R, Baldwin D, Hammer M, Moniz C. Substitution of aluminum salts by magnesium salts in control of dialysis hyperphosphataemia. *Lancet* 1986;1:880–882.
75. Moriniere P, Vinatier I, Westeel P, et al. Magnesium hydroxide as a complementary aluminium-free phosphate binder to moderate doses of oral calcium in uraemic patients on chronic haemodialysis: lack of deleterious effect on bone mineralisation. *Nephrol Dial Transplant* 1988;3:651–656.
76. Delmez JA, Kelber J, Norword KY, Giles KS, Slatopolsky E. Magnesium carbonate as a phosphorus binder: a prospective, controlled, crossover study. *Kidney Int* 1996;49:163–167.
77. Parsons VA, Baldwin D, Moniz C, Marsden J, Ball E, Rifkin I. Successful control of hyperparathyroidism in patients on continuous ambulatory peritoneal dialysis using magnesium carbonate and calcium carbonate as phosphate binders. *Nephron* 1993;63:379–383.
78. Plekta P, Bernstein DS, Hampers CL, Merill JP, Sherwood LM. Relationship between magnesium and secondary hyperparathyroidism during long term hemodialysis. *Metabolism* 1974;23:619–624.

79. Oe PL, Lips P, Wan der Meulen J, De Bries PMJM, Van Bronswijk H, Donker ASM. Long term use of magnesium hydroxide as a phosphate binder in patients on hemodialysis. *Clin Nephrol* 1987;28:180–185.
80. Parsons V, Papapoulos E, Weston MJ, Tomlinson S, O'Riordan JLH. The long term effect of lowering dialysate magnesium on circulating parathyroid hormone in patients on regular hemodialysis therapy. *Acta Endocrinol (Copenhagen)* 1980; 93:455–460.
81. Nilsson P, Johansson SG, Danielson BG. Magnesium studies in hemodialysis patients before and after treatment with low dialysate magnesium. *Nephron* 1984;37:25–29.
82. McGonigle R, Weston M, Keenan J, Jackson D, Parsons V. Effect of hypermagnesemia on circulating plasma parathyroid hormone in patients on regular hemodialysis therapy. *Magnesium* 1984;3:1–7.
83. Navarro JF, Macía ML, Gallego E, et al. Serum magnesium concentration and PTH levels. Is long-term chronic hypermagnesemia a risk factor for acynamic bone disease? *Scand J Urol Nephrol* 1997;31:275–280.
84. Saha H, Harmoinen A, Pietilä K, Mörsky P, Pasternack A. Measurement of serum ionized versus total levels of magnesium and calcium in hemodialysis patients. *Clin Nephrol* 1996;46:326–331.

26
Magnesium in Hemodialysis Patients

Senji Okuno and Masaaki Inaba

Magnesium (Mg) is the second most abundant intracellular cation and the fourth most abundant cation of the human body. Magnesium plays an essential role as a cofactor for a variety of enzymes, including those involved in several key steps of intermediary metabolism and phosphorylation. In addition, Mg is required for protein and nucleic acid synthesis, the cell cycle progression, cytoskeletal and mitochondrial integrity, and the binding of substances to the plasma membrane. Disorders of Mg homeostasis may lead to profound changes in the function and well being of the organism. Serum Mg concentration is normal in patients with early renal failure, but hypermagnesemia usually occurs in the advanced stage of renal failure due to the reduced urinary Mg excretion. Following the introduction of chronic hemodialysis or continuous ambulatory peritoneal dialysis (CAPD) treatment, the major factor to determine Mg balance is Mg levels in the dialysate. Patients with end-stage renal disease (ESRD) who are receiving dialysis may develop various complications including hypertension, atherosclerosis, dyslipidemia, and renal osteodystrophy. The disturbance of Mg balance in those patients maintained on dialysis may affect the development of these complications.

Serum Magnesium Level

The total body Mg content is approximately 2000 mEq or 25 g. Like calcium (Ca), only a small fraction (about 1%–2%) of whole-body Mg store exists in the extracellular fluid compartment. Approximately 60% and 20% of total body Mg are found in bone and muscle, respectively. Circulating level of Mg is tightly controlled within narrow ranges between 1.7 to 2.6 mg/dL in healthy individuals. Although no single mechanism has been demonstrated for Mg homeostasis, the cellular availability of Mg is closely regulated by the kidney, gastrointestinal tract, and bone. On a normal dietary intake of Mg, urinary Mg excretion averages 100 to 150 mg/day. The kidney is able to respond rapidly to changes in Mg level in the extracellular fluid by modulating tubular Mg re-absorption. In patients receiving Mg-containing antacids, urinary Mg

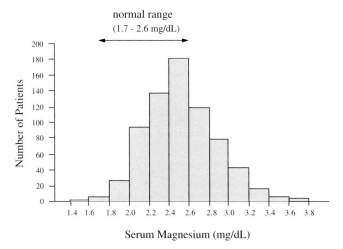

FIGURE 26.1. Serum magnesium concentration in patients with maintained on hemodialysis ($n = 714$). In 714 patients of our unit dialyzed with a standard dialysate Mg (1.0 mEq/L) the mean serum Mg concentration was 2.5 mg/dL, and moreover 35% of subjects had hypermagnesemia.

excretion can increase to 500 to 600 mg/day or more. Conversely, when dietary intake of Mg is restricted, urinary Mg excretion decreases to as low as 10 to 12 mg/day, indicating kidney as the major organ to maintain Mg homeostasis. Taken collectively, Mg disturbance may occur more easily as renal dysfunction progresses. However, the derangement of Mg homeostasis in chronic kidney disease has received less attention than that of Ca or phosphorus. Serum Mg level is usually normal in patients with early stages of renal failure, but hypermagnesemia becomes common in patients with progression of renal impairment (Figure 26.1). Hypermagnesemia becomes apparent when the glomerular filtration rate falls below 30 mL/min. Hypermagnesemia of chronic renal failure is usually asymptomatic, probably due to the gradual increase of serum Mg, although abrupt increase of serum Mg may occur when ESRD patients take Mg-containing antacids or laxatives.

Metabolic balance studies indicate that 25% to 60% of dietary Mg is absorbed. The preferential site of Mg absorption in the intestine may be different depending on the animal species studied. In humans, the temporal pattern of the appearance of radiolabeled Mg into the plasma following ingestion suggests that most of Mg absorption occurs in the small intestine. Although Mg shares with Ca the common pathways for intestinal absorption, most of the evidence suggests that Mg is absorbed mainly by ionic diffusion and solvent drag resulting from the bulk flow of water, in contrast to the fact that most of Ca absorption occurs by vitamin D–induced active transport at the gastrointestinal tract. The data available are not consistent with regard to whether the intestine is capable of modifying Mg absorption according to the Mg content of the diet. Although a few studies suggested that vitamin D and parathyroid hormone (PTH) may increase the intestinal absorption of Mg, other studies found no effect. Whether intestinal absorption of Mg may be altered in uremia is controversial. Studies of net intestinal absorption of Mg in patients with uremia have yielded results that are either normal or only marginally decreased.[1,2] However, data from perfusion of different intestinal segments indicate that jejunal and ileal transport of Mg is reduced in uremic patients.[3] The reasons for the discrepancies between results obtained from intestinal perfusion studies and metabolic balance studies have yet to be determined.

The major factor in determining serum Mg level is a dietary Mg intake in ESRD patients who are not receiving dialysis, and Mg level in the dialysate in uremic patients undergoing dialysis. Mg readily crosses the dialysis membrane with its movement mainly determined by the gradient of diffusible Mg concentrations between blood and dialysate. The usual Mg level in the dialysate is 0.5 to 1.0 mEq/L. With the use of dailysate containing 1.5 mEq/L of Mg, serum Mg levels is increased up to 3.0 to 4.8 mg/dL. The use of dialysate containing 0.5 to 0.8 mEq/L of Mg results in serum Mg concentration near the normal upper limit or only slightly elevated. In the patients maintained on CAPD, the removal rate of Mg through peritoneal membrane can also be increased by lowering Mg level in the peritoneal dialysate. It was originally reported that CAPD patients showed mild hypermagnesemia when Mg level in peritoneal dialysate was 1.5 mEq/L. Because dialysate containing Mg level of 0.5 mEq/L are now commonly used, serum Mg level has become normalized with the body burden of Mg decreased.

Approximately 75% to 80% of the Mg in serum is ultrafiltrable, and the rest is bound to protein, mostly albumin. As the ultrafiltrable form of Mg in serum, Mg existed as ionized Mg, and in the form of complexes with small anions such as bicarbonate, citrate, and inorganic phosphate. It has been suggested that the relationship between ionized Mg and total circulating Mg is altered in dialysis patients, because Mg bind more to anionic compounds that accumulate in uremic conditions. However, the studies on the metabolism of the ionized Mg were conflicting in uremic patients. While some described a lower fraction of ionized Mg in uremic patients than in normal controls,[4,5] others did not.[6,7] This discrepancy could, in part, be attributed to analytical problems associated with the measurement of both total and ionized Mg.[8]

Intracellular Level of Magnesium

Almost 99% of the body Mg exists inside the cells. It has been suggested that serum Mg levels poorly reflect the intracellular Mg content and that, in patients with renal failure, hypermagnesemia may be accompanied with normal or even depleted tissue Mg content. Muscle normally has 76 mEq of Mg per kilogram of fat-free solids, and much of this exists in a complex form with intracellular organic phosphate and protein. Erythrocytes contain normally about 4.6 mEq/L of Mg, of which 84% is thought to be complexed with adenosine triphosphate (ATP). As noted, bone is the principal source of Mg. The normal

Ca/Mg ratio in bone is 50, with the ratio consistently higher in trabecular bone than in cortical bone. Most of the Mg in bone exits in the form associating with apatite crystal rather than bone matrix. Approximately 30% of Mg in bone is present as a surface-limited ionized form on the bone crystal and thus is freely exchangeable. Investigation of the intracellular content of total Mg in uremic patients has given variable results in various tissues. In patients maintained on hemodialysis, Mg level in erythrocytes was found to be increased. The increase of Mg content in erythrocytes could solely reflect Mg retention due to high serum Mg, although other factors may contribute. An inverse correlation was found between erythrocytes Mg content and hematocrit in dialysis patients. Mg contents of erythrocytes appeared to be inversely related to the age of the cell, with the reticulocytes containing about two times more Mg than older erythrocytes. Therefore, the erythrocyte Mg content may not provide a reliable index of Mg status in dialysis patients. Magnesium content of muscle in ESRD patients has been reported to be either low, normal, or elevated. Normal Mg levels have been noted in lymphocytes of hemodialysis patients despite elevated serum Mg levels. In contrast, increased Mg content in bone has been a consistent finding in uremic patients.[9]

Nearly all intracellular Mg is bound or complexed with proteins or nucleic acids, or is confined within intracellular organelles. Intracellular ionized Mg, which constitutes only 1% to 10% of total intracellular Mg content, plays an important role in a wide range of cellular processes. These include energy production and utilization, protein and DNA synthesis, and membrane transport. Accordingly, under normal circumstances the intracellular levels of ionized Mg appears to be strictly regulated under tight homeostatic control. The main factors affecting intracellular levels of ionized Mg are the concentration of nucleotides and operation of transport systems in the plasma membrane and mitochondria.

Intracellular level of ionized Mg in dialysis patients is controversial. Kaupke and colleagues measured with molecular fluorescent probes intracellular level of ionized Mg in platelets from ESRD patients maintained on hemodialysis.[10] They found that ESRD patients exhibited significantly lower intracellular level of ionized Mg, despite higher serum Mg level, than the normal control group. Hemodialysis reduced serum Mg level without affecting intracellular level of ionized Mg. Nishida and colleagues reported that intracellular ionized Mg, measured with ^{31}P magnetic resonance spectroscopy, was normal in skeletal muscle of uremic patients.[11] Kister and coworkers determined cytosolic ionized Mg in lymphocytes from uremic patients and normal controls.[12] Serum total and ionized Mg concentrations were greater in uremic patient population, whereas the concentration of ionized Mg in lymphocytes was comparable in both groups. Huijgen and coworkers demonstrated that intracellular ionized Mg in mononuclear blood cells from hemodialysis patients was greater than that from the healthy population.[13] The reason for the discrepancy of these results is not well understood, but at least it may be attributed to different methods and tissues analyzed.

uated the progression of atherosclerotic lesions in aortas and the intimal thickening.[30] Epidemiological studies suggested an inverse relationship between Mg content in drinking water and cardiovascular mortality. Lower level of tissue Mg is associated with greater occurrence of coronary heart disease. Gartside and colleaguesd reported an inverse relation between serum Mg and incidence of coronary heart disease.[31] Liao and coworkers also reported an inverse association in healthy middle-aged adults; relative risk of coronary hear disease across quartiles of serum Mg was 1.00, 0.92, 0.48, and 0.44 among women and 1.00, 1.32, 0.95, and 0.73 among men.[32] Amighi and colleagues found that low serum Mg levels were associated with an increased risk of neurological events in patients with symptomatic peripheral artery disease.[33] A few studies reported a correlation between Mg metabolism and atherosclerosis in dialysis patients. Tzanakis and colleagues showed that there was a negative linear correlation existed between intra- or extracellular Mg contents and intima–media thickness of common carotids in hemodialysis patients.[34] This observation suggested that Mg may play a protective role against the development and/or acceleration of arterial atherosclerosis in dialysis patients.

There is evidence enough to draw a conclusion on how low Mg is implicated in the pathogenesis of atherosclerosis. Magnesium is a biological competitor of Ca and inhibits its entry into the cells. Intracellular accumulation of Ca in smooth muscle as well as in vascular endothelial cells may induce vasospasm, endothelial dysfunction, arterial wall thickening, and, finally, atherosclerosis. Experimental and clinical data indicate that high serum Mg may result in decreased intracellular Ca and in attenuation of the above-mentioned consequences. In addition, Mg interferes with other mechanisms involved in atherosclerosis. It is noteworthy that Mg deficiency also stimulates the peroxidation of lipoprotein. In Mg-deficient rats, oxidation of very-low-density lipoprotein (VLDL) and low-density lipoprotein (LDL) becomes greater than in control rats, partly because the lipids of Mg-deficient rats were more susceptible to oxidative damage following iron incubation.[35] Similarly, when drinking water was fortified with Mg, serum levels of lipid peroxides were lower in mice fed an atherogenic diet.[36] Oxidized LDL is more atherogenic than the wild type of lipoprotein, and high-density lipoprotein (HDL), after oxidative modifications, loses its effect to stimulate efflux of cholesterol from foam cells. The role of Mg deficiency in inflammatory processes should also be considered. Magnesium deficiency in rats led to the sequence of events characteristic for the inflammatory response, that is, a significant increase in plasma interleukin-6 levels.[37] Inflammatory process was accompanied with an increase in plasma levels of acute-phase proteins. Increases in circulating levels of pro-inflammatory cytokines may activate macrophage and endothelial cells. The inflammatory response in Mg-deficient rats may contribute to atherogenesis by modifications of lipoprotein metabolism, oxidation of lipoproteins, inflammatory cell recruitment, and release of cytokine and growth factors that induce cell migration and proliferation.[38]

Hypertension

Hypertension occurs in approximately 80% of ESRD patients. In ESRD patients, the relationship between blood pressure and cardiovascular events seems to be U shaped. Low predialysis systolic blood pressure is associated with decreased survival. Likewise, systolic blood pressure greater than 180 mm Hg is associated with poor clinical outcome. The pathogenesis of hypertension in ESRD patients is complex and probably multifactorial, including intravascular volume, renin–angiotensin system, vascular stiffness, nitric oxide formation, and uremic toxin such as PTH.

Magnesium deficiency may contribute to the development of hypertension in ESRD patients. Epidemiological studies have shown an inverse relationship between dietary Mg intake and blood pressure. Magnesium supplementation decreases blood pressure in several, but not in all, clinical studies. A few studies have evaluated the serum concentrations of Mg in hypertensive patients and have show a significant decrease or an inverse relationship between serum Mg and blood pressure. An abnormality of intracellular ionic homeostasis was proposed as a contributory pathophysiological factor for essential hypertension. Intracellular free Mg in vascular smooth muscle or circulating red blood cells has been found to be decreased in hypertensive patients. Irish and colleagues reported that there was a significant inverse correlation of intracellular ionized Mg in skeletal muscle with systolic and diastolic blood pressure in the uremic population.[39] This study suggested that intracellular ionized Mg played an important role for the hypertension in ESRD patients. End-stage renal disease patients are frequently associated with secondary hyperparathyroidism, leading to increased intracellular Ca in almost any organs. A relationship between intracellular Ca and blood pressure has been demonstrated in patients with essential hypertension as well as ESRD, suggesting that increased serum levels of PTH may be responsible for the increase of blood pressure in ESRD patients.

Lipid Metabolism

End-stage renal disease is associated with dyslipidemia due to characteristic alterations of lipoprotein metabolism. The dyslipidemia characteristic for dialysis patients is increases of VLDL, VLDL remnant, and lipoprotein (a) [Lp(a)] as well as a decrease of HDL. Although both increased synthesis and decreased catabolism of VLDL contribute to increased VLDL-remnant level in dialysis patients, decreased catabolism of VLDL is the predominant mechanism. Plasma lipoprotein lipase, hepatic triglyceride lipase, and adipose tissue lipase activities are all impaired in dialysis patients. Decrease of plasma HDL levels in ESRD patients are probably caused by decreased synthesis that impairs the ability of HDL to esterfy and transport free cholesterol from tissues and Apo B-100 containing lipoproteins. Although the mechanism of increased Lp(a) in uremia is unknown, increased synthesis is the most likely explanation.

Magnesium deficiency has a possible role in the perturbation of lipid metabolism in the non-uremic population. In rat, dietary Mg restriction increases plasma levels of LDL and VLDL and reduces that of HDL. The lower concentration of HDL is accounted for by a decreases in the availability of apo E and apo A-I, the major apoproteins of HDL. In cholesterol-fed animals, oral Mg supplementation lowers serum cholesterol and triglyceride and attenuates the development of atherosclerotic lesions. Delva and colleagues reported that the ionized Mg inside lymphocytes from hypertriglyceridemic patients was decreased compared with normotriglyceridemic patients and that negative correlation was observed between ionized intralymphocyte Mg and plasma triglycerides levels.[40]

The association between abnormalities of Mg and lipid metabolism was also suggested in uremia. In a study by Inagaki and colleagues, Mg deficiency in five-sixths nephrectomized uremic rats resulted in an increase of serum triglyceride levels and a decrease of HDL cholesterol levels.[41] Robles and coworkers reported that there was a positive and significant correlation between serum levels of Mg and total cholesterol, LDL cholesterol, VLDL cholesterol, and apo-B in hemodialyzed uremic patients.[42] Nasri and coworkers also reported that there were a positive correlation between serum Mg and Lp(a) and also between serum Mg and triglyceride levels in hemodialysis patients. Kirsten and coworkers showed that Mg supplementation in patient with chronic renal failure resulted in a decrease of serum total and LDL cholesterols, and an increase of HDL cholesterol.

Several mechanisms have been proposed to explain the relationship between hyperlipidemia and Mg depletion. Magnesium-deficient rats developed hyperlididemia as a result of reduction in plasma lipoprotein lipase activity and lecithin cholesterol acyltransferase activity. It has been considered that these enzymes require the Mg ion as an important cofactor. It is reported that, in Mg-deficient rats, there is increased HMG-CoA reductase activity and increased cholesterol synthesis. These factors may contribute of hyperlipidemia in Mg deficiency.[43]

References

1. Clarkson EM, McDonald SJ, Dewardener HE, et al. Magnesium metabolism in chronic renal failure. *Clin Sci* 1965;28:107–115.
2. Kopple JD, Coburn JW. Metabolic studies of low protein diets in uremia. II. Calcium, phosphorus, and magnesium. *Medicine* 1973;52:597–607.
3. Brannan PGP, Verne-Marini P, Pak CYC, et al. Magnesium absorption in the human small intestine. Results in normal subjects, patients with chronic renal disease and patients with absorptive hypercalciuria. *J Clin Invest* 1976;57:1412–1418.
4. Truttmann AC, Faraone R, von Vigier RO, et al. Maintenance hemodialysis and circulating ionized magnesium. *Nephron* 2002;92:616–621.

5. Markell MS, Altura BT, Sarn Y, et al. Deficiency of serum ionized magnesium in patients receiving hemodialysis or peritoneal dialysis. *ASAIO J* 1993;39:M801–M804.
6. Saha H, Harmoinen A, Pietila K, et al. Measurement of serum ionized versus total levels of magnesium and calcium in hemodialysis patients. *Clin Nephrol* 1996; 46:326–331.
7. Huijgen HJ, Sanders R, van Olden RW, et al. Intracellular and extracellular blood magnesium fractions in hemodialysis patients; is the ionized fractions a measure of magnesium excess? *Clin Chem* 1998;44:639–648.
8. Dewitte K, Dhondt A, Lameire N, et al. The ionized fraction of serum total magnesium in hemodialysis patients: is it really lower than in healthy subjects? *Clin Nephrol* 2002;58:205–210.
9. Mountokalakis TD. Magnesium metabolism in chronic renal failure. *Magnes Res* 1990;3:121–127.
10. Kaupke CJ, Zhou XJ, Vaziri ND. Cytosolic ionized magnesium concentration in end-stage renal disease. *ASAIO J* 1993;39:M614–M617.
11. Hishida A, Shapiro JI, Chan L. Effect of uremia on cytosolic free magnesium [Mg++]i and energy metabolism in skeletal muscle. *J Am Soc Nephrol* 1992;3:687.
12. Kisters K, Spieker C, Tepel M, et al. Plasma, cytosolic and membrane magnesium content in renal insufficiency. *Magnes Res* 1995;8:167–172.
13. Huijgen HJ, Sanders R, van Olden RW, et al. Intracellular and extracellular blood magnesium fractions in hemodialysis patients; is the ionized fraction a measure of magnesium excess? *Clin Chem* 1998;44:639–648.
14. Block GA, Hulbert-Shearon TE, Levin NW, et al. Association of serum phosphorus and calcium X phosphate product with mortality risk in chronic hemodialysis patients: a national study. *Am J Kidney Dis* 1998;31:607–617.
15. Malluche HH, Mawad H. Management of hyperphosphataemia of chronic kidney disease: lessons from tha past and future directions. *Nephrol Dial Transplant* 2002;17:1170–1175.
16. Fine KD, Santa Ana CA, Porter JL, et al. Intestinal absorption of magnesium from food and supplements. *J Clin Invest* 1991;88:394–402.
17. Guillot AP, Hood VL, Runge CF, et al. The use of magnesium-containing phosphate binders in patients with end-stage renal disease on maintenance hemodialysis. *Nephron* 1982;30:114–117.
18. O'Donovan R, Baldwin D, Hammer M, et al. Substitution of aluminium salts by magnesium salt in control of dialysis hyperphosphataemia. *Lancet* 1986;1:880–882.
19. Delmez JA, Kelber J, Norword KY, et al. Magnesium carbonate as a phosphorus binder: a prospective, controlled, crossover study. *Kidney Int* 1996;49:163–167.
20. Pletka P, Bernstein DS, Hampers CL, et al. Relationship between magnesium and secondary hyperparathyroidism during long-term hemodialysis. *Metabolism* 1974;23:619–624.
21. Navarro JF, Macía ML, Gallego E, et al. Serum magnesium concentration and PTH levels. Is long-term chronic hypermagnesemia a risk factor for adynamic bone disease? *Scand J Urol Nephrol* 1997;31:275–280.
22. Navarro JF, Mora C, Jiménez A, et al. Relationship between serum magnesium and parathyroid hormone levels in hemodialysis patients. *Am J Kidney Dis* 1999;34:43–48.

23. Gonella M, Ballanti P, Della RC, et al. Improved bone morphology by normalizing serum magnesium in chronically hemodialyzed patients. *Miner Electrolyte Metab* 1988;13:240–245.
24. Morinière P, Vinatier I, Westeel PF, et al. Magnesium hydroxide as a complementary aluminum-free phosphate binder to moderate doses of oral calcium in uraemic patients on chronic haemodialysis: lack of deleterious effect on bone mineralization. *Nephrol Dial Transplant* 1988;3:651–656.
25. Ng AHM, Hercz G, Kandel R, et al. Association between fluoride, magnesium, aluminum and bone quality in renal osteodystrophy. *Bone* 2004;34:216–224.
26. Meema HE, Oreopoulos G, Rapoport A. Serum magnesium level and arterial calcification in end-stage renal disease. *Kidney Int* 1987;32:388–394.
27. Tzanakis I, Pras A, Kounali D, et al. Mitral annular calcification in haemodialysis patients: a possible protective role of magnesium. *Nephrol Dial Transplant* 1997;12:2036–2037.
28. Okuno S, Ishimura E, Maeno Y, et al. Relationship between serum magnesium and vascular calcification in hemodialysis patients. *J Jap Soc Magnes Res* 2005;24:59–67.
29. Shoji T, Emoto M, Tabata T, et al. Advanced atherosclerosis in predialysis patients with chronic renal failure. *Kidney Int* 2002;61:2187–2192.
30. Altura BT, Brust M, Bloom S, et al. Magnesium dietary intake modulates blood lipid levels and atherogenesis. *Proc Natl Acad Sci U S A* 1990;87:1840–1844.
31. Gartside PS, Glueck CJ. The important role of modifiable dietary and behavioral characteristics in the causation and prevention of coronary heart disease hospitalization and mortality: the Prospective NHANES I Follow-up Study. *J Am Coll Nutr* 1995;14:71–79.
32. Liao F, Folsom AR, Brancati FL. Is low magnesium concentration a risk factor for coronary heart disease? The atherosclerosis risk in communities (ARIC) study. *Am Heart J* 1998;136:480–490.
33. Amighi J, Sabeti S, Schlager O, et al. Low serum magnesium predicts neurological events in patients with advanced atherosclerosis. *Stroke* 2004;35:22–27.
34. Tzanakis I, Virvidakis K, Tsomi A, et al. Intra- and extracellular magnesium levels and atheromatosis in haemodialysis patients. *Magnes Res* 2004;17:102–108.
35. Rayssiguier Y, Gueux E, Bussiere L, et al. Dietary magnesium affects susceptibility of lipoproteins and tissues to peroxidation in rats. *J Am Coll Nutr* 1993;12:133–137.
36. Yamaguchi Y, Kitagawa S, Kunitomo M, et al. Preventive effects of magnesium on raised serum lipid peroxide levels and aortic cholesterol deposition in mice fed an atherogenic diet. *Magnes Res* 1994;7:31–37.
37. Malpuech-Brugere C, Nowacki W, Daveau M, et al. Inflammatory response following acute magnesium deficiency in the rat. *Biochem Biophys Acta* 2000;1501:91–98.
38. Maier JAM. Low magnesium and atherosclerosis: an evidence-based link. *Mol Aspects Med* 2003;24:137–146.
39. Irish AB, Thompson CH, Kemp GJ, et al. Intracellular free magnesium concentrations in skeletal muscle in chronic uraemia. *Nephron* 1997;76:20–25.
40. Delva PT, Pastori C, Degan M, et al. Intralymphocyte free magnesium in a group of subjects with essential hypertension. *Hypertension* 1996;28:433–439.

41. Inagaki O, Shono T, Nakagawa K, et al. Effect of magnesium deficiency on lipid metabolism in uremic rats. *Nephron* 1990;55:176–180.
42. Robles NR, Escola JM, Albarran L, et al. Correlation of serum magnesium and serum lipid levels in hemodialysis patients. *Nephron* 1998;78:118–119.
43. Mitsopoulos E, Griveas I, Zanos S, et al. Increase in serum magnesium level in haemodialysis patients receiving sevelamer hydrochloride. *Int Urol Nephrol* 2005;37:321–328.

Neurology

27
Neurology Overview

Keiichi Torimitsu

The neurological effect of the magnesium ion (Mg^{2+}) has long been well recognized as a blocker of the N-methyl-D-aspartate (NMDA) receptor. Recent reports indicate that Mg^{2+} plays an important role in brain disease and injury. Mg^{2+} deficiency causes severe brain function damage. Although the mechanism is unclear with respect to neurological functions, neuroprotective effects have been widely investigated. The neurological requirements of Mg^{2+} are still unknown, unlike those of Ca^{2+}. The recent development of analytical technologies has made it possible to investigate the effect of Mg^{2+} in real time. This chapter discusses a wide variety of neurological effects induced by Mg^{2+} and includes fundamental and clinical research.

Mg^{2+} is an important ion with respect to various biological functions (see Figure 27.1). It has been long identified as an essential ion in cell and body functions because of the significant effects of magnesium deficiency. The consequences of following a diet deficient in Mg^{2+} have been widely investigated. The maintenance of a homeostatic condition is important for the prevention of various diseases, including brain disease. A low dietary magnesium intake leading to a hypomagnesium condition increases the risk of stroke, ischemia, and other cardiovascular diseases.[1] There have been many reports indicating the crucial role of Mg^{2+} in brain injuries and neurodegenerative disease. The neuroprotective effect of Mg^{2+} is also attracting considerable attention. The Mg blockade of the NMDA receptor ion channel that prevents ischemic damage to neurons has been well discussed. A pharmacological approach to neuroprotection using Mg^{2+} as a medicine has been extensively tested. In addition to Mg, glycine, zinc, and other antagonists are believed to be possible agents for reducing brain damage by keeping the NMDA receptor ion channels closed.[2–4]

As described above, a considerable amount is known about Ca^{2+}. However, the role of Mg^{2+} and the mechanism of the Mg^{2+} effects are not well understood. This is because of a lack of research, and a lack of a technique for measuring Mg^{2+} activity, such as fluorescent dye. Fura-2, Fluo-3, and Fluo-4 are widely used calcium fluorescent dyes that could be employed to investigate changes in intracellular calcium concentration. The role of Ca^{2+} has been well identified

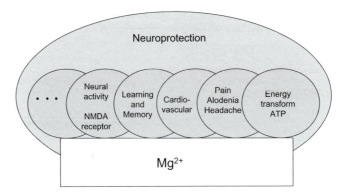

FIGURE 27.1. Role of magnesium in the biological system.

using these dyes. In contrast, until recently there were only a few methods for investigating Mg^{2+} behavior in cells. Microdialysis is a commonly used measurement technique, especially with in vivo experiments. Although real-time measurement is possible if a selective detection system is employed, the target will be diluted 10 times or more because of the circulation of the medium inside the microdialysis tube. Chemical analysis in an external fluid also suggests the idea of determining magnesium content with a sampling procedure. A magnesium-selective electrode has also been used. However, the recent development of fluorescent dye to measure the intracellular magnesium concentration provides us with a great advantage when investigating Mg^{2+} activity. Magfura-2, magnesium green, and KMG-104 are the dyes that can be used for Mg^{2+} measurement.[5]

Role of Magnesium in Neuronal Cells

As described above, the importance of magnesium ions can be seen in a wide variety of fields. The role of magnesium can be categorized as followings: (1) therapeutic importance; (2) neuroprotection; and (3) ion regulation (Figure 27.2).

Therapeutic Importance

Magnesium deficiency causes various symptoms, such as cardiovascular problems, stroke, and headache. It may also induce cell death, resulting in a reduced neuroprotective effect as described below.

Headache is a typical symptom caused by the NMDA receptor–related hyperactivity of neurons. It is reported that magnesium reduces the pain of patients who suffer from headaches according to a dose-dependent manner. An increment in learning ability has also been observed in tests on animals.[6]

Recent research has identified that a secondary process is essential after primary brain injury or damage if cells are to survive. Cerebral ischemia and traumatic brain injury are representative examples of this phenomenon. Cerebral ischemia induces a decrease in energy supply because of a lack of oxygen. A breakdown of homeostatic conditions is induced by oxidative stress, the depolarization of membrane potential, and energy transduction failure. These phenomena increase the release of neurotransmitters, such as glutamate, which results in an intracellular calcium increase, and they finally induce apoptosis and/or necrosis leading to neuronal death.

An analysis of magnesium supplementation using an experimental model of animal cerebral ischemia indicates that reducing apoptotic neurons, and infarct volume, increases neuronal survival in CA1 region of the hippocampus.

Neuroprotection

The neuroprotective effect of magnesium in relation to brain damage is widely recognized.[4] One interesting factor is the supportive function of magnesium during the process of recovery from injury and/or damage. Neurons are very sensitive to any kind of stress compared with other supportive cells, such as glial cells and Schwann cells. Nerve growth factor (NGF) and brain-derived neurotrophic factor (BDNF) are well-known neuroprotective agents in neurons, and they use the tyrosine kinase pathway to provide damaged protection. Transient neurotransmitter release and intracellular calcium increase are also induced by these agents.[7,8]

On the other hand, magnesium compounds are essential in various enzyme processes, including DNA and protein synthesis. The adenosine triphosphate (ATP) energy process also depends on magnesium, but it does not require a specialized target, and this indicates its complexity.

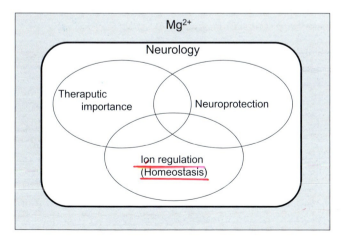

FIGURE 27.2. Role of magnesium in neuronal cells.

In neural systems, the blockade of the NMDA receptor is a well-known phenomenon that regulates Ca^{2+} transport through a plasma membrane. The blockade is dependent both on membrane potential and on ionic strength and species. Neural activity depends on the magnesium concentration when glutamate is released under conditions of anoxia. In this respect, magnesium protects neurons from hyperactive situations as a neuromodulator.

The modulation of neural activity by magnesium may be related to memory and learning. Although increases in animal learning ability have been reported, further study, including an investigation of synaptic regulation, is required if we are to understand the effect of magnesium on memory and learning system. Magnesium could be effective for both neuroprotection and synaptic regulation.

Ion Regulation

Although the intracellular magnesium concentration is regulated at a certain level, its mechanism is still not well understood. Intracellular store site, such as mitochondria, and endoplastic reticulum (ER), play a role in regulating the intracellular magnesium level together with ion channels and the ion exchange mechanism. The sodium (Na)—Mg exchange system[9–11] and transient receptor potential (TRPM)7[12,13] are reported to be regulators of magnesium transport through plasma membranes. Another important role is Ca transport. An inhibitory effect on voltage-gated calcium channels might be involved in the neuroprotection system as described above.[14]

It is very common for cells to have a Na—Ca exchange system for controlling Ca transport. The Na—Mg system plays an important role in the regulation of magnesium transport. The system is considered to work along with the Na—Ca system to regulate magnesium homeostasis. However, it is unclear whether Na—Ca and Na—Mg systems are different independent systems. As the Na—Mg exchange system has yet to be identified and purified, the role of the system, especially its stoichiometry, is still under investigation.

There are some reports indicating a relationship between neurotransmitter release and intracellular magnesium concentration. Glutamate has been reported to increase intracellular magnesium in cultured cortical neurons.

Conclusion

This chapter described the effect of magnesium on the nervous system, particularly that of the brain. Although magnesium has been recognized as a very important substance in various nervous systems whose deficiency causes severe damage, its role and mechanism are still not well understood. The recent development of measurement methods has provided us with more information on neuronal disease and protection. As the research is strongly related to clinical applications, we can expect significant pharmaceutical developments in the near future. The following chapters will discuss the important role played by magnesium in the neural system.

References

1. Altura BT, Memon ZI, Zhang A, et al. Low levels of serum ionized magnesium are found in patients early after stroke which result in rapid elevation in cytosolic free calcium and spasm in cerebral vascular muscle cells. *Neurosci Lett* 1997;230:37–40.
2. Vink R, Cernak I. Regulation of brain intracellular free magnesium following traumatic injury to the central nervous system. *Front Biosci* 2000;5:656–665.
3. Turner RJ, Van Den Heuvel C, Vink R. Amiloride increases neuronal damage after traumatic brain injury in rats. *Magnes Res* 2003:16;315.
4. Suzer T, Coskun E, Islekel H, et al. Neuroprotective effect of magnesium on lipid peroxidation and axonal function after experimental spinal cord injury. *Spinal Cord* 1999;37:480–484.
5. Suzuki Y, Komatsu H, Ikeda T. Design and synthesis of Mg^{2+}-selective fluoroionophores based on a coumarin derivative and application for Mg^{2+} measurement in a living cell. *Anal Chem* 2002;74:1423–1428.
6. Nomura M. Magnesium deficiency—view point of "Life salt nuti ma-su." *Clin Calcium* 2004;14:71–75.
7. Torimitsu K, Furukawa Y, Kasai N, et al. Effect of neurotrophins on intracellular Ca concentration and electrical activity in cultured rat cortex. *Am Soc Cell Biol* 1999;1330.
8. Torimitsu K, Furukawa Y, Kasai N. Real-time detection of neurotransmitter release and its spatial distribution. *Folia Pharma Japonica* 2003;121:349–356.
9. Brocard JB, Rajdev S, Reynolds IJ. Glutamate-induced increases in intracellular free Mg^{2+} in cultured cortical neurons. *Neuron* 1993;11:751–757.
10. Stout AK, Li-Smerin Y, Johnson JW, Reynolds IJ. Mechanisms of glutamate-stimulated Mg^{2+} influx and subsequent Mg^{2+} efflux in rat forebrain neurons in culture. *J Physiol* 1996;492:641–657.
11. Tashiro M, Konishi M. Na^+ gradient-dependent Mg^{2+} transport in smooth muscle cells of guinea pig tenia cecum. *Biophys J* 1997;73:3371–3384.
12. Takezawa R, Schmitz C, Demeuse P, Scharenberg AM, Penner R, Fleig A. Receptor-mediated regulation of the TRPM7 channel through its endogenous protein kinase domain. *Proc Natl Acad Sci U S A* 2004;101:6009–6014.
13. Nadler MJ, Hermosura MC, Inabe K, et al. LTRPC7 is a Mg-ATP-regulated divalent cation channel required for cell viability. *Nature* 2001;411:590–595.
14. Zhang A, Fan SH, Cheng TP, et al. Extracellular Mg^{2+} modulates intracellular Ca^{2+} in acutely isolated hippocampal CA1 pyramidal cells of the guniea-pig. *Brain Res* 1996;728:204–208.

28
Magnesium in the Central Nervous System

Renee J. Turner and Robert Vink

It has been almost a century since a role for magnesium in the central nervous system (CNS) was first proposed. Despite intensive efforts, the subsequent 75 years saw few advances in our understanding of magnesium's precise role in brain function or the mechanisms by which the cation influences these functions. More recently, the advent of noninvasive techniques to measure intracellular free magnesium concentration, plus the recognition that magnesium plays a critical role in regulating neurotransmitter receptor function, have ushered in a new era for magnesium research in neuroscience. The result has been thousands of published studies describing various effects of magnesium in the CNS, ranging from effects on normal physiology and biochemistry to modulation of pathological events at the molecular level. The current review critically examines the evidence suggesting that alterations in intracellular free magnesium concentration may be an injury factor in acute and chronic CNS injury, as well as the potential for magnesium administration to be neuroprotective under these conditions. Finally, the reasons for contradictory results in the literature regarding therapeutic efficacy are discussed, with an emphasis on cellular energy state and how it may affect treatment, as well as dosage strategies and the potential for adverse side effects.

A role for magnesium in CNS function has been recognised since 1916, when magnesium sulphate was first described as a clinical anesthetic.[1] Despite this early recognition, an understanding of magnesium's role in cell physiology and metabolism has not been readily forthcoming largely because of the inherent difficulty in measuring free concentration of the ion. In contrast, calcium determinations have been possible for several decades and a critical role for calcium has now been well established. Despite several early reports supporting a similar role for magnesium, the evidence was difficult to substantiate and accordingly, magnesium became known as the forgotten cation. More recently, there has been a renewed interest in the role of magnesium in cell function, particularly in the obstetric, cardiovascular, and neural disciplines. With respect to this resurgence in interest in the neurosciences, the impetus can be arguably attributed to the ability to readily measure free magnesium concentration in the CNS, and to the finding that free magnesium regulates

N-methyl-D-aspartate (NMDA) channel activity. Over the past decade, substantial evidence has accumulated suggesting that magnesium changes may occur in the CNS and that these changes in magnesium concentration may have considerable affects on outcome. This recent evidence has never been summarized and consequently, the importance of the magnesium ion in CNS function is still poorly appreciated. The present review therefore examines the recent literature concerning magnesium's regulatory role in neuronal metabolism, particularly the evidence suggesting a role in acute and chronic injury.

Basal Concentration

A number of methods exist to measure total magnesium concentration and these have been used clinically for several years.[2] These include spectrophotometric, atomic absorption, and mass spectrometer techniques. While these techniques are highly accurate, the diagnostic use of the values obtained is limited because clinical hypermagnesemia and hypomagnesemia overlap extensively with normal values. Moreover, it should be recognized that the free ion concentration is the critical physiological value and this value does not always correlate with the total concentration. Several methods have been used to determine free magnesium concentration.[3,4] Fluorometric techniques were first used to measure enzyme equilibria and thereby indirectly calculate free magnesium concentration. Similarly, ionophores were used in a null point titration technique to do likewise. The difficulty is that both these techniques were highly invasive and tissue destruction was inevitable. The development of ion-selective electrodes was far less destructive, and indeed could be used very successfully on body fluids to determine free magnesium concentration. However, it was still extremely difficult to determine free intracellular levels with these electrodes. The development of fluorescent probes for the measurement of free magnesium concentration has opened the way for intracellular magnesium determinations, however, the technique was limited to cell culture. The only truly noninvasive technique used to date is nuclear magnetic resonance (NMR) spectroscopy.[5] This technique takes advantage of an intracellular magnesium-binding ligand to provide information that can be used to calculate free magnesium concentration. Various ligands have been used with the most widely used being the endogenous ligand, adenosine triphosphate (ATP). The discussion of these techniques is beyond the scope of the present review and readers are referred to some excellent reviews of these NMR-based techniques.[4,6]

The use of NMR to demonstrate that brain intracellular free magnesium concentration can vary in response to an insult was pioneered by Vink and colleagues in the late 1980s.[7,8] Subsequently, a number of laboratories have demonstrated that both total and free brain magnesium concentration declines following injury and that these declines are associated with the development of functional deficits. Reported values for brain intracellular free magnesium concentration range from 0.3 mM to 1.0 mM, depending on the species. Although some variability in this value can be attributed to small differences in the method of calculation from the NMR data, this can be largely avoided by using the specific equations available on the internet to calculate magnesium concentration by NMR.[9] In the rodent brain, the accepted mean resting value is now considered to be between 0.5 and 0.6 mM, while values on the order of 0.25 to 0.3 mM are considered to represent injured tissue. There is some evidence to suggest that magnesium is not homogeneously distributed throughout the brain with the concentration of the ion varying between cortex and subcortex and between structures.[10,11] Use of transgenic animals with magnesium deficiency may clarify why such variation exists.[10]

Changes in Disease States

Trauma

Changes in brain magnesium concentration that could be shown to have functional consequences were first demonstrated in traumatic brain injury.[7,8] These authors used NMR to show that traumatic injury to the brain resulted in a 50% to 60% decline in the free intracellular concentration of the ion, and that this decline was associated with the development of neurological motor deficits. Subsequent studies by the same group demonstrated that depleting tissue magnesium concentration exacerbated neurological deficits while attenuation of the decline with parenteral magnesium administration prior to the injury significantly improved neurological outcome.[12] The decline in free magnesium after trauma was reflected in a much smaller decline in tissue total magnesium concentration that was limited to the injury zone[7] and did not extend to non-injured tissue.[13] Such a relationship between free and total magnesium changes was consistent with the previously described buffering role that total magnesium plays in the control of free concentration.[14] Thus, decline in magnesium concentration was thought to be a specific indicator of brain cellular damage.

Since those initial studies, a number of other laboratories have reported similar results suggesting that such declines in magnesium concentration were not limited to any specific model of brain trauma or to any particular species.[15–18] In all species and models, the declines reported in brain intracellular free magnesium concentration ranged between 40% and 60% (Figure 28.1). The fact that magnesium does decline following trauma irrespective of the injury model suggests that magnesium decline is a ubiquitous feature of CNS trauma. However, it is notable that the free magnesium decline never exceeds a minimum value of approximately 0.2 mM, suggesting that this concentration may be a threshold below which the ion cannot fall under physiological conditions. Because free magnesium changes were related to much smaller changes in total tissue magnesium,[7] buffering by intracellular ligands

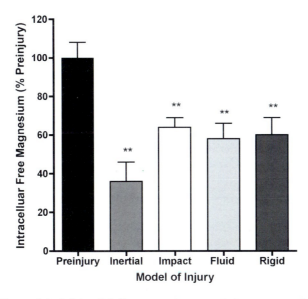

FIGURE 28.1. Changes in brain intracellular free magnesium concentration as measured at 4 to 6 h after experimental brain injury induced using a variety of models. Inertial, inertial injury; Impact, impact acceleration injury; Fluid, fluid percussion injury; Rigid, rigid indentation injury. **$p < 0.01$ versus pre-injury.

presumably prevents any further decline. The declines in brain free magnesium have been shown to persist for at least 4 days following traumatic brain injury and, with increasing severity of injury, out to 7 days.[15,19] Thus, it is not the magnitude of decline that is exclusively related to outcome but also the time period for which magnesium is depressed.

The availability of ion-selective electrodes for magnesium has meant that blood free magnesium concentration can also be readily measured after brain injury. Similar to brain free magnesium concentration, blood levels of free magnesium in rats significantly decline from a mean pre-injury level of approximately 0.50 mM to 0.35 mM by 30 min after traumatic brain injury.[20] This decline in magnesium persisted for at least 4 days following trauma recovering to pre-injury levels by 6 days after injury. This temporal profile of blood free magnesium change was similar to that observed in rat brain using magnetic resonance techniques. As with the brain determinations, there was a pattern whereby milder levels of injury resulted in transient declines in blood free magnesium while more severe injury caused persistent declines. However, unlike the brain determinations, a relationship between blood free magnesium and outcome only existed for the more severe injury levels. While the experimental evidence suggests that trauma to the brain specifically results in a decline in plasma magnesium concentration, there is little evidence to support the view that plasma magnesium reflects brain magnesium concentration, or that a clear relationship exists between blood free magnesium concentration and motor outcome following less severe injury. This is in contrast to findings in clinical trauma where blood hypomagnesia is common[21] and is related to the severity of injury.[22,23] Nonetheless, contradictory reports do exist, with Frankel and colleagues reporting no correlation between blood magnesium and outcome in trauma patients.[24]

The declines in blood free concentration reported in experimental models have been associated with declines in total magnesium concentration,[25] although this is in contrast with observations in clinical brain trauma where there was always a decline in blood free magnesium concentration that was not associated with any decline in blood total levels.[22] Given that increased urinary excretion of magnesium occurs after clinical brain injury,[26] it would be expected that some loss of total blood concentration of the cation would be observed. A recent comprehensive study of total and free magnesium concentration in stress, mild, moderate, and severe brain injury[23] has confirmed that declines in blood total magnesium concentration are present in chronic or severe conditions but not in acute stress or following mild-to-moderate trauma. In contrast, declines in free magnesium concentration occur at all levels of injury.

Finally, extracellular magnesium changes have not been detected in trauma by microdialysis,[27] although the technique has been effective in detecting significant declines following ischemia.[28] This difference may be due to the detection methods used in the study, with the earlier study using ion chromatography as opposed to the more sensitive graphite furnace atomic absorption spectroscopy used in the later study.

Ischemia/Stroke

A number of studies have now reported changes in free magnesium concentration following stroke. Intracellular brain levels have been shown to increase and then decline following stroke,[29] with a number of studies implying that glutamate excitotoxicity may contribute to the initial increase. However, a rise in magnesium in response to glutamate exposure was shown to depend on high extracellular magnesium concentration,[30] and it is now appreciated that the reported magnesium increase in the in vitro studies was primarily a change in calcium. Therefore, any increase in free magnesium in stroke is now thought to be associated with declines in ATP concentration. The initial increase in intracellular levels following stroke is temporally correlated to approximately a 60% fall in extracellular magnesium levels as measured by microdialysis,[28] as well as by magnesium declines in blood and cerebrospinal fluid (CSF).[31] While a significant correlation was reported between CSF magnesium and infarct size and clinical outcome following acute stroke,[32] no correlation was observed between CSF magnesium and serum magnesium. Finally, magnesium deficiency has been noted to exacerbate injury from stroke[33,34] in a manner similar to that reported following traumatic brain injury.[12]

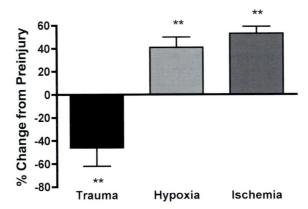

FIGURE 28.2. Change in brain intracellular free magnesium concentration as measured in the first 24 h following experimental trauma, hypoxia, or ischemia. **$p < 0.01$ from pre-injury.

Related to ischemia is the condition known as hypoxia–ischemia. There are a number of similarities between ischemia and hypoxia–ischemia, particularly in relation to energy depletion with profound hypoxia–ischemia. However, with milder levels of insult, it has been possible to maintain ATP levels and determine free magnesium levels using magnetic resonance spectroscopy. Under these conditions, intracellular free magnesium declines during hypoxia–ischemia in neonatal rats.[35] Declines in brain magnesium by 18% over 4 h have also been reported after in utero hypoxia.[36] Finally, there is a deficiency in cord-blood magnesium following hypoxia–ischemia.[37]

Although there are many similarities in stroke and traumatic brain injury, there is the fundamental difference in stroke in that ATP levels decline and that this is associated with an increase in intracellular magnesium concentration, whereas following traumatic brain injury, ATP levels remain stable and that magnesium levels decline after injury (Figure 28.2). At the extracellular level, both injuries produce declines in CSF and serum magnesium concentration.

Neurodegenerative Disease

There is a long history of reports associating low magnesium levels to neurodegenerative disease. While it is recognized that magnesium deficiency alone cannot account for neurodegenerative disease,[38] there is increasing evidence to suggest that in combination with other predisposing factors, altered magnesium homeostasis may play some part in the neurodegenerative condition. For example, in Parkinson's disease, lower concentrations of magnesium and copper were observed in specific brain regions associated with the disease.[39] Phosphorus magnetic resonance spectroscopy confirmed that the brain free magnesium concentration was significantly decreased in patients affected with Parkinson's disease,[40] and that this low magnesium concentration was not apparent in multiple system atrophy (including the olivopontocerebellar atrophy and stiatonigral degeneration variants). A number of studies have also reported an association between low magnesium and high aluminium concentration in areas of brains affected by Parkinson's disease,[41] with recent studies in animal models demonstrating that high aluminium exacerbates tissue magnesium depletion,[42] and that reduced dietary magnesium intake plays a critical role in disease onset.[43] While the mechanism associated with low magnesium facilitating disease onset is unknown, Golts and coworkers[44] have recently shown that the reduced magnesium concentration may affect the conformation of α-synuclein and thereby increase its aggregation in Lewy bodies.

Another neurodegenerative disease demonstrating an association with low magnesium levels is Alzheimer's disease. Early reports demonstrated a low plasma magnesium concentration in severe dementia of the Alzheimer's type.[45] Subsequently, decreased magnesium concentration in specific brain structures of Alzheimer's patients has been identified.[11] This magnesium deficit is particularly apparent in the hippocampus,[46] whose degeneration has been associated with short-term memory loss. Of interest is the fact that survivors of traumatic brain injury have an increased risk of Alzheimer's disease.[47] Part of the pathology of traumatic brain injury is decline in brain intracellular magnesium concentration and activation of amyloidogenic pathways that result in formation of the toxic β amyloid protein.[48] Administration of magnesium after brain trauma is thought to alter the post-translational processing of amyloid precursor protein away from the toxic form and toward the neuroprotective α forms.[48]

Finally, magnesium has been shown to be increased in schizophrenia.[49] While increases in free magnesium concentration can be attributed to a decrease in ATP ligands that bind the cation, no decrease in ATP has been reported in schizophrenia. On the other hand, phosphorus magnetic resonance spectroscopy has shown decreased phosphomonoester/phosphodiester peaks in schizophrenia, implying changes in membrane structure. Because it is well known that the membrane is the major buffer for free magnesium concentration,[14] it could be hypothesized that the altered membrane structure would decrease the number of available binding sites for magnesium.

Therapeutic Efficacy

Therapies targeting the restoration of magnesium homeostasis have demonstrated beneficial effects in a wide variety of CNS pathologies including focal[50,51] and diffuse traumatic brain injury,[52-55] sensorimotor lesions,[56] ischemia,[57-59] hypoxia,[60] subarachnoid hemorrhage,[61,62] and fetal brain damage following maternal seizures,[63] amongst others.

Nonetheless, there are reports that do not support a protective role for magnesium. $MgSO_4$ at doses of 300 to 400 mg/kg is not protective in hypoxia-ischemia[64,65] or in asphyxia in lambs.[66] When administered pre-injury, magnesium at a dose of 270 mg/kg is protective in hypoxia–ischemia, but not when administered postinjury.[67] Similarly, MgCl$_2$ does not protect in ischemia when used at doses of 235 mg/kg.[68] It is significant that dose optimization studies in both traumatic brain injury[69] and stroke[70] suggest that the optimal magnesium dose is between 30 and 90 mg/kg, and certainly less than 100 mg/kg. Higher doses of magnesium have the clear potential to cause hypotension and other cardiovascular effects which may adversely affect outcome. Our conclusion is therefore that studies that use more than 100 mg/kg IV run the risk of being negative.

A second factor when considering whether magnesium will be efficacious in the treatment of CNS injury is whether ATP declines during the insult. As stated above, ATP does decline in ischemia and in severe hypoxia. Under these conditions, the intracellular free magnesium concentration will increase as more of the cation is released from the binding ligand. Hence, the observations that magnesium initially increases in ischemia. This raises the issue of whether the increase in magnesium is an injury factor or a consequence of injury. In our opinion, any magnesium increase that accompanies energy depletion is more likely a consequence of injury rather than an injury factor in itself. Magnesium administration at this point may have little effect, particularly at the intracellular level where magnesium is already elevated. This does not preclude the fact that magnesium treatment may have beneficial effects at the extracellular level. However, in the absence of ATP, increasing the intracellular concentration of the cation will have little effect.

Neuroprotective Properties

As stated above, magnesium administration can potentially be protective at both the intracellular and the extracellular level depending upon the ATP concentration during the insult. Where ATP levels fall, and free magnesium increases, there can only be a benefit from the extracellular effects of the cation. In contrast, where intracellular magnesium declines, there can be benefits from the combined intracellular and extracellular effects of the cation. Clearly, this will impact on the therapeutic efficacy.

Extracellular Effects

Magnesium has been shown to cross the blood–brain barrier and enter the CSF,[71] a process that will be enhanced after acute CNS injury given that the blood–brain barrier is usually disrupted for extended periods of time. Having gained entry to the CNS, magnesium has a number of neuroprotective actions at the extracellular level, the most widely characterized being its effects on glutamate release and excitotoxicty. Increased extracellular magnesium is known to reduce glutamate release,[72,73] as well as the capacity of the neurotransmitter to bind to the NMDA receptor.[74] In addition, intracellular and extracellular magnesium is a voltage-dependent blocker of this NMDA channel[75,76] whose activity is associated with entry of calcium into neurons and the initiation of an autodestructive cascade.[72] By blocking the NMDA channel, magnesium reduces calcium influx into the cell and protects against this injury cascade. This neuroprotective property of magnesium has now been extensively demonstrated both in vitro and in vivo.[72,74,75,77–79] Interestingly, injury reduces the magnesium blockade of the NMDA channel[80] and the block of the NMDA channel increases substantially with age,[79] the latter suggesting that the importance of the magnesium block increases as the complexity of the CNS increases.

The effects of magnesium on calcium entry into cells are not limited to the NMDA channel. Indeed, the magnesium ion has been described as nature's physiological calcium antagonist, having direct effects on calcium uptake.[81] Moreover, the modulation of ion fluxes is not limited to calcium, with the K+ channel also being directly regulated by magnesium concentration.[81] In subarachnoid hemorrhage, the ability of magnesium to modulate ion fluxes and reduce depolarization time is thought to account, in part, for smaller lesion volumes.[82] Finally, with respect to neurotransmitters, magnesium has been shown to decrease the dopamine surge during anoxia,[83] suggesting that neurotransmitter release in general may be inhibited by extracellular magnesium, perhaps by inhibiting presynaptic calcium channels.

This combined effect on presynaptic neurotransmitter release and postsynaptic ion fluxes may therefore account for the observed ability of magnesium to inhibit membrane depolarization and cortical spreading depression.[84]

Magnesium's extracellular effects are not limited to neurotransmission, with a number of reports suggesting actions at the level of the vasculature. Magnesium can reverse vasoconstriction and increase cerebral blood flow in rats.[85,86] As a vasodilator, magnesium has the ability to inhibit cerebrovasospasms,[87] perhaps accounting for, in part, the reduced incidence of vasospastic episodes following subarachnoid hemorrhage, and the reduced lactate production after brain trauma.[88] Finally, magnesium has recently been shown to attenuate blood–brain barrier permeability following traumatic brain injury[89] and to reduce associated vasogenic edema formation.[90,91]

Intracellular Effects

Not only does magnesium enter the cerebrospinal fluid following parenteral administration,[71] it has also been shown to increase brain intracellular free magnesium concentration[53] and to influence a number of intracellular processes, including energy metabolism, free radical formation, membrane structure, apoptosis, and receptor activity.[92] For example, the effects of extracellular magnesium on the excitatory amino acids or ligand-gated ionotropic receptors described above are also mediated by intracellular magnesium, and in the case of the NMDA channel, intracellular magnesium is the main regulator of channel activity.[75] Modulatory effects on receptor binding and activity have also been described for seven metabotropic, transmembrane receptors including serotonergic and opioid receptors. In these receptors, intracellular magnesium is required for the activation of G proteins and therefore the coupling of G protein–linked receptors to their effector mechanisms.[93]

In terms of energy metabolism, magnesium is an essential cofactor in all energy-producing and -consuming reactions, and therefore changes in the cation concentration will have profound consequences for all bioenergetic reactions, including those of oxidative phosphorylation.[94] After ischemia, magnesium is thought to reduce infarct size, in part, by preserving energy metabolism.[73] Specifically, magnesium preserves glucose and pyruvate levels after ischemia, while attenuating elevation of glutamate, lactate,[73] and malondialdehyde levels.[95] At the level of the mitochondrion, magnesium preserves mitochondrial membrane potential,[96] and after traumatic brain injury, improves mitochondrial ultrastructure.[97] It has also been shown to inhibit the opening of the mitochondrial permeability transition pore,[98] whose activity is thought to be integral to apoptosis.

Since the initial observations that intracellular magnesium deficiency enhances apoptosis[99] and that decline in blood magnesium causes an increase in pro-apoptotic intracellular ceramide,[100] a role for magnesium in apoptosis has become increasingly apparent.[101] Magnesium decreases nuclear oxidative damage and DNA fragmentation,[102] possibly through its ability to inhibit the induction of DNA fragmentation factor.[103] The cation also attenuates induction of p53 after trauma,[104] and reduces hippocampal apoptosis following hypoxia–ischemia,[105] although both of these studies suggest differential effects depending on the brain region. Whether this is dependent on the basal levels of magnesium and adenylates, which vary according to brain region,[10,11] is unknown. The work of Delivoria-Papadopolous and colleagues[106] has demonstrated that magnesium alters the balance between pro-apoptotic bax and the anti-apoptotic bcl-2, thus inhibiting the activation of caspases. These findings were consistent with others in hypoxia–ischemia, demonstrating magnesium inhibition of caspase-3 activation and MAP-2 immunostaining.[107] Clearly, magnesium has multiple effects on the apoptotic cell death pathway, although further studies are required to fully characterize the implications.

At the level of the cell membrane, magnesium is an integral component, with the membrane actually serving as a free magnesium buffer.[14] By reducing membrane lipolysis,[108] magnesium is thought to inhibit lipid peroxidation and generation of free radicals.[109,110] The reduction in free-radical production can also be mediated by magnesium's ability to inhibit reduced nicotinamide-adenine dinucleotide phosphate oxidase, an enzyme that produces superoxide radicals.[111] This inhibition is equivalent to other known inhibitors of this enzyme, such as copper. Magnesium also reduces free radicals by significantly increasing the levels of endogenous antioxidants after injury,[88] specifically activating superoxide dismutase and catalase, as well as the concentration of glutathione.[112] Through its effects on the Na$^+$/K$^+$ ATPase, magnesium can be considered as having an indirect effect on membrane potential.

While the critical role of magnesium in RNA aggregation, protein synthesis and the replication of DNA itself has been known for some time,[113] recent results suggest that the cation also reduces acute cytoskeletal alterations following traumatic brain injury,[114] perhaps through modulation of calpain. Finally, the cation is thought to regulate the processing of amyloid precursor protein (APP) away from the neurotoxic β amyloid form and toward the neuroprotective soluble APPα form.[48] This potentially has implications for Alzheimer's disease.

Conclusion

A role for magnesium in both physiological and pathological states of the CNS has become increasingly evident, with considerable evidence now supporting the view that the intracellular free ion concentration is critical in normal cell functioning, and that the ion concentration can change in response to various stimuli. The change in magnesium concentration in pathological states can be considered as either an injury factor, or an injury marker, depending on the energy status of the cell. Where the cell does not have energy failure, the change is considered an injury factor and therapeutic approaches that restore magnesium homeostasis have shown considerable success in improving outcome. Because magnesium is involved in so many physiological and biochemical processes, administering a dose that restores homeostasis without producing adverse side effects is critical to therapeutic efficacy. While our understanding of magnesium's role in this multitude of processes has grown considerably over the last three decades, there is still a need to identify and characterize the transporter mechanisms that are responsible for maintaining normal magnesium concentration. Only then can the critical role for magnesium in biological processes, and in particular the central nervous system, be fully appreciated.

References

1. Peck CH, Meltzer SJ. Anesthesia in human beings by intravenous injection of magnesium sulfate. *JAMA* 1916;67:1131–1133.
2. Birch NJ, ed. *Magnesium and the Cell*. London: Academic Press; 1993.
3. Alvarez-Leefmans FJ, Giraldez F, Gamino SM. Intracellular free magnesium in excitable cells: its measurements and its biological significance. *Can J Physiol Pharmacol* 1987;65:915–925.
4. London RE. Methods for measurement of intracellular magnesium: NMR and fluorescence. *Annu Rev Physiol* 1991;53:241–258.
5. Gupta RK, Gupta P, Yushok WD, Rose ZB. On the noninvasive measurements of intracellular free magnesium by ^{31}P NMR spectroscopy. *Physiol Chem Physics Med NMR* 1983;15:265–280.

6. Gupta RK, Gupta P. NMR studies of intracellular metal ions in intact cells and tissues. *Ann Rev Biophys Bioeng* 1984;13:21–43.
7. Vink R, McIntosh TK, Demediuk P, Faden AI. Decrease in total and free magnesium concentration following traumatic brain injury in rats. *Biochem Biophys Res Commun* 1987;149:594–599.
8. Vink R, McIntosh TK, Demediuk P, Weiner MW, Faden AI. Decline in intracellular free magnesium concentration is associated with irreversible tissue injury following brain trauma. *J Biol Chem* 1988;263:757–761.
9. Iotti S, Frassineti C, Alderighi L, Sabatini A, Vacca A, Barbiroli B. In vivo (31)P-MRS assessment of cytosolic [Mg(2+)] in the human skeletal muscle in different metabolic conditions. *Magn Reson Imaging* 2000;18:607–614.
10. Chollet D, Franken P, Raffin Y, Malafosse A, Widmer J, Tafti M. Blood and brain magnesium in inbred mice and their correlation with sleep quality. *Am J Physiol Regul Integr Comp Physiol* 2000;279:R2173–R2178.
11. Andrasi E, Igaz S, Molnar Z, Mako S. Disturbances of magnesium concentrations in various brain areas in Alzheimer's disease. *Magnes Res* 2000;13:189–196.
12. McIntosh TK, Faden AI, Yamakami I, Vink R. Magnesium deficiency exacerbates and pretreatment improves outcome following traumatic brain injury in rats: ^{31}P magnetic resonance spectroscopy and behavioural studies. *J Neurotrauma* 1988;5:17–31.
13. McIntosh TK, Vink R, Soares H, Hayes RL, Simon RP. Effect of noncompetitive blockade of N-methyl-D-aspartate receptors on the neurochemical sequelae of experimental brain injury. *J Neurochem* 1990;55:1170–1179.
14. Corkey BE, Duszynski J, Rich TL, Matschinsky B, Williamson JR. Regulation of free and bound magnesium in rat hepatocytes and isolated mitochondria. *J Biol Chem* 1986;261:2567–2574.
15. Heath DL, Vink R. Traumatic brain axonal injury produces sustained decline in intracellular free magnesium concentration. *Brain Res* 1996;738:150–153.
16. Cernak I, Radosevic P, Malicevic Z, Savic J. Experimental magnesium depletion in adult rabbits caused by blast overpressure. *Magnes Res* 1995;8:249–259.
17. Suzuki M, Nishina M, Endo M, et al. Decrease in cerebral free magnesium concentration following closed head injury and effects of VA-045 in rats. *Gen Pharmacol* 1997;28:119–121.
18. Smith DH, Cecil KM, Meaney DF, et al. Magnetic resonance spectroscopy of diffuse brain trauma in the pig. *J Neurotrauma* 1998;15:665–674.
19. Vink R, Heath DL, McIntosh TK. Acute and prolonged alterations in brain free magnesium following fluid percussion induced brain trauma in rats. *J Neurochem* 1996;66:2477–2483.
20. Heath DL, Vink R. Blood free magnesium concentration declines following graded experimental traumatic brain injury. *Scand J Clin Lab Invest* 1998;58:161–166.
21. Polderman KH, Bloemers FW, Peerdeman SM, Girbes AR. Hypomagnesemia and hypophosphatemia at admission in patients with severe head injury. *Crit Care Med* 2000;28:2022–2025.
22. Memon ZI, Altura BT, Benjamin JL, Cracco RQ, Altura BM. Predictive value of serum ionized but not total magnesium levels in head injuries. *Scand J Clin Lab Invest* 1995;55:671–677.

23. Cernak I, Savic VJ, Kotur J, Prokic V, Veljovic M, Grbovic D. Characterization of plasma magnesium concentration and oxidative stress following graded traumatic brain injury in humans. *J Neurotrauma* 2000;17:53–68.
24. Frankel H, Haskell R, Lee SY, Miller D, Rotondo M, Schwab CW. Hypomagnesemia in trauma patients. *World J Surg* 1999;23:966–999.
25. Bareyre FM, Saatman KE, Helfaer MA, et al. Alterations in ionized and total blood magnesium after experimental traumatic brain injury: relationship to neurobehavioral outcome and neuroprotective efficacy of magnesium chloride. *J Neurochem* 1999;73:271–280.
26. Cavaliere F, Sciarra M, Crea MA, Rossi M, Proietti R. Variazioni del magnesio sierico ed urinario in pazienti tramatizzati cranici. *Recent Prog Med* 1985;76: 561–566.
27. Goodman JC, Valadka AB, Gopinath SP, Uzura M, Grossman RG, Robertson CS. Simultaneous measurement of cortical potassium, calcium, and magnesium levels measured in head injured patients using microdialysis with ion chromatography. *Acta Neurochir Suppl (Wien)* 1999;75:35–37.
28. Lee MS, Wu YS, Yang DY, Lee JB, Cheng FC. Significantly decreased extracellular magnesium in brains of gerbils subjected to cerebral ischemia. *Clin Chim Acta* 2002;318:121–125.
29. Helpern JA, Van de Linde AMQ, Welch KMA, et al. Acute elevation and recovery of intracellular Mg2+ following human focal cerebral ischemia. *Neurol* 1993; 43:1577–1581.
30. Cheng C, Reynolds IJ. Subcellular localization of glutamate-stimulated intracellular magnesium concentration changes in cultured rat forebrain neurons using confocal microscopy. *Neuroscience* 2000;95:973–979.
31. Borowik H, Pryszmont M. Concentration of magnesium in serum and cerebrospinal fluid in patients with stroke. *Neurol Neurochir Pol* 1998;32:1377–1383.
32. Lampl Y, Geva D, Gilad R, Eshel Y, Ronen L, Sarova-Pinhas I. Cerebrospinal fluid magnesium level as a prognostic factor in ischaemic stroke. *J Neurol* 1998;245: 584–588.
33. Altura BM, Gebrewold A, Zhang AM, Altura BT, Gupta RK. Magnesium deficiency exacerbates brain injury and stroke mortality induced by alcohol—a P-31-NMR in vivo study. *Alcohol* 1998;15:181–183.
34. Demougeot C, Bobillier-Chaumont S, Mossiat C, Marie C, Berthelot A. Effect of diets with different magnesium content in ischemic stroke rats. *Neurosci Lett* 2004;362:17–20.
35. Williams GD, Smith GD. Application of the accurate assessment of intracellular magnesium and pH from the 31P shifts of ATP to cerebral hypoxia-ischemia in neonatal rat. *Magn Reson Med* 1995;33:853–857.
36. Powell SR, Wahezi SE, Maulik D. The effect of in utero hypoxia on fetal heart and brain trace elements. *J Trace Elem Med Biol* 2002;16:245–248.
37. Ilves P, Blennow M, Kutt E, et al. Concentrations of magnesium and ionized calcium in umbilical cord blood in distressed term newborn infants with hypoxic-ischemic encephalopathy. *Acta Paediatr* 1996;85:1348–1350.
38. Durlach J, Bac P, Durlach V, Durlach A, Bara M, Guiet-Bara A. Are age-related neurodegenerative diseases linked with various types of magnesium depletion? *Magnes Res* 1997;10:339–353.
39. Uitti RJ, Rajput AH, Rozdilsky B, Bickis M, Wollin T, Yuen WK. Regional metal concentrations in Parkinson's disease, other chronic neurological diseases, and control brains. *Can J Neurol Sci* 1989;16:310–314.

40. Barbiroli B, Martinelli P, Patuelli A, et al. Phosphorus magnetic resonance spectroscopy in multiple system atrophy and Parkinson's disease. *Mov Disord* 1999;14:430–435.
41. Yasui M, Kihira T, Ota K. Calcium, magnesium and aluminum concentrations in Parkinson's disease. *Neurotoxicology* 1992;13:593–600.
42. Yasui M, Ota K. Aluminum decreases the magnesium concentration of spinal cord and trabecular bone in rats fed a low calcium, high aluminum diet. *J Neurol Sci* 1998;157:37–41.
43. Oyanagi K. The nature of the parkinsonism-dementia complex and amyotrophic lateral sclerosis of Guam and magnesium deficiency. *Parkinsonism Relat Disord* 2005;11(suppl. 1):S17–S23.
44. Golts N, Snyder H, Frasier M, Theisler C, Choi P, Wolozin B. Magnesium inhibits spontaneous and iron-induced aggregation of alpha-synuclein. *J Biol Chem* 2002; 277:16116–16123.
45. Lemke MR. Plasma magnesium decrease and altered calcium/magnesium ratio in severe dementia of the Alzheimer type. *Biol Psychiatry* 1995;37:341–343.
46. Durlach J. Magnesium depletion and pathogenesis of Alzheimer's disease. *Magnes Res* 1990;3:217–218.
47. Starkstein SE, Jorge R. Dementia after traumatic brain injury. *Int Psychogeriatr* 2005;17(suppl. 1):S93–S107.
48. Van Den Heuvel C, Finnie JW, Blumbergs PC, et al. Upregulation of neuronal amyloid precursor protein (APP) and APP mRNA following magnesium sulphate (MgSO4) therapy in traumatic brain injury. *J Neurotrauma* 2000;17:1041–1053.
49. Hinsberger AD, Williamson PC, Carr TJ, et al. Magnetic resonance imaging volumetric and phosphorus 31 magnetic resonance spectroscopy measurements in schizophrenia. *J Psychiatry Neurosci* 1997;22:111–117.
50. McIntosh TK, Vink R, Yamakami I, Faden AI. Magnesium protects against neurological deficit after brain injury. *Brain Res* 1989;482:252–260.
51. Bareyre FM, Saatman KE, Raghupathi R, McIntosh TK. Postinjury treatment with magnesium chloride attenuates cortical damage after traumatic brain injury in rats. *J Neurotrauma* 2000;17:1029–1039.
52. Heath DL, Vink R. Neuroprotective effects of MgSO4 and MgCl2 in closed head injury: a comparative phosphorus NMR study. *J Neurotrauma* 1998;15:183–189.
53. Heath DL, Vink R. Delayed therapy with magnesium up to 24h following traumatic brain injury improves motor outcome. *J Neurosurg* 1999;90:504–509.
54. Vink R, O'Connor CA, Nimmo AJ, Heath DL. Magnesium attenuates persistent functional deficits following diffuse traumatic brain injury in rats. *Neurosci Lett* 2003;336:41–44.
55. Turner RJ, Dasilva KW, O'Connor C, van den Heuvel C, Vink R. Magnesium gluconate offers no more protection than magnesium sulphate following diffuse traumatic brain injury in rats. *J Am Coll Nutr* 2004;23:541S–544S.
56. Hoane MR, Irish SL, Marks BB, Barth TM. Preoperative regimens of magnesium facilitate recovery of function and prevent subcortical atrophy following lesions of the rat sensorimotor cortex. *Brain Res Bull* 1998;45:45–51.
57. Yang Y, Qiu L, Fayyaz A, Shuaib A. Survival and histological evaluation of therapeutic window of post-ischemia treatment with magnesium sulfate in embolic stroke model of rat. *Neurosci Lett* 2000;285:119–122.
58. Lampl Y, Gilad R, Geva D, Eshel Y, Sadeh M. Intravenous administration of magnesium in acute stroke: a randomized double-blind study. *Clin Neuropharmacol* 2001;24:11–15.

59. Miles AN, Majda BT, Meloni BP, Knuckey NW. Postischemic intravenous administration of magnesium sulfate inhibits hippocampal CA1 neuronal death after transient global ischemia in rats. *Neurosurgery* 2001;49:1143–1150.
60. Hallak M, Hotra JW, Kupsky WJ. Magnesium sulfate protection of fetal rat brain from severe maternal hypoxia. *Obstet Gynecol* 2000;96:124–128.
61. Pyne GJ, Cadoux-Hudson TA, Clark JF. Magnesium protection against in vitro cerebral vasospasm after subarachnoid haemorrhage. *Br J Neurosurg* 2001;15:409–415.
62. Veyna RS, Seyfried D, Burke DG, et al. Magnesium sulfate therapy after aneurysmal subarachnoid hemorrhage. *J Neurosurg* 2002;96:510–514.
63. Hallak M, Kupsky WJ, Hotra JW, Evans JB. Fetal rat brain damage caused by maternal seizure activity: prevention by magnesium sulfate. *Am J Obstet Gynecol* 1999;181:828–834.
64. Galvin KA, Oorschot DE. Postinjury magnesium sulfate treatment is not markedly neuroprotective for striatal medium spiny neurons after perinatal hypoxia/ischemia in the rat. *Pediatr Res* 1998;44:740–745.
65. Greenwood K, Cox P, Mehmet H, et al. Magnesium sulfate treatment after transient hypoxia–ischemia in the newborn piglet does not protect against cerebral damage. *Pediatr Res* 2000;48:346–350.
66. de Haan HH, Gunn AJ, Williams CE, Heymann MA, Gluckman PD. Magnesium sulfate therapy during asphyxia in near-term fetal lambs does not compromise the fetus but does not reduce cerebral injury. *Am J Obstet Gynecol* 1999;176:18–27.
67. Sameshima H, Ota A, Ikenoue T. Pretreatment with magnesium sulfate protects against hypoxic-ischemic brain injury but postasphyxial treatment worsens brain damage in seven-day-old rats. *Am J Obstet Gynecol* 1999;180:725–730.
68. Milani H, Lepri ER, Giordani F, Favero-Filho LA. Magnesium chloride alone or in combination with diazepam fails to prevent hippocampal damage following transient forebrain ischemia. *Braz J Med Biol Res* 1999;32:1285–1293.
69. Heath DL, Vink R. Optimization of magnesium therapy following severe diffuse axonal brain injury in rats. *J Pharmacol Exp Ther* 1999;288:1311–1316.
70. Muir KW. New experimental and clinical data on the efficacy of pharmacological magnesium infusions in cerebral infarcts. *Magnes Res* 1998;11:43–56.
71. Hallak M, Berman RF, Irtenkauf SM, Evans MI, Cotton DB. Peripheral magnesium sulphate enters the brain and increases the threshold for hippocampal seizures in rats. *Am J Obstet Gynecol* 1992;167:1605–1610.
72. Rothman SM, Olney JW. Glutamate and the pathophysiology of hypoxic-ischemic brain damage. *Ann Neurol* 1986;19:105–111.
73. Lin JY, Chung SY, Lin MC, Cheng FC. Effects of magnesium sulfate on energy metabolites and glutamate in the cortex during focal cerebral ischemia and reperfusion in the gerbil monitored by a dual-probe microdialysis technique. *Life Sci* 2002;71:803–811.
74. Hallak M, Berman RF, Irtenkauf SM, Janusz CA, Cotton DB. Magnesium sulphate treatment decreases N-Methyl-D-Aspartate receptor binding in the rat brain—an autoradiographic study. *J Soc Gynecol Invest* 1994;1:25–30.
75. Mayer ML, Westbrook GL, Guthrie PB. Voltage-dependent block by Mg^{2+} of NMDA responses in spinal cord neurons. *Nature* 1984;309:261–263.
76. Johnson JW, Ascher P. Voltage-dependent block by intracellular Mg^{2+} of N-methyl-D-aspartate activated channels. *J Biophys* 1990;57:1085–1090.

77. Hoffman DJ, Marro PJ, McGowan JE, Mishra OP, Deliveria Papadopoulos M. Protective effect of MgSO4 infusion on NMDA receptor binding characteristics during cerebral cortical hypoxia the newborn piglet. *Brain Res* 1994;644: 144–149.
78. Frandsen A, Schousboe A. Effect of magnesium on NMDA mediated toxicity and increases in [Ca2+]i and cGMP in cultured neocortical neurones: evidence for distinct regulation of different responses. *Neurochem Int* 1994;25:303–308.
79. Strecker GJ. Blockade of NMDA activated channels by magnesium in the immature rat hippocampus. *J Neurophysiol* 1994;72:1538–1548.
80. Zhang L, Rzigalinski BA, Ellis EF, Satin LS. Reduction of voltage-dependent Mg^{2+} blockade of NMDA current in mechanically injured neurons. *Science* 1996; 274:1921–1923.
81. Agus ZS, Morad M. Modulation of cardiac ion channels by magnesium. *Annu Rev Physiol* 1991;53:299–307.
82. van den Burgh WM, Zuur JK, Kamerling NA, et al. Role of magnesium in the reduction of ischemic depolarization and lesion volume after experimental subarachnoid hemorrhage. *J Neurosurg* 2002;97:416–422.
83. Nakajima W, Ishida A, Takada G. Magnesium attenuates a striatal dopamine increase induced by anoxia in the neonatal rat brain: an in vivo microdialysis study. *Pediatr Res* 1997;41:809–814.
84. van der Hel WS, van den Bergh WM, Nicolay K, Tulleken KA, Dijkhuizen RM. Suppression of cortical spreading depressions after magnesium treatment in the rat. *Neuroreport* 1998;9:2179–2182.
85. Kemp PA, Gardiner SM, Bennnett T, Rubin PC. Magnesium sulphate reverses the carotid vasoconstriction caused by endothelin-I, angiotensin-II and neuropeptide Y, but not that caused by N(G)-nitro-L-arginine methyl ester, in conscious rats. *Clin Sci* 1993;85:175–181.
86. Kemp PA, Gardiner SM, March JE, Rubin PC, Bennett T. Assessment of the effects of endothelin-1 and magnesium sulphate on regional blood flows in conscious rats, by the coloured microsphere reference technique. *Br J Pharmacol* 1999; 126:621–626.
87. Altura BM, Altura BT, Gupta RK. Alcohol intoxication results in rapid loss in free magnesium in brain and disturbances in brain bioenergetics: relation to cerebrospasm, alcohol-induced strokes, and barbiturate anesthesia-induced deaths. *Magnes Trace Elem* 1992;10:122–135.
88. Ustun ME, Duman A, Ogun CO, Vatansev H, Ak A. Effects of nimodipine and magnesium sulfate on endogenous antioxidant levels in brain tissue after experimental head trauma. *J Neurosurg Anesthesiol* 2001;13:227–232.
89. Kaya M, Kucuk M, Kalayci RB, et al. Magnesium sulfate attenuates increased blood-brain barrier permeability during insulin-induced hypoglycemia in rats. *Can J Physiol Pharmacol* 2001;79:793–798.
90. Feldman Z, Gurevitch B, Artru AA, et al. Effect of magnesium given 1 hour after head trauma on brain edema and neurological outcome. *J Neurosurg* 1996;85: 131–137.
91. Esen F, Erdem T, Aktan D, et al. Effects of magnesium administration on brain edema and blood-brain barrier breakdown after experimental traumatic brain injury in rats. *J Neurosurg Anesthesiol* 2003;15:119–125.
92. Vink R, Cernak I. Regulation of brain intracellular free magnesium following traumatic injury to the central nervous system. *Front Biosci* 2000;5:656–665.

93. Birnbaumer L, Abramowitz J, Brown AM. Receptor-effector coupling by G proteins. *Biochim Biophys Acta* 1990;1031:163–224.
94. Vink R, Golding EM, Headrick JP. Bioenergetic analysis of oxidative metabolism following traumatic brain injury in rats. *J Neurotrauma* 1994;11:265–274.
95. Ustun ME, Gurbilek M, Ak A, Vatansev H, Duman A. Effects of magnesium sulfate on tissue lactate and malondialdehyde levels in experimental head trauma. *Intensive Care Med* 2001;27:264–268.
96. Sharikabad MN, Ostbye KM, Brors O. Increased [Mg2+]o reduces Ca2+ influx and disruption of mitochondrial membrane potential during reoxygenation. *Am J Physiol Heart Circ Physiol* 2001;281:H2113–H2123.
97. Xu M, Dai W, Deng X. Effects of magnesium sulfate on brain mitochondrial respiratory function in rats after experimental traumatic brain injury. *Chin J Traumatol* 2002;4:361–364.
98. Halestrap AP, Kerr PM, Javadov S, Woodfield KY. Elucidating the molecular mechanism of the permeability transition pore and its role in reperfusion injury of the heart. *Biochim Biophys Acta* 1998;1366:79–94.
99. Malpuech-Brugere C, Nowacki W, Gueux E, et al. Accelerated thymus involution in magnesium-deficient rats is related to enhanced apoptosis and sensitivity to oxidative stress. *Br J Nutr* 1999;81:405–411.
100. Morrill MA, Gupta RK, Kostellow AB, et al. Mg2+ modulates membrane shingolipid and lipid second messenger levels in vascular smooth muscle cells. *FEBS Lett* 1998;440:167–171.
101. Wolf FI, Cittadini A. Magnesium in cell proliferation and differentiation. *Front Biosci* 1999;4:D607–D617.
102. Maulik D, Qayyum I, Powell SR, Karantza M, Mishra OP, Delivoria-Papadopoulos M. Post-hypoxic magnesium decreases nuclear oxidative damage in the fetal guinea pig brain. *Brain Res* 2001;26:130–136.
103. Zhang C, Raghupathi R, LaPlaca MC, Bareyre FM, McIntosh TK. Changes in DNA fragmentation factor (DFF) following experimental brain trauma in the rat: effect of posttraumatic magnesium treatment. *J Neurotrauma* 1998;15:904.
104. Muir JK, Raghupathi R, Emery DL, Bareyre FM, McIntosh TK. Postinjury magnesium treatment attenuates traumatic brain injury-induced cortical induction of p53 mRNA in rats. *Exp Neurol* 1999;159:584–593.
105. Turkyilmaz C, Turkyilmaz Z, Atalay Y, Soylemezoglu F, Celasun B. Magnesium pretreatment reduces neuronal apoptosis in newborn rats in hypoxia-ischemia. *Brain Res* 2002;955:133–137.
106. Ravishankar S, Ashraf QM, Fritz K, Mishra OP, Delivoria-Papadopoulos M. Expression of Bax and Bcl-2 proteins during hypoxia in cerebral cortical neuronal nuclei of newborn piglets: effect of administration of magnesium sulfate. *Brain Res* 2001;18:23–29.
107. Sameshima H, Ikenoue T. Effect of long-term, postasphyxial administration of magnesium sulfate on immunostaining of microtubule-associated protein-2 and activated caspase-3 in 7-day-old rat brain. *J Soc Gynecol Invest* 2002;9:203–209.
108. Gasparovic C, Berghmans K. Ca2+- and Mg2+-modulated lipolysis in neonatal rat brain slices observed by one- and two-dimensional NMR. *J Neurochem* 1998;71:1727–1732.
109. Gunther T, Hollriegl V, Vormann J, Bubeck J, Classen HG. Increased lipid peroxidation in rat tissues by magnesium deficiency and vitamin E depletion. *Magnes Bull* 1994;16:38–43.

110. Maulik D, Zanelli S, Numagami Y, Ohnishi ST, Mishra OP, Delivoria-Papadopoulos M. Oxygen free radical generation during in-utero hypoxia in the fetal guinea pig brain: the effects of maturity and of magnesium sulfate administration. *Brain Res* 1999;817:117–122.
111. Afanas'ev LB, Suslova TB, Chermisina ZP, Abramova NE, Korkina LG. Study of antioxidant properties of metal aspartates. *Analyst* 1995;120:859–862.
112. Matkovics B, Kiss I, Kiss SA. The activation by magnesium treatment of antioxidants eliminating the oxygen free radicals in *Drosophila melanogaster* in vivo. *Magnes Res* 1997;10:33–38.
113. Terasaki M, Rubin H. Evidence that intracellular magnesium is present in cells at a regulatory concentration for protein synthesis. *Proc Natl Acad Sci USA* 1985;82:7324–7326.
114. Saatman KE, Bareyre FM, Grady MS, McIntosh TK. Acute cytoskeletal alterations and cell death induced by experimental brain injury are attenuated by magnesium treatment and exacerbated by magnesium deficiency. *J Neuropathol Exp Neurol* 2001;60:183–194.

Magnesium in Dental Medicine

29
Tooth and Magnesium

Masayuki Okazaki

Inorganic tooth materials are composed of hydroxyapatite. However, biological apatites contain many trace elements and have different crystallographic properties. Tooth enamel is well crystallized, contrary to poorly crystallized dentin and bone. Magnesium is richer in dentin than in enamel, and especially significantly affects the physicochemical properties of teeth. In general, magnesium decreases the crystallinity of hydroxyapatite and promotes solubility. Magnesium also has important roles in the relationship to cells. Divalent ions, such as magnesium, promote cell adhesion and contribute to the metabolism of hard tissues. This may affect the formation of teeth and be related to regulation during maturation.

Tooth Structure and Its Chemical Composition

Teeth are hard tissues vital for human life because we maintain our health by chewing food. This biting action is also very important in maintaining brain activity through the impulse of the biting force. These teeth are sometimes destroyed by caries or periodontal diseases. Teeth are structured with enamel, dentin and cementum supported by alveolar bone, and periodontal membrane (ligament). Enamel is composed of almost 95 wt % well-crystallized hydroxyapatite containing a few percentages of noncollagenous proteins and water.[1] Dentine is composed of approximately 60 wt % to 70 wt % of poorly crystallized hydroxyapatite and 30 wt % to 40 wt % of collagen. These chemical compositions are shown in Table 29.1.[2] Magnesium is contained as a trace element and is rich in dentin.

Although the calcium and phosphate molar ratio approximately reflects their content in hydroxyapatite for enamel, it shows a higher value for dentin because several percentage of CO_3^{2-} ions are substituted for PO_4^{3-} positions in the apatite crystals of dentin.[3] The variety of these chemical compositions in hydroxyapatite crystals affects the physicochemical properties, especially the crystallographic and mechanical properties of teeth.

TABLE 29.1. Chemical compositions of human enamel, dentin, and bone (dry wt %).

Composition	Enamel	Dentin	Bone
Ash	95.7	70.0	57.1
Ca	37.9	25.9	22.5
P	17.0	12.6	10.3
Mg	0.42	0.82	0.26
Na	0.55	0.25	0.52
K	0.17	0.09	0.09
CO_2	2.4	3.2	3.5
Cl	0.27	0.0	0.1
F	0.01	0.02	0.05

Crystallographic Aspect

Apatites and related compounds are of fundamental importance in several areas, including the structural chemistry of hard tissue. Hydroxyapatite is the first compound to be examined in a program which calls, in part, for the detailed refinement of several of these materials.[4] The space group and unit cell dimensions are, respectively, $P6_3/m$ and $a = b = 9.432$, $c = 6.881$.

With constantly improving knowledge of structural detail now being provided by precision techniques in X-ray and neutron diffraction, the study of the structure–property relationships at the atomic level of organization seems to be especially promising. Atomic-scale mechanisms and the effects of substitution are particularly important in biological apatites which, apparently, always show considerable substitution and other crystalline defects.[5]

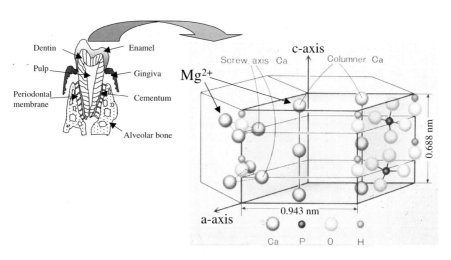

FIGURE 29.1. Tooth structure and hydroxyapatite crystal model.

Apatites are generally formulated as $A_{10}(BO_4)6C2$.[6] Hydroxyapatite is a kind of calcium phosphate and is the main inorganic component of hard tissues such as teeth and bones. Hydroxyapatite has a chemical formula of OH-containing apatites, and is an ionic compound in the hexagonal crystal group, theoretically expressed as $Ca_{10}(PO_4)_6(OH)_2$ (Figure 29.1).[4] Recent computer graphics programs can draw the accurate hydroxyapatite structure.[7]

However, biological apatites contain many of trace elements such as magnesium.[2] The incorporation of magnesium in apatites is very limited unless carbonate or fluoride ions are simultaneously incorporated.[8,9] The amount of magnesium incorporated in apatite is proportional to the magnesium concentration in the solution. The incorporation of magnesium is increased by the simultaneous incorporation of fluoride or carbonate. In particular, the incorporation of magnesium accelerated with the increase of fluoride.[10] The limited incorporation of magnesium is also observed in apatites prepared at temperatures above 900°C by solid-state reaction.[11]

Trace Elements in Teeth

It is well known that a number of elements can be substituted into the apatite crystal structure.[6] The behavior of many trace elements has been reported. It is said that most elements on the periodic table can be substituted into apatite crystals. Both cationic and anionic ions can substitute into the Ca^{2+} position and PO_4^{3-} or OH^- position. Furthermore, some of these substituted ions contribute to metabolism in the human body and cell adhesion. From the viewpoint of ionic charge balance, divalent ions are easy to substitute. Of course, this depends on the ionic radius.[12]

The most interesting ion is Mg^{2+}. Magnesium exists in human minerals, especially as a phosphorus complex in bone, and in chlorophyll in plants, and is known to be closely related to the metabolism and physiology of the human body.[13] Hexokinase, for example, which transfers a phosphoryl group from adenosine triphosphate (ATP) to a variety of six-carbon sugars, requires Mg^{2+} for activity in human cells.[14] In that human hard tissues contain certain amounts of magnesium, it may play an important role in the initial formation of tooth apatites and have a significant effect on their physicochemical properties. However, because magnesium is a minor constituent of human enamel, dentin, and bone, the nature of its association with the mineral phase, especially hydroxyapatite, and its contribution to the properties of biological apatites are still not clear. Because this ion has a smaller ionic radius (0.064 nm) than that of Ca^{2+} ion (0.099 nm),[17] it was previously thought that Mg^{2+} ion is hard to substitute into apatite crystals. Mostly, it was considered that it was adsorbed at the surface of the crystal[15] or combined with organic compounds.[16] However, LeGeros first suggested that some Mg^{2+} ions could substitute into apatite crystals.[8] Okazaki and colleagues[9] also demonstrated this considering the change of the c-axis dimension of magnesium-containing apatites. It is estimated that the maximum magnesium content maintaining the apatite structure may be on fifth of calcium ions. In addition to the magnesium content, tricalium phosphate [TCP; $Ca_3(PO_4)_2$] with magnesium, which is a kind of calcium phosphate, is formed. By incorporating with magnesium, TCP is stabilized.

The presence of Mg in apatites decreases the a-axis dimension, decreases crystallinity, increases HPO_4^{2-} incorporation, and increases the extent of dissolution. Magnesium inhibits remineralization. In vitro, Mg suppresses the crystallization of apatite, stabilizes dicalcium phophate dihydrate (DCPD; $CaHPO_4·2H_2O$), and promotes its formation even in neutral pH, stabilizes amorphous calcium phosphate (ACP), and increases the solubility of synthetic apatites.

Magnesium Distribution in Teeth

Some trace elements such as magnesium, sodium, zinc, fluoride, and chloride have characteristic distribution patterns in human matured enamel and dentin.[17] For example, at the narrow surface layer of enamel, magnesium and sodium show a very steep decrease in concentration, whereas zinc, chloride, and fluoride show a peak concentration much higher than in the rest of enamel. It seems likely that these specific patterns of distribution are related to the progressive mineralization pattern observed during the maturation stage, which is characteristic of the layers. Suga and colleagues[17] indicated by electron microprobe analysis that in the matured enamel of cow, dog, rat, and guinea pig, magnesium, sodium, and chloride have almost the same distribution patterns as in human matured enamel.

Magnesium is distributed in dentin at a much higher level than in enamel.[2,18] Although the magnesium distribution difference between dentin and enamel has not been clarified, it can be speculated that this phenomena may be related to the association of magnesium with organic compounds such as collagen in dentin. Magnesium may also be related to caries susceptibility of teeth the same as carbonate.[19] Both magnesium and carbonate decrease crystallinity and increased the solubility of apatite crystals.

Magnesium Incorporation with Cells

Although enamel develops from epithelial cells, dentin and bone develop from mesenchymal cells by epithelial–mesenchymal interaction.[20] The chemical composition and crystallinity of dentin and bone are relatively similar. Therefore, osteoblast behavior may be partially reflected in odontoblasts, which form dentin. Recently, adhesion molecules, such as those of the integrin family, were examined concerning cell structure and function. Divalent ions affect cell adhesion in relation to the integrin molecule as an adhesion molecule at the cell surface,[21] and it has been especially, reported that Mg^{2+} ions promote cell adhesion.[22] Integrins are crucially important receptor proteins because they are the main way in which cells both bind to and respond to the extracellular matrix. They are composed of two noncovalently associated transmembrane glycoprotein subunits called α and β, both of which contribute to the binding of the matrix protein. The binding of integrins to their ligands depends on extracellular divalent cations, reflecting the presence of three or four divalent cation-binding domains in the large extracellular part of the α chain (Figure 29.2).[21] Mg^{2+} ions also play some role in cell adhesion. Thus, magnesium seems to be an important factor even in controlling in vivo bone metabolism because it plays a part in both bone formation and resorption. Mg^{2+} ions may contribute to the bone metabolism of osteoclasts and osteoblasts with integrins at their cell surfaces.

Recently, scaffold biomaterials have been studied in the tissue engineering field. In a continuation of those studies, functionally graded CO_3apatite containing magnesium, producing a negative gradient of magnesium concentration from the surface toward the core, was synthesized.[23] Second, the degree of cell adhesion to a composite made by mixing $FGMgCO_3$–apatite and collagen to facilitate bonding and processing was investigated. Furthermore, the biocompatibility and effect of magnesium on bone formation by implanting the same $FGMgCO_3Ap$–collagen composite into rabbit femur was examined.[24]

Electron Spectroscopy for Chemical Analysis (ESCA) analysis clearly showed a negative gradient distribution of Mg1s intensity (atomic concentration) of magnesium from the crystal surface toward the inner core. The WST cell growth assay of mouse MC3T3-E1 osteoblast cells incubated on the surface of the composites showed similar optical densities between the control

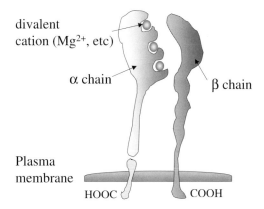

FIGURE 29.2. Schema of integrin as an adhesion molecule.

CO_3Ap–collagen composite and the $FGMgCO_3Ap$–collagen. This suggests that cell growth was good and that there was no significant inhibition in relation to the components of each sample. In the cell adhesion assay, after the nonadhering cells were rinsed off, the optical density of the $FGMgCO_3Ap$–collagen composite was higher than that of the CO_3Ap–collagen composite. Furthermore, the cell adhesion assay by radioisotope showed that the adhesion fraction of the $FGMgCO_3Ap$–collagen composite was higher than that of the CO_3Ap–collagen composite and much higher than that of the titanium plate as a control. After 4 weeks of incubation, many more osteoblasts adhered to the $FGMgCO_3Ap$–collagen composite than to the CO_3Ap–collagen composite and the layer they formed was thicker.

After 1 week of implantation into rabbit femurs, both the $FGMgCO_3Ap$–collagen composite and the CO_3Ap–collagen composite showed no clear difference from the control in histological results. However, hematoxylin–eosin stains of magnified new bone revealed the existence of many osteoblasts after 2 weeks of implantation. After 4 weeks of implantation, both the $FGMgCO_3Ap$–collagen composite and the CO_3Ap–collagen composite showed a clear bone formation, although the control hole with no implantation appeared to also have been repaired with a thinner layer of new bone. The bone density of the $FGMgCO_3Ap$–collagen composite was higher than that of the CO_3Ap–collagen composite.

Thus, the modification of gradational Mg^{2+} ions on apatite crystals was successfully performed. The apatite crystals showed no special function on the surface but did promote cell adhesion. Furthermore, as a scaffold material, the $FGMgCO_3Ap$–collagen composite was suggested to contribute to bone formation. Because this composite is easy to process into any form, it is extremely useful in the reconstruction of bone defects. Thus, trace elements in apatite crystals are strongly related to biological metabolism and physiology. Bone apatites contribute to store trace elements in addition to the skeletal structure.

These scaffold biomaterials are useful to replace the jawbone when it is lost by accident or maxillofacial disease and adsorbed after the loss of a tooth, and during the implantation of artificial teeth. Furthermore, it is speculated that Mg^{2+} ions may contribute to the formation and maturation of enamel.

The latest researches have gradually clarified the unknown phenomena inside and outside cells with magnesium. Mg^{2+} ions regulate numerous cellular functions, serving as a cofactor in many different enzymatic pathways. The cytoplasmic Mg^{2+} ion, $[Mg^{2+}]_c$, is regulated by both passive (influx driven by the electrochemical gradient of this ion) and active (active efflux) transport.[25] It has been clarified that $[Mg^{2+}]_c$ is tightly regulated by Mg^{2+} efflux, depending on extracellular $[Na^+]$.[26] Therefore, we speculate that during the formation and maturation of teeth, $[Mg^{2+}]_c$ in ameloblasts and odontoblasts may have significant roles. In conclusion, it can be speculated that magnesium is strongly related to the maturation of teeth and their physicochemical properties.

References

1. Miles AEW. *Structural and Chemical Organization of Teeth.* Vol. II. New York: Academic Press; 1967.
2. Nizel AE. *The Science of Nutrition and Its Application in Clinical Dentistry.* 2nd ed. Philadelphia: Sauders; 1966.
3. LeGeros RZ. Apatite crystallites: effects of carbonate on morphology. *Science* 1967;155:1409–1411.
4. Kay MI, Young RA, Posner AS. Crystal structure of hydroxyapatite. *Nature* 1964;204:1050–1052.
5. Young RA. Dependence of apatite properties on crystal structural details. *Trans N Y Acad Sci* 1967;29:949–959.
6. Van Wazer JR. *Phosphorus and Compounds.* New York: Interscience; 1958.
7. Okazaki M, Sato M. Computer graphics of hydroxyapatite and β-tricalcium phosphate. *Biomaterials* 1990;11:573–578.
8. LeGeros RZ. Incorporation of magnesium in synthetic and in biological apatites. In: Fearnhead RW, Suga S, eds. *Tooth Enamel IV.* Amsterdam: Elsevier; 1984:23–36.
9. Okazaki M, Takahashi J, Kimura H. Unstable behavior of magnesium-containing hydroxyapatites. *Caries Res* 1986;20:324–331.
10. Okazaki M. Mg^{2+}-F^- interaction during hydroxyapatite formation. *Magnesium* 1987;6:296–301.
11. LeGeros RZ. Calcium phosphates in oral biology and medicine. In: Myers HM, ed. *Monographs in Oral Science.* Basel: Karger; 1991.
12. Pauling L. *The Nature of the Chemical Bond.* 3rd ed. Ithaca, NY: Cornell University Press; 1960.
13. Itokawa Y, Durlach J. *Magnesium in Health and Disease.* London: John Libbey; 1989.
14. Stryer L. *Biochemistry.* San Francisco: Freeman; 1981.
15. Neuman MF, Mulryan BJ. Synthetic hydroxyapatite crystals. IV. Magnesium incorporation. *Calcif Tissue Res* 1971;7:133–138.
16. Brudevold F, Steadman LT, Smith FA. Inorganic and organic components of tooth structure. *Ann N Y Acad Sci* 1960;85:110–132.
17. Suga S. *Mechanisms of Tooth Enamel Formation.* Tokyo: Quintessence; 1983.
18. Hallsworth AS, Robinson C, Weatherell JA. Mineral and magnesium distribution within the approximal carious lesion of dental enamel. *Caries Res* 1972;6:156–158.
19. Thiradilok S, Feagin F. Effects of magnesium and fluoride on acid resistance of remineralized enamel. *Ala J Med Sci* 1978;15:144–148.
20. Markwald R, Eisenberg C, Eisenberg L, Trusk T, Sugi Y. Epithelial-mesenchymal transformations in early avian heart development. *Acta Anat (Basel)* 1996;156:173–186.
21. Albert B, Bray D, Lewis J, Raff M, Roberts K, Watson JD. *Molecular Biology of the Cell.* 3rd ed. New York: Garland Publishing; 1994:949–1010.
22. Lange TS, Bielinsky AK, Kirchberg K, et al. Mg^{2+} and Ca^{2+} differentially regulate β_1 integrin-mediated adhesion of dermal fibroblasts and keratinocytes to various extracellular matrix proteins. *Exp Cell Res* 1994;214:381–388.
23. Yamasaki Y, Yoshida Y, Okazaki M, et al. Synthesis of functionally graded $MgCO_3$apatite accelerating osteoblast adhesion. *J Biomed Mater Res* 2002;62:99–105.

24. Yamasaki Y, Yoshida Y, Okazaki M, et al. Action of FGMgCO$_3$Ap-collagen composite in promoting bone formation. *Biomaterials* 2003;24:4913–4920.
25. Flatman PW. Mechanisms of magnesium transport. *Ann Rev Physiol* 1991;53: 259–171.
26. Tursun P, Tashiro M, Konishi M. Modulation of Mg^{2+} efflux from rat ventricular myocytes studied with the fluorescent indicator furaptra. *Biophys J* 2005;88: 1911–1924.

Magnesium in Neurology and Psychiatry

30
Magnesium in Psychoses

Mihai Nechifor

Psychoses are severe psychiatric diseases. They affect a significant number of patients and have a various symptoms. The main psychoses are schizophrenia, major depression, and bipolar disorders. There are also other types of psychoses, such as interictal psychosis and postictal psychosis in epileptic patients.[1] These psychoses were less studied regarding the influence on the changing concentrations of magnesium and other cations. Magnesium and other bivalent cations play multiple roles in the central nervous system (CNS). Misbalances in intra- or extracellular concentrations of these cations are met in some neurological and psychiatric diseases and are involved sometimes in the pathogenic mechanisms of these diseases.

Bipolar Disorder (BD)

This disease (also called manic–depressive psychosis) is included in the affective psychosis category. The clinical aspect of bipolar disorder is extremely different from patient to patient and this creates a difficult classification of this disease. There are patients with predominant manic manifestations and there are also patients with predominant depressive symptoms.

There are changes in the plasmatic and cellular concentrations of magnesium and certain other cations in case of BD patients.

Data existing about magnesium plasmatic concentration levels in this affective psychosis are heterogeneous. Certain authors detected an increase in magnesium intra- and extracellular concentrations, but others have reported results suggesting the contrary. It was found that erythrocyte Mg^{2+} level is significantly increased in case of BD nonhospitalized patients: this increase is not seen after hospitalization.[2] Some authors have reported that there are no changes in plasmatic and erythrocyte Na^+ and Mg^{2+} levels in patients suffering from affective psychoses (including BD).[3] Also, other authors consider that in patients with BD there are not significant changes of Mg^{2+} levels in cerebrospinal fluid (CSF) and there are no correlations between magnesium concentrations and disease evolution.[4] Some differences between data obtained by different authors might originate into the existence of more than one subtype of BD.[5] Other differences are due to distinct phases in disease evolution at the moment of taking cation samples. There are few data about the influence of pharmacotherapy on patient's magnesium concentrations in BD. Mood stabilizers represent the main pharmacotherapy for these patients. Beside lithium, there are used nowadays new mood stabilizers as lamotrigine, sodium valproate, carbamazepine, etc.[6,7]

This class of drugs is very heterogeneous and includes drugs used from a long time (such as lithium) as well as drugs recently introduced in the treatment of BD (such as some anti-epileptic drugs). In 1988, the possibility was dicovered that a part of therapeutic effect of Li^+ in BD is due to competition between Li^+ and Mg^{2+} for the intracellular binding site of magnesium, because Li^+ and Mg^{2+} have some similar physicochemical characteristics.[8] It was shown by nuclear magnetic resonance (NMR) spectroscopy that Li^+ competes with Mg^{2+} for phosphates groups in GTP and GDP.[9] In this way, the transduction of biological signal is modified through second messenger inositol-1,4,5-triphosphate (IP3). There were identified some intracellular molecules, where Li^+ competes for magnesium binding site (inositol, monophosphatase, inositol polyphosphate 1–phosphatase, etc.).[10] Lithium increases magnesium intracellular concentration; the intake of 1 to $2\,nM$ lithium in the medium of culture significantly increases magnesium concentration in SH-SYSY neuroblastoma cells.[11] It is considered that the mechanism of increasing this magnesium concentration is represented by a competition between Li^+ and Mg^{2+} for magnesium intracellular binding sites. Increasing binding of lithium to these molecular sites decreases magnesium binding and raises the concentration of free intracellular magnesium.[9] An intracellular lithium concentration of 15 mm/L determines an important increase in Mg^{2+} concentration in neuroblastoma cells (approximately 158%).[12] The lithium/magnesium competition can also be noticed in case of therapeutic lithium concentration in BD patients. Only 72 h incubation at 1 to $2\,mM$ extracellular Li^+ concentrations determines a significant increase in magnesium intracellular concentration. Intracellular Li^+ was found to be at therapeutic levels (between $1–1.5\,nM$) and its level may reach $2.5\,nM$ in the neurons of Li^+-treated patients.

In BD and normal subjects, sodium (Na)–potassium (K)–adenosine triphosphatase (ATPase) activity is similar, but Mg^+ dependent activity of membrane Na-K-ATPase in erythrocytes is higher in BD patients than in normal patients.[13] Lithium treatment does not influence significantly activity of Mg^{2+}-dependent ATPase.

Carbamazepine therapy did not modify CSF concentration of this cation. The positive correlation between Na-K-ATPase activity and lithium concentration was emphasized in BD and also in the positive correlation between lithium and magnesium concentration in these patients.[14]

In neuroblastoma cells, an increase was observed in intracellular free Mg^{2+}, from $0.39 \pm 0.04\,nM$ to $0.6 \pm 0.04\,nM$ during Li^+ incubation (when intracellular Li^+ concentration increased from $0–5.5\,nM$).[15]

Verapamil is calcium-blocking channel drug, which proved to be useful as maintenance therapy of manic patients. The association between verapamil and magnesium was found to be significantly more effective than single verapamil in reducing manic symptoms in BD patients.[16] Administration of magnesium sulfate infusion in patients with severe therapy-resistant maniacal agitation improved their symptoms.[17]

In patients with severe maniacal agitation, the co-administration of Mg^{2+} allows the decrease of benzodiazepines or neuroleptic doses. Therapy with magnesium aspartate (40 mEq/day) leads to improvement in condition of BD patients.[18]

The fact that magnesium salts improved the symptoms of some BD patients argues for a role of the intracellular Mg^{2+} increment in BD drug mechanism. Changes in glutamatergic brain neurotransmission are involved in pathophysiology of psychoaffective psychoses, including BD.[19] We consider that symptoms in BD patients are alleviated by decreasing presynaptic release of glutamate, by partially blocking Ca^{2+} channel coupled with N-methyl-D-aspartate (NMDA) receptors, and by increasing intracellular concentrations of Mg^{2+}.

Calcium is an antagonist on carbamazepine effect used in the therapy of affective and schizoaffective psychoses, including BD.[20] This pleads for a magnesium role in supporting therapeutic effect of carbamazepine by decreasing calcium entrance through neuronal membrane. Decreasing magnesium concentrations could exacerbate the anxiety, weakness, and other symptoms in BD patients.[21] Increasing intracellular concentrations of magnesium in BD patients treated with lithium could not be the consequence of the effect on urinary elimination because Li^+ increases Mg^{2+} urinary elimination. The Li^+ effect is not the same in relation with all bivalent cations. For example, urinary elimination of Ca^{2+} is decreased.[22] Plasmatic concentrations of Mg^{2+} are not correlated with serum concentration other than Li^+.[23] Our data[24] have shown that treatment with carbamazepine and sodium valproate significantly increases Mg^{2+} intracellular concentration and plasmatic level of zinc. It is not known the effect of therapy in BD with atypic antipsychotics (quetiapine, ziprasidone, risperidone, olanzapine) on intracellular concentration of Mg^{2+}.[25]

Major Depression (MD)

One of the difficulties linked with assessment of extra- and intracellular concentrations of bivalent cations in depression is the existence of substantial differences between MD and depressive states associated with other diseases.

Regarding only MD, data are heterogenic. In MD, in drug-free patient treatment low levels of plasmatic magnesium was found.[26] In patients with severe MD that committed suicide or attempted suicide, Mg^{2+} concentrations in CSF were significantly lower.[27] Decreased Mg^{2+} and increased Ca^{2+} levels in the neocortex of the patients with severe MD was identified.[28] Decreasing plasmatic or erythrocyte magnesium concentration in patients with MD were accompanied by increasing plasmatic concentrations of calcium.[29] An increased Ca/Mg ratio in CSF was identified in patients with MD before treatment.[30] On contrary, other authors have found in drug-free depressed patients a higher plasma and erythrocyte magnesium concentration.[2,31] In patients with MD, changes in plasmatic or erythrocyte magnesium concentration were found, compared to healthy subjects.[32–34] A relationship between magnesium status and mood disorders is possible, but evidence remains inconsistent.[35]

There are few data referring to the influence of anti-depressant therapy on plasmatic and cellular concentration of magnesium and other cations. The effect of chronic treatment with imipramine was shown to increase plasmatic concentrations of zinc on experimentally induced depression in rats.[36] Difference in plasmatic Mg^{2+} concentrations in patients with MD or changes in concentrations after the drug therapy were not found.[37] It was shown that tricyclic anti-depressants inhibit voltage-dependent calcium channels and decrease intracellular concentrations of calcium.[38] Increasing magnesium concentration may contribute to reducing Ca^{2+} entrance in neurons. In our studies of drug-free patients diagnosed with MD (Hamilton score > 23), the intraerythrocytic magnesium level is significantly lower compared to control subjects. A decrease in plasmatic zinc level (0.68 ± 0.09 mg/L in MD subjects vs. 0.99 ± 0.11 mg/L in control group; $p < 0.01$) is associated with decreasing of magnesium intraerythrocytic level. Also, a significant increase in copper plas-

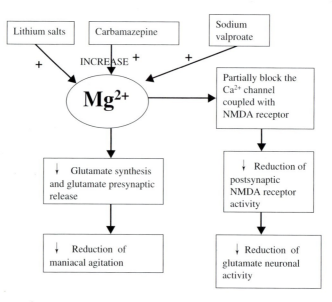

FIGURE 30.1. Mg^{2+} involvement in mechanism of action of mood stabilizers used in BD treatment.

TABLE 30.1. Influence of amitryptiline and sertraline on intraerythrocytic Mg^{2+} level in MD patients.

Group	Control	MD patients before treatment	MD patients + amitriptyline	MD patients + sertraline
Intraerythrocytic magnesium (mg/L)	59.1 ± 3.2*	44 ± 2.7	57.6 ± 4.5*	56.9 ± 5.32*

*$p < 0.01$ versus MD group before treatment.

matic level can be noticed in MD patients. Magnesium plasmatic concentration suffered no significant variations in MD. Calcium plasmatic concentrations are not influenced in MD. Our data show that amitriptyline therapy (tricyclic anti-depressant drug), 3 × 25 mg/day per os, 4 weeks, as well as sertraline (Zoloft®) (serotonine re-uptake inhibitor), 150 mg/day per os, 4 weeks, determine a significant increase in magnesium intraerythrocytic levels in two different groups of patients with MD, the patients being in a drug-free state before the initialization of this treatment.[39] This effect is shown in Table 30.1. We consider that the increasing magnesium intraerythrocytic and zinc plasmatic level effect of amitriptyline and sertraline represents a component of its anti-depressant mechanism of action. Some anti-depressant drugs decrease Ca^{2+} concentrations. It is considered that this might be a mechanism of action for these drugs.[40] The fact that anti-depressant drugs (with different chemical structures and different mechanism of action) has a similar influence on the concentration of bivalent cations (significantly increases intraerythrocytic Mg^{2+} and plasmatic Zn^{2+}) pleads for including changes in some bivalent cations concentrations in the anti-depressant mechanism of action of these drugs.

Schizophrenia (SCH)

There are various and sometimes divergent data about plasmatic and intracellular concentrations of Mg^{2+} and other cations in patients with schizophrenia. In some studies, the schizophrenic patients shown lower Mg^{2+} plasmatic levels compared to normal subjects.[21,41] Plasmatic level of magnesium and calcium were not found modified in patients with untreated schizophrenia in other studies.[14,42,43] On the contrary, other authors found an increased magnesium level in CSF in patients with schizophrenia.[27,31]

Our research[43] show that both typical (haloperidol, 8 mg/kg/day, 3 weeks) and atypical (risperidone, 6 mg/kg/day, 3 weeks) anti-psychotic drugs produce a significantly increase in intracellular concentrations of magnesium in adult patients with paranoid schizophrenia. Our data are in agreement with authors that shown haloperidol (typical anti-psychotic) increases intraerythrocytic level of magnesium.[45] There are several hypotheses regarding schizophrenia pathophysiology. One of these hypothesis is the glutamatergic hypothesis, which suggests that schizophrenia is determined by an imbalance between glutamatergic and dopaminergic mediation in some brain areas. Mg^{2+} ions block the Ca^{2+} ion channel of NMDA receptors[46] and reduces the activation of glutamatergic systems in the brain. It is considered by some authors that a high level of free radicals is involved in the pathogeny of acute psychoses. Other possible way that increased intracellular Mg^{2+} and plasmatic Zn^{2+} might influence SCH evolution is decreasing formation of peroxidic radicals.

We consider that increasing intracellular magnesium concentration is an important part of the mechanism of action of some typical and atypical antipsychotics. Intra- and extracellular concntrations of magnesium and other bivalent cations may be considered as a biological marker for some psychoses. We consider that a low level of intracellular Mg^{2+} correlated with a raised plasmatic ratio (Cu^{2+}/Zn^{2+}) might be a severity indicator of MD and SCH.

In all of the psychoses described above, we have observed a good correlation between normalization of intra-erythrocytic magnesium level/plasmatic zinc level and clinical evolution of the patient during drug therapy.[47] Periodical assesment of intraerythrocytic magnesium level correlated with plasmatic zinc and copper concentrations may offer a guidance for the efficiency of applied pharmacotherapy.

References

1. Adachi N, Onuma T, Nishiwaki S, et al. Inter-ictal and post-ictal psychoses in frontal lobe epilepsy: a retrospective comparison with psychoses in temporal lobe epilepsy. *Seizure* 2000;9:328–335.
2. Frazer A, Ramsey TA, Swann A, et al. Plasma and erythrocyte electrolytes in affective disorders. *J Affect Disord* 1983;5:103–113.
3. Ramsey TA, Frazer A, Mendels J. Plasma and erythrocyte cations in affective illness. *Neuropsychobiology* 1979;5:1–10.
4. George MS, Rosenstein D, Rubinow DR, Kling MA, Post RM. CSF magnesium in affective disorder: lack of correlation with clinical course of treatment. *Psychiatry Res* 1994;51:139–146.
5. Mick E, Biederman J, Faraone SV, Murray K, Wozniak J. Defining a developmental subtype of bipolar disorder in a sample of nonreferred adults by age at onset. *J Child Adolesc Psychopharmacol* 2003;13:453–462.
6. Schatzberg AF. Employing pharmacologic treatment of bipolar disorder to greatest effect. *J Clin Psychiatry* 2004;65(suppl. 15):15–20.
7. Arban R, Maraia G, Brackenborough K, et al. Evaluation of the effects of lamotrigine, valproate and carbamazepine in a rodent model of mania. *Behav Brain Res* 2005;158:123–132.
8. Avissar S, Schreiber G, Danon A, Belmaker RH. Lithium inhibits adrenergic and cholinergic increases in GTP binding in rat cortex. *Nature* 1988;331:440–442.
9. Amari L, Layden B, Nikolakopoulos J, et al. Competition between Li^+ and Mg^{2+} in neuroblastoma SH-SY5Y cells: a fluorescence and 31P NMR study. *Biophys J* 1999;76:2934–2942.
10. Gould TD, Quiroz JA, Singh J, Zarate CA, Manji HK. Emerging experimental therapeutics for bipolar disorder: insights from the molecular and cellular actions of current mood stabilizers. *Mol Psychiatry* 2004;9:734–755.

11. Abukhdeir AM, Layden BT, Minadeo N, Bryant FB, Stubbs EB Jr, Mota de Freitas D. Effect of chronic Li+ treatment on free intracellular Mg^{2+} in human neuroblastoma SH-SY5Y cells. *Bipolar Disord* 2003;5:6–13.
12. Layden BT, Abukhdeir AM, Williams N, et al. Effects of Li(+) transport and Li(+) immobilization on Li(+)/Mg(2+) competition in cells: implications for bipolar disorder. *Biochem Pharmacol* 2003;66:1915–1924.
13. Thakar JH, Lapierre YD, Waters BG. Erythrocyte membrane sodium-potassium and magnesium ATPase in primary affective disorder. *Biol Psychiatry* 1985;20:734–740.
14. Alexander DR, Deeb M, Bitar F, Antun F. Sodium-potassium, magnesium, and calcium ATPase activities in erythrocyte membranes from manic-depressive patients responding to lithium. *Biol Psychiatry* 1986;21:997–1007.
15. Layden B, Diven C, Minadeo N, Bryant FB, Mota de Freitas D. Li+/Mg2+ competition at therapeutic intracellular Li+ levels in human neuroblastoma SH-SY5Y cells. *Bipolar Disord* 2000;2:200–204.
16. Giannini AJ, Nakoneczie AM, Melemis SM, Ventresco J, Condon M. Magnesium oxide augmentation of verapamil maintenance therapy in mania. *Psychiatry Res* 2000;93:83–87.
17. Heiden A, Frey R, Presslich O, Blasbichler T, Smetana R, Kasper S. Treatment of severe mania with intravenous magnesium sulphate as a supplementary therapy. *Psychiatry Res* 1999;89:239–246.
18. Chouinard G, Beauclair L, Geiser R, Etienne P. A pilot study of magnesium aspartate hydrochloride (Magnesiocard) as a mood stabilizer for rapid cycling bipolar affective disorder patients. *Prog Neuropsychopharmacol Biol Psychiatry* 1990;14:171–180.
19. Levine J, Panchalingam K, Rapoport A, Gershon S, McClure RJ, Pettegrew JW. Increased cerebrospinal fluid glutamine levels in depressed patients. *Biol Psychiatry* 2000;47:586–593.
20. Walden J, Grunze H, Olbrich H, Berger M. Importance of calcium ions and calcium antagonists in affective psychoses. *Fortschr Neurol Psychiatr* 1992;60:471–476.
21. Kirov GK, Tsachev KN., Magnesium, schizophrenia and manic-depressive disease. *Neuropsychobiology* 1990;23:79–81.
22. Plenge P, Rafaelsen OJ. Lithium effects on calcium, magnesium and phosphate in man: effects on balance, bone mineral content, faecal and urinary excretion. *Acta Psychiatr Scand* 1982;66:361–373.
23. Rybakowski J, Szajnerman Z. Lithium-magnesium relationship in red blood cells during lithium prophylaxis. *Pharmakopsychiatr Neuropsychopharmakol* 1976;9:242–246.
24. Nechifor M, Vaideanu C, Mindreci I, Palamaru I, Boisteanu P. Research on plasmatic and tissular concentration of some bivalent cations in patients with bipolar disorders. In: Ermidou Pollet S, Pollet S, eds. *Abstract Book of 5th International Symposium on Trace Elements in Human: New Perspectives*. Athens: Entypossis Athens Publishing House; 2005:18.
25. Bowden CL. Atypical antipsychotic augmentation of mood stabilizer therapy in bipolar disorder. *J Clin Psychiatry* 2005;66(suppl. 3):12–19.
26. Rybakowski J, Jankowiak E. Plasma and erythrocyte electrolytes in endogenous depression—neurophysiologic implications. *Psychiatr Pol* 1989;23:265–270.
27. Banki CM, Vojnik M, Papp Z, Balla KZ, Arato M. Cerebrospinal fluid magnesium and calcium related to amine metabolites, diagnosis, and suicide attempts. *Biol Psychiatry* 1985;20:163–171.

28. Basarsky TA, Duffy SN, Andrew RD, MacVicar BA. Imaging spreading depression and associated intracellular calcium waves in brain slices. *J Neurosci* 1998;18:7189–7199.
29. Bowden CL, Huang LG, Javors MA, et al. Calcium function in affective disorders and healthy controls. *Biol Psychiatry* 1988;23:367–376.
30. Levine J, Stein D, Rapoport A, Kurtzman L. High serum and cerebrospinal fluid Ca/Mg ratio in recently hospitalized acutely depressed patients. *Neuropsychobiology* 1999;39:63–70.
31. Widmer J, Stella N, Raffin Y, Bovier P, Gaillard JM, Hilleret H, Tissot R. Blood magnesium, potassium, sodium, calcium and cortisol in drug-free depressed patients. *Magnes Res* 1993;6:33–41.
32. Kamei K, Tabata O, Muneoka K, Muraoka SI, Tomiyoshi R, Takigawa M. Electrolytes in erythrocytes of patients with depressive disorders. *Psychiatry Clin Neurosci* 1998;52:529–533.
33. Young LT, Robb JC, Levitt AJ, Cooke RG, Joffe RT. Serum Mg^{2+} and Ca^{2+}/Mg^{2+} ratio in major depressive disorder. *Neuropsychobiology* 1996;34:26–28.
34. Nechifor M, Văideanu C, Boişteanu P, Mindreci I, Cuciureanu R, Nechifor C. Changes in Mg^{2+} and other cations plasmatic concentration in patients with major depression. In: Escanero JF, Alda JO, Guerra M, Durlach J, eds. *Advances in Magnesium Research Physiology, Pathology and Pharmacology*. Zaragoza: Prensas Universitarias de Zaragoza; 2003:177–181.
35. Singewald N, Sinner C, Hetzenauer A, Sartori SB, Murck H. Magnesium-deficient diet alters depression- and anxiety-related behavior in mice—influence of desipramine and *Hypericum perforatum* extract. *Neuropharmacology* 2004;47:1189–1197.
36. Schlegel-Zawadzka M, Papp M, Krosniak M, Nowak G. Effect of chronic imipramine treatment on serum and plasma zinc levels in chronic mild stress model of depression in rats. In: *Mengen und Spurenelemente*. Leipzig: Verlag Harald Schubert; 1998:616–620.
37. Imada Y, Yoshioka S, Ueda T, Katayama S, Kuno Y, Kawahara R. Relationships between serum magnesium levels and clinical background factors in patients with mood disorders. *Psychiatry Clin Neurosci* 2002;56:509–514.
38. Lavoie PA, Beauchamp G, Elie R. Tricyclic antidepressants inhibit voltage-dependent calcium channels and Na(+)-Ca^{2+} exchange in rat brain cortex synaptosomes. *Can J Physiol Pharmacol* 1990;68:1414–1418.
39. Nechifor M. Involvement of some cations in major depression. In: Cser MA, Sziklai L, eds. *Metal Ions in Biology and Medicine*. Paris: John Libbey; 2004:518–521.
40. Jagadeesh SR, Subhash MN. Effect of antidepressants on intracellular Ca^{++} mobilization in human frontal cortex. *Biol Psychiatry* 1998;44:617–621.
41. Kanofsky JD, Sandyk R. Magnesium deficiency in chronic schizophrenia. *Int J Neurosci* 1991;61:87–90.
42. Athanassenas G, Papadopoulos E, Kourkoubas A, et al. Serum calcium and magnesium levels in chronic schizophrenics. *J Clin Psychopharmacol* 1983;3:212–216.
43. Yassa R, Nair NP, Schwartz G. Plasma magnesium in chronic schizo-phrenia. A preliminary report. *Int Pharmacopsychiatry* 1979;14:57–64.
44. Nechifor M, Vaideanu C, Palamaru I, Borza C, Mindreci I. The influence of some antipsychotics on erythrocyte magnesium and plasma magnesium, calcium, copper and zinc in patients with paranoid schizophrenia. *J Am Coll Nutr* 2004;23:549S–551S.

45. Jabotinsky-Rubin K, Durst R, Levitin LA, et al. Effects of haloperidol on human plasma magnesium. *J Psychiatr Res* 1993;27:155–159.
46. Kahn DA, Sachs GS, Printz DJ, Carpenter D, Docherty JP, Ross R. Medication treatment of bipolar disorder 2000: a summary of the expert consensus guidelines. *J Psychiatr Pract* 2000;6:197–211.
47. Nechifor M, Văideanu C, Mândeci I, Borza C. Variations of magnesium concentrations in psychoses. *Magnes Res* 2004;17:226.

Veterinary Medicine

31
Significance of Magnesium in Animals

Tohru Matsui

Magnesium (Mg) metabolism differs among animal species because the digestive system and feeds are different. The diseases related to Mg nutrition are rare in pigs and poultry under practical conditions because their diets are formulated as containing an appropriate level of Mg. On the other hand, Mg deficiency is not rare in grazing animals because Mg in pasture is affected by several factors such as soil and plant species and maturity, and thus Mg concentration is largely varied in pasture. Grass tetany in ruminants is induced by the reduction of Mg absorption resulting from low Mg intake with high potassium and nitrogenous compounds, and with the reduction of ruminal fermentation. Additionally, cold stress stimulates the incidence of tetany through decreasing Mg concentration in the cerebrospinal fluid. Excess Mg is one of the factors inducing urolithiasis in cats and cattle, and enterolithiasis in horses. However, Mg level in the practical diets alone cannot induce these diseases. Cat urolithiasis is developed in combination with alkaline urine, and cattle urolithiasis and horse enterolithiasis are developed in combination with high phosphorus intake. The diseases related to Mg nutrition are mainly developed in combination with other dietary factors and/or environmental factors in ruminants, horses, and cats.

Comparative Aspects of Magnesium Metabolism

The digestive system is widely different among animal species. Herbivores consume high fibrous feeds (rich in cellulose) originated from plants and their digestion is largely owing to microbial fermentation in the rumen (a forestomach) of ruminants, such as cattle and sheep, or in the large intestine of monogastric herbivores, such as horses and rabbits. These sites of microbial digestion are large because longer transit time is necessary for sufficient fermentation. Carnivores natively obtain most of their feed by eating other animals, and their digestion is mainly owing to digestive enzymes and microbial digestion is minimal. Thus, the alimentary tract is short and simple in carnivores. In omnivores such as dogs, pigs, and rats, both enzymatic and microbial digestion are important and the structure of digestive tract is intermediate between carnivores and herbivores.

Apparent absorption of magnesium (Mg) is different among animal species (Table 31.1). Magnesium absorption is generally lower in ruminants than in other animals. In ruminants, sheep absorb Mg 1.7-times more efficiently than cattle because the ratio of surface area to content is higher in the major site of Mg absorption of sheep.[1] The efficiency of Mg absorption is higher in horses than in ruminants. Magnesium absorption is less in pigs than in rats. Dietary phytate (inositol hexaphosphate) is known to decrease Mg solubility in intestinal digesta of monogastric animals and suppresses Mg absorption.[2] Practical diets of pig contain whole grains, oilseed meals, and bran that are rich in phytate, but rats are usually given semipurified diets without containing phytate. Thus, Mg absorption may be different between these omnivore species. Because cats have a relatively short intestine resulting in a rapid passage rate of digesta, the digestibility for many natural feedstuffs is generally lower in cats than in rats and dogs.[3] The absorption of Mg also may be less in cats than in rats.

Ruminants mainly absorb Mg from the rumen.[4] The major site of Mg absorption is the distal small intestine[5] or the ileum and the colon[6] in rats. There have been few reports showing the major site of Mg absorption in other domestic animals. Pigs were reported to absorb Mg in the ileum and the colon.[7] Cats and dogs predominantly absorbed Mg from the large intestine.[8] The major site of Mg absorption was the small intestine in horses[9,10] and rabbits.[11]

There are two pathways for Mg absorption, that is, paracellular route and transcellular route.[6] The paracellular route consists of tight junctions and intercellular space between the epithelial cells, which depends on the passive driving force and the permeability of this route for Mg. The transcellular route consists of the influx through the apical membrane of epithelial cells and the

TABLE 31.1. Magnesium absorption in some species.

	Apparent absorption (% of intake)	Major site of absorption
Cattle	23.3 ± 7.8^a	Forestomach (rumen)
Sheep	32.8 ± 9.0^b	Forestomach (rumen)
Horse	51.2 ± 5.6^c	Small intestine
Pig	39.2 ± 9.7^d	Ileum and colon
Rat	60.8 ± 14.5^e	Ileum and colon
Cat	39.3 ± 12.9^f	Large intestine

Values are mean ± standard deviation (SD).
[a]Calculated from 35 publications in dairy cattle given diets containing appropriate amounts of magnesium and potassium.
[b]Calculated from 37 publications in lambs given diets containing appropriate amounts of magnesium and potassium.
[c]Calculated from seven publications in horses given diets containing sufficient amounts of magnesium.
[d]Calculated from 12 publications in growing pigs given diets containing sufficient amounts of magnesium.
[e]Calculated from 27 publications in growing rats given AIN diets.
[f]Calculated from 11 publications in cats given diets containing sufficient amounts of magnesium.

efflux across the basolateral membrane. The Mg concentration was 5 mM in the liquid phase of ileal digesta of pigs given a conventional diet[12] and the luminal Mg concentration ranged between 4 and 13 mM in the rumen.[13] The intracellular concentration of ionized Mg ranged between 0.5 and 1.0 mM in ruminal epithelial cells[14] and between 0.4 and 0.7 mM in Caco-2 intestinal cells.[15] Additionally, ionized Mg concentration is 0.4 to 0.6 mM in blood. Therefore, the intracellular concentration of ionized Mg is generally considered lower than its luminal concentration and is close to its concentration in blood. The entry of ionized Mg into the epithelial cells does not require energy but the efflux is energy dependent.[8]

Coudray and colleagues[16] suggested that the active transport of Mg was important only under conditions of extremely low dietary Mg in rats because the amount of absorbed Mg linearly increased with increasing dietary Mg up to the requirement level. Some researchers also reported that Mg was primarily absorbed by a passive diffusion at usual Mg intake in rats.[17,18] Additionally, metabolic inhibitors and an adenosine triphosphatase (ATPase) inhibitor did not affect transepithelial Mg transport, which also supported passive transport as the major route of Mg absorption.[19]

As reviewed by Schweigel and Martens,[8] ruminants mainly absorb Mg through transcellular route across the rumen epithelium by the secondary active transport. Ruminants obtain dietary energy as volatile fatty acids produced by ruminal microbes. Additionally, ruminal microbes degrade dietary nitrogenous components (protein and nonprotein nitrogen) to ammonia and they reconstitute protein from ammonia. The microbial protein largely contributes to protein nutrition of ruminants. Magnesium absorption increases with increasing readily fermentable carbohydrates in diets. The ingestion of readily fermentable carbohydrates rapidly raises ruminal concentration of volatile fatty acids and CO_2/HCO_3^-, which stimulate directly Mg uptake by the epithelial cells. The high intake of nitrogenous substances increases ruminal ammonia concentration through the fermentation because the degradation of dietary protein is higher than microbial protein synthesis in this condition, which transiently reduces Mg absorption. Additionally, the reduction of ruminal pH increases Mg absorption through rising Mg solubility. The concentrations of volatile fatty acids and ammonia affect ruminal pH and Mg absorption. The high concentration of potassium (K) in the rumen largely and directly suppresses Mg uptake by the epithelial cells.

Magnesium is endogenously excreted in both urine and feces. Urinary Mg excretion was largely more than the endogenous fecal excretion in sheep[20] and cats[21] given Mg at the requirement level. On the other hand, the endogenous fecal excretion was as much as urinary excretion in rats.[16] Horses secreted a half of absorbed Mg into the large intestine[10] or the endogenous fecal loss of Mg was more than urinary excretion in horses given Mg at its requirement level.[22] Urinary Mg excretion increases with dietary Mg in most animals and it is known that there is a good correlation between Mg absorption and its urinary excretion. The endogenous excretion into feces was reported to increase with increasing dietary Mg in rats[16] and in sheep.[23] However, the relationship between dietary Mg and its endogenous excretion into feces is still controversial.

The concentration of Mg is approximately 5 mM in cow's milk. A cow producing 30 kg of milk would loss 150 mmol Mg/day into the milk, which approximately corresponds to half of absorbed Mg.[24] The Mg concentration in milk was relatively stable in Mg-deficient cows but the milk production decreased.[25] Rats secreted Mg into milk at approximately 40% of apparently absorbed Mg.[26] The Mg concentration was approximately 1.6 mM in mare's milk and mares lost Mg at 12 to 40 mmol/day into milk.[27] Lactating mares absorb 160 mmol Mg/day when dietary Mg is satisfied with its requirement.[28] Therefore, the absorbed Mg is fourfold more than its secretion into milk in horses. Lactation stimulates Mg absorption due to increasing feed intake and efficiency of Mg absorption. The positive balance of Mg may be maintained even in lactating cows when a sufficient amount of Mg is given.

Magnesium Deficiency

The main manifestation of Mg deficiency induces retarded growth, hyperirritability and tetany, peripheral vasodilation, anorexia, muscular incoordination, and convulsion.[29] Typical ingredients of feeds contain sufficient amounts of Mg and thus practical diets usually contain adequate Mg in many species and Mg deficiency is rare. Although the Mg concentration in forages generally satisfies its requirement of ruminants, the Mg concentration varies largely with plant species, maturity of plants, and with the soil and climate in which plants are grown. Therefore, Mg in forages is occasionally low and hypomagnesemia is observed in grazing herbivores. Hypomagnesemia in ruminants is classified into a rapidly developing type and a slowly developing type.[30]

Acute Type of Hypomagnesemia (Grass Tetany)

Grass tetany results from hypomagnesemia that occurs suddenly in early spring just after the initiation of grazing.[29] Grass tetany is seldom developed in horses grazing pastures that develop grass tetany in cattle.

The reduction of Mg absorption is considered as a major factor of the pathogenesis, which results from low Mg in diets and factors reducing Mg bioavailability. Grass tetany is found in areas where dairying or beef production is highly developed. Pastures in areas with intensive livestock production are generally rich in K and nitrogenous components due to frequent fertilization with manure. Grazing cattle on such pastures entails the risk of hypomagnesemia, primarily due to K-suppressing Mg absorption.[31] Additionally, excess nitrogenous components raise ruminal ammonia concentration, which reduces Mg absorption.[31]

The feed intake is reduced by the rapid change of environment after the initiation of grazing, which decreases Mg absorption. Lactating cows are more susceptible to development of grass tetany because of Mg secretion into milk. Furthermore, the susceptibility to grass tetany is increased in order ruminants. Bone Mg concentration is lower in order animals. The reduction of available Mg in bone was possibly related to the higher incidence of grass tetany in older cows.[32] The administration of a pyrophosphate analogue suppressed bone resorption in sheep, which did not affect plasma Mg concentration in sheep given an adequate amount of Mg[33] but the pyrophosphate analogue enhanced the reduction of plasma Mg concentration in sheep given a Mg-deficient diet.[34] On the other hand, Robson nd colleagues[35] reviewed the relationship between bone resorption and the plasma Mg concentration and they suggested that bone Mg was not important for maintaining the plasma Mg concentration. Therefore, the relationship between bone resorption and the incidence of hypomagnesemia is not clear in old ruminants. The Mg absorption was low in old ruminants,[36] which may be related to higher incidence of grass tetany in older animals.

The onset of grass tetany is more closely associated with the Mg concentration in the cerebrospinal fluid (CSF) than with blood Mg. The Mg concentration in CSF was lower in clinically affected cows than in nontetanic cows but plasma Mg concentration was almost similar between them.[37] The low Mg concentration in CSF was associated with alterations in monoamine concentrations in the central nervous systems that played an important role in both voluntary and involuntary motor function.[38] Therefore, the disturbance of monoamine concentrations was considered to play a role in the etiology of hypomagnesemic tetany.

The Mg concentration is higher in CSF than in plasma and the difference of Mg concentration is generated by its active transport. Mild hyperkalemia lowered Mg concentration in CSF of sheep.[39] Therefore, Mg influx into CSF may be inhibited by high concentration of K in blood. The initiation of grazing in early spring stresses animals through the rapid changes in environment. A stress reaction involving the adrenal–glucocorticoid axis increased circulating K concentration and lowered Mg transport across the choroidal plexus, which was one of causes of this disease.[35] Thus, dietary and environmental factors synergistically develop grass tetany.

Slow Type of Hypomagnesemia

Subclinical hypomagnesemia is observed in cattle for several months especially during winter. When a plasma Mg concentration reaches critically low level, it is accompanied by clinical symptoms such as moderate incoordination, tetany, and hyperirritability. This disease is called winter tetany.[40] Winter tetany is developed by low dietary Mg, low quality of feeds, and environmental stresses such as extremely cold and wet weather.[40] The reduction of energy intake suppresses ruminal fermentation, which decreases volatile fatty acids and CO_2 concentrations, and elevates ammonia concentration in the rumen. Additionally, the reduction of energy intake increases ruminal pH because of high ammonia and low volatile fatty acid concentrations in the rumen. These changes decrease Mg absorption in the rumen.

Hypomagnesemia is observed in calves consuming whole milk for an extended period, particularly calves suckling cows that are subclinically hypomagnesemic. This disease is called milk tetany.[40] The etiology of this disease is a simple deficiency of Mg. The Mg requirement was calculated as 1.3 g/kg dry matter in a 75-kg suckling calf gaining at 1 kg/day.[41] Magnesium concentration was 1 g/kg dry matter in milk.[42] Therefore, the fast-growing calves may be susceptible to development of milk tetany. Additionally, Mg absorption decreases with growth, that is, 70% for a 50-kg calf and 30% for 75-kg calf[41] and hypomagnesemia occurs in older suckling calves.

Spontaneous atherosclerosis (AS) is considered to occur in almost all animal species, including wild ruminants. Although AS is rare in adult ruminants, AS is developed in calves consuming whole milk for an extended period. A sclerosis of arteries was also found in calves given artificial magnesium-deficient diets. The margarine-fed calves developed hypomagnesemia and severe arteriosclerosis that could be prevented with Mg supplementation. Thus, Mg deficiency is considered as a trigger of AS in calves.[43]

Excess Magnesium

Magnesium toxicosis has not been reported and does not appear in many animals given natural feedstuffs but would be most likely to occur using excess supplementation with Mg. On the other hand, excess Mg may induce urolithiasis of ruminants (Figure 31.1) and cats, and enterolithiasis of horses (Figure 31.2) in practical conditions. Excess Mg-induced uroliths are com-

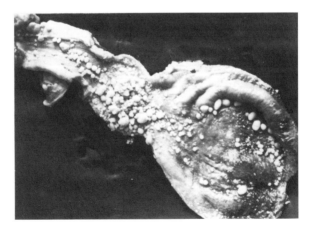

FIGURE 31.1. Uroliths in the bladder and urinary duct of beef cattle. (Courtesy of H. Yano, Kyoto University, Kyoto, Japan.)

FIGURE 31.2. Enteroliths from the large intestine of a horse. Transverse section of an enterolith. Scale bar = 2 cm. (Courtesy of Y. Tajima and T. Ueno, Equine Research Institute, Japan Racing Association, Utsunomiya, Japan.)

posed of Mg ammonium phosphate (P) and are determined as struvite (Mg ammonium P hexahydrate) in cats and dogs. Crystallization of Mg ammonium P depends on the urinary concentration of its components, that is, the product of $[Mg^{2+}] \times [NH_4^+] \times [PO_4^{3-}]$ and urinary pH affecting NH_4^+ and PO_4^{3-} concentrations.[44]

Urolithiasis is an important clinical problem, especially in dogs and cats. Uroliths are mainly composed of struvite or calcium oxalate in these animals. Although the formation of struvite stone in dogs usually results from a urinary tract infection with urea-splitting microbes that increase urinary ammonium concentration, the majority of struvite uroliths are observed in sterile cats forming sterile urine.[45] Thus, the diet is the major causal factor inducing urolithiasis in cats. Since the cat evolved as a desert animal, it has the capacity to produce highly concentrated urine in order to conserve water. Cats usually consume diets rich in animal protein, which produces net amounts of acid from sulfur amino acids and acidifies urine. Urinary P and ammonia concentrations are higher in cats than in other animals, although urinary pH and Mg may be lower in cats than in other animals (Table 31.2).

Urinary Mg excretion increases with Mg intake in animals including cats. Excess Mg is linked to struvite urolithiasis. The dietary Mg over 0.35%

TABLE 31.2. Urinary concentrations of factors affecting struvite formation in some species.

	Magnesium mM	Phosphorus mM	Ammonia mM	pH	References
Cow	19	0.9	10.5	8.3	Vagnoni et al.[46]
Rat	19	12	45	6.97	Amanzadeh et al.[47]
Cat	3.1	81	118	6.37	Cottam et al.[48]

produced struvite urolithiasis.[49] Pet food industries reduce Mg content in commercial cat foods and Mg concentration ranges from 0.05% and 0.3% in many commercial cat foods. Thus, dietary Mg is not critical at present. Nevertheless, excess Mg should be avoided and low-Mg diets are considered advantageous in the prevention of struvite urolithiasis. Cats produce acidic urine with a pH of 6.0 to 7.0 in the normal situation. Struvite remains largely in solution below pH 6.6, while the crystallization may occur spontaneously if the urinary pH rises above 7.1.[50] High levels of dietary Mg did not induce urolithiasis when acidic urine was produced.[51] Therefore, urinary pH may be more important factor than the urinary Mg concentration in cats. However, urine acidification together with a low Mg intake increases the risk of calcium oxalate urolithiasis in cats.[52]

Ruminants ingest plant materials that contain large amounts of K and organic anions. Organic anions are protonated during catabolism and then oxidized to water and CO_2 (base forming), leaving K and HCO_3^- to be excreted into urine. Therefore, urinary pH is higher in ruminants than in some other animals (Table 31.2). The alkaline urine may also stimulate the development of urolithiasis in ruminants. Although the urinary Mg concentration is not high in cattle, excess Mg increases its concentration. Urinary P and ammonia concentrations are low in ruminants because they excreted mainly P and ammonia into the digestive tract via saliva (P and ammonia) or via transport across ruminal wall (ammonia). However, excess P and protein increase urinary P and ammonia. A field survey indicated that dietary Mg was positively correlated with both the morbidity and mortality rates due to urolithiasis in fattening lambs.[53] However, some researchers suggested that high dietary Mg per se did not develop urolithiasis and that high dietary P was required for the urolith formation in calves[54] and lambs.[55]

Enterolithiasis is a serious problem in horses. The enteroliths consist primarily of Mg ammonium P. Prominent clinical features were recurrent mild abdominal pain, gaseous distension, and minimal intestinal motility. Most obstructing enteroliths were found near the beginning of the small colon. Horses with enterolithiasis represented 15.1% of patients admitted for treatment of colic, and 27.5% of patients undergoing celiotomy for treatment of colic.[56] Enteroliths ranged from 200 g to 9 kg, but generally weighed 450 g to 3 kg.[57] Horses secrete Mg and P into the large intestine, which may stimulate the incidence of enterolithiasis. Wheat bran was reported as a dietary factor inducing enterolithiasis because of its high concentration of P and Mg.[58]

References

1. Shockey WL, Conrad HR, Reid RL. Relationship between magnesium intake and fecal magnesium excretion of ruminants. *J Dairy Sci* 1984;67:2594–2598.
2. Hirabayashi M, Matsui T, Ilyas A, Yano H. Fermentation of soybean meal by Aspergillus usami increases magnesium availability in rats. *Jpn J Magnes Res* 1995;14:45–53.

3. Kendall PT, Blaza SE, Smith PM. Comparative digestible energy requirements of adult beagles and domestic cats for body weight maintenance. *J Nutr* 1983;113:1946-1955.
4. Tomas FM, Potter BJ. The site of magnesium absorption from the ruminant stomach. *Br J Nutr* 1976;36:37-45.
5. Hardwick LL, Jones MR, Brautbar N, Lee DB. Site and mechanism of intestinal magnesium absorption. *Miner Electrolyte Metab* 1990;16:174-180.
6. Kayne LH, Lee DB. Intestinal magnesium absorption. *Miner Electrolyte Metab* 1993;19:210-217.
7. Partridge IG. Studies on digestion and absorption in the intestines of growing pigs. 3. Net movements of mineral nutrients in the digestive tract. *Br J Nutr* 1978;39:527-537.
8. Schweigel M, Martens H. Magnesium transport in the gastrointestinal tract. *Front Biosci* 2000;5:d666-d677.
9. Hintz HF, Schryver HF. Magnesium metabolism in the horse. *J Anim Sci* 1972;35:755-759.
10. Matsui T, Murakami Y, Yano H. Magnesium in digesta of horses fed diets containing different amounts of phytate. In: Theophanides T, Anastassopoulou J, eds. *Magnesium: Current Status and New Developments—Theoretical, Biological and Medical Aspects*. New York: Kluwer; 1997:143-144.
11. Aikawa JK, Rhoades EL, Harmas DR, Readon JZ. Magnesium metabolism in rabbits using ^{28}Mg as a tracer. *Am J Physiol* 1959;197.99-101.
12. Matsui T, Yano H. Magnesium ligands in ileal digesta of piglets fed skim milk and soybean flour. In: Theophanides T, Anastassopoulou J, eds. *Magnesium: Current Status and New Developments—Theoretical, Biological and Medical Aspects*. New York: Kluwer; 1997:71-76.
13. Jittakhot S, Schonewille JT, Wouterse H, Yuangklang C, Beynen AC. Apparent magnesium absorption in dry cows fed at 3 levels of potassium and 2 levels of magnesium intake. *J Dairy Sci* 2004;87:379-385.
14. Schweigel M, Lang I, Martens H. Mg^{2+} transport in sheep rumen epithelium: evidence for an electrodiffusive uptake mechanism. *Am J Physiol* 1999;277:G976-G982.
15. Quamme GA. Intracellular free Mg^{2+} with pH changes in cultured epithelial cells. *Am J Physiol* 1993;264:G383-G389.
16. Coudray C, Feillet-Coudray C, Grizard D, Tressol JC, Gueux E, Rayssiguier Y. Fractional intestinal absorption of magnesium is directly proportional to dietary magnesium intake in rats. *J Nutr* 2002;132:2043-2047.
17. Hardwick LL, Jones MR, Buddington RK, Clemens RA, Lee DB. Comparison of calcium and magnesium absorption: in vivo and in vitro studies. *Am J Physiol* 1990;259:G720-G726.
18. Karbach U, Schmitt A, Saner FH. Different mechanism of magnesium and calcium transport across rat duodenum. *Dig Dis Sci* 1991;36:1611-1618.
19. Phillips JD, Davie RJ, Keighley MR, Birch NJ. Brief communication: magnesium absorption in human ileum. *J Am Coll Nutr* 1991;10:200-204.
20. Larvor P. ^{28}Mg kinetics in ewes fed normal or tetany prone grass. *Cornell Vet* 1976;66:413-429.
21. Matsui T, Kawashima Y, Yano H. True absorption, and endogenous excretion of magnesium in cats given dry-type food and wet-type food [abstract]. *Jpn J Magnes Res* 2001;1:88-89.

22. Hintz HF, Schryver HF. Magnesium, calcium and phosphorus metabolism in ponies fed varying levels of magnesium. *J Anim Sci* 1973;37:927–930.
23. Allsop TF, Rook JAF. The effect of diet and blood-plasma magnesium concentration on the endogenous faecal loss of magnesium in sheep. *J Agric Sci Camb* 1979;92:403–408.
24. Georgievskii VI. The physiological role of macroelements. In: Georgievskii VI, Annenkov BN, Samokhin VI, eds. *Mineral Nutrition of Animals*. London: Butterworths; 1982:91–170.
25. Lucey S, Rowlands GJ, Russell AM. Short-term associations between disease and milk yield of dairy cows. *J Dairy Res* 1986;53:7–15.
26. Brommage R. Magnesium fluxes during lactation in the rat. *Magnes Res* 1989;2: 253–255.
27. Asai Y, Matsui A, Osawa T, et al. Nutrient intake from milk in Thoroughbred foals. *Proc AAAP Anim Sci Cong* 1996;1:527–532.
28. Equine Research Institute, Japan Racing Association. *Japanese Feeding Standard for Horses*. Tokyo: Animal Media; 2004.
29. McDowell LR. Minerals in animal and human nutrition. Amsterdam: Elsevier; 2003.
30. Allcroft R. Hypomagneseamia in cattle. *Vet Rec* 1954;66:517–522.
31. Fontenot JP, Wise MB, Webb KE Jr. Interrelationships of potassium, nitrogen, and magnesium in ruminants. *Fed Proc* 1973;32:1925–1928.
32. Blaxter KL, McGill RF. Magnesium metabolism in cattle. *Vet Rev Annot* 1956;2: 35–55.
33. Matsui T, Kawabata T, Harumoto T, Yano H. The effect of a synthetic analogue of pyrophosphate on calcium, magnesium and phosphorus homeostasis in sheep. *Asian-Austral J Anim Sci* 1992;5:303–308.
34. Matsui T, Yano H, Harumoto T. The effect of suppressing bone resorption on Mg homeostasis in sheep. *Comp Biochem Physiol* 1994;107A:233–236.
35. Robson AB, Sykes AR, McKinnon AE, Bell ST. A model of magnesium metabolism in young sheep: transactions between plasma, cerebrospinal fluid and bone. *Br J Nutr* 2004;91:73–79.
36. Garcia-Gomez F, Williams PA. Magnesium metabolism in ruminant animals and its relationship to other inorganic elements. *Asian Austral J Anim Sci* 2000;13: 158–170.
37. Allsop TF, Pauli JV. Cerebrospinal fluid magnesium concentrations in hypomagnesaemic tetany. *Proc N Z Soc Anim Prod* 1975;35:170–174.
38. McCoy MA, Young PB, Hudson AJ, Davison G, Kennedy DG. Regional brain monoamine concentrations and their alterations in bovine hypomagnesaemic tetany experimentally induced by a magnesium-deficient diet. *Res Vet Sci* 2000;69:301–307.
39. Parkinson GB, Leaver DD. The effect of experimental hyperkalaemia on cerebrospinal fluid magnesium. *Anim Prod Aust* 1980;13:447.
40. Hunt E. Disorder of magnesium metabolism. In: Smith BB, ed. *Large Animal Internal Medicine*. St. Louis: Mosby; 1996:1474–1480.
41. Agricultural Research Council. *The Nutrient Requirements of Ruminant Livestock*. Slough UK: Commonwealth Agricultural Beureaux; 1980.
42. Agriculture, Forestry and Fisheries Research Council Secretariat. *Standard Tables of Feed Composition in Japan*. Tokyo: Japan Livestock Industry Association; 1997.

43. Haaranen S. Does high plant feed magnesium and potassium protect healthy ruminants from atherosclerosis? A review. *Pathophysiology* 2003;10:1–6.
44. Buffington CA, Rogers QR, Morris JG. Effect of diet on struvite activity product in feline urine. *Am J Vet Res* 1990;51:2025–2030.
45. Bovee KC. Urolithiasis. In: Bovee KC, ed. *Canine Nephrology*. Philadelphia: Harwell; 1984:355–379.
46. Vagnoni DB, Oetzel GR. Effects of dietary cation-anion difference on the acid-base status of dry cows. *J Dairy Sci* 1998;81:1643–1652.
47. Amanzadeh J, Gitomer WL, Zerwekh JE, et al. Effect of high protein diet on stone-forming propensity and bone loss in rats. *Kidney Int* 2003;64:2142–2149.
48. Cottam YH, Caley P, Wamberg S, Hendriks WH. Feline reference values for urine composition. *J Nutr* 2002;132:1754S–1756S.
49. Kallfelz FA, Bressett JD, Wallace RJ. Urethral obstruction in random source SPF male cats introduced by dietary magnesium. *Feline Pract* 1980;10:25–35.
50. National Research Council. *Nutrient Requirements of Cats*. rev. ed. Washington, DC: National Academy Press; 1986.
51. Buffington CA, Rogers QR, Morris JG, Cook NE. Feline struvite urolithiasis–magnesium effect depends on urinary pH. *Feline Pract* 1985;15:29–33.
52. Buffington CA, Chew D. Intermittent alkaline urine in a cat fed an acidifying diet. *J Am Vet Med Assoc* 1996;209:103–104.
53. Malone F, Goodall E, O'Hagan J. Factors associated with disease in intensive lamb fattening units. *Irish Vet J* 1998;51:78–82.
54. Kallfelz FA, Ahmed AS, Wallace RJ, et al. Dietary magnesium and urolithiasis in growing calves. *Deut Tierarztl Woch* 1985;92:407–411.
55. Cuddeford D. Role of magnesium in the aetiology of ovine urolithiasis in fattening store lambs and intensively fattened lambs. *Vet Rec* 1987;121:194–197.
56. Hassel DM, Langer DL, Snyder JR, Drake CM, Goodell ML, Wyle A. Evaluation of enterolithiasis in equids: 900 cases (1973–1996). *J Am Vet Med Assoc* 1999;214: 233–237.
57. Butters AL. Intestinal calculi in the horse. *Vet J* 1894;18:348–352.
58. Lloyd K, Hintz HF, Wheat JD, Schryver HF. Enteroliths in horses. *Cornell Vet* 1987;77:172–186.

Index

A

ACDP 2 (ancient conserved domain protein 2), 22, 23, 24, 25
Acetylcholine, 128–129
Acid-base balance, effect on renal magnesium reabsorption, 59, 305
Adenosine, cardioprotective effects of, 232
Adenosine diphosphate (ADP). *See also* Magnesium-adenosine diphosphate (ADP) complex
in magnesium mitochondrial uptake, 51
Adenosine triphosphate (ATP). *See also* Magnesium-adenosine triphosphate (ATP) complex
magnesium/calcium-mediated production of, 127
in magnesium mitochondrial uptake, 51
in sodium/magnesium exchange, 25
β-Adrenergic desensitization, 234–235
Albumin, magnesium-bound, 55, 293
Alcohol abuse
as hepatic cancer cause, 166
as hypomagnesemia cause, 157
as magnesium deficiency and depletion cause, 69, 75, 159, 166, 243
Aldosterone, effect on magnesium renal re-absorption, 276–277
Alkalosis, ionized and total magnesium levels in, 59
Aluminum
as Parkinson's disease risk factor, 344
weight of, 4

Alzheimer's disease, 105, 344
Amiloride, 28, 290
Aminoglycosides, as renal magnesium wasting cause, 60, 281
Amitriptyline, 373
Amyotrophic lateral sclerosis/parkinsonism/dementia complex, 8–9, 23
Ancient conserved domain protein 2 (ACDP), 22, 23, 24, 25
Animals, magnesium in, 5, 381–391
magnesium absorption, 382–383
magnesium deficiency, 384–386
magnesium excess, 386–388
Annexin 1, as TRPM7 kinase substrate, 41–42
Antacids, magnesium-containing, 134, 316–317
Antidepressants, 372–373
Antidiuretic hormone, effect on magnesium renal reabsorption, 276–277
Antioxidants, 177, 230, 347–348, 373
Anxiety, photosensitive magnesium depletion-related, 119
Apoptosis, 347
Arachidonic acid, 108, 129, 130
Arrhythmias
ionized and total magnesium levels in, 58
magnesium as treatment for, 131, 132, 231–232
magnesium deficiency-related, 176, 236
Asthma, magnesium supplementation treatment of, 81, 133

393

Atherogenesis/atherosclerosis
  definition of, 239
  extracellular magnesium in, 239
  in hemodialysis patients, 323–324
  leukocytes in, 246–247
  magnesium deficiency-related,
      244–245, 247
    cellular signaling mechanisms in,
        239, 252–254
    fatty acid saturation in, 253–254
    ferrylmyoglobin radicals in,
        251–252
    free radicals in, 247–249
    growth factors in, 246
    leukocytosis in, 247
    lipid peroxidation in, 247–249
    membrane phospholipids in, 239,
        252–254
    platelet aggregation in, 247
    pro-inflammatory agents in,
        245–246, 254
    sphingolipids in, 239, 254
  magnesium deficiency-related
      acceleration of, 193
  magnesium-related amelioration of,
      249–251
  pathogenesis of, 239, 324
  spontaneous, in ruminants, 386
Athletes
  immunosuppression in, 178
  magnesium status of, 174–176
  magnesium supplementation in, 176
Atrial fibrillation, 236–237

B
Balance studies, of minerals, 94–103
Bartter-like phenotype, 272, 280
Bartter syndrome, 76, 277
Basolateral membrane, sodium/
    magnesium exchange in, 295–300
Biological clock, dysregulation of, 9, 117,
    118
  scototherapy for, 121–124
Bipolar disorder, 369–371
Blood-brain barrier, magnesium passage
    across, 345
Blood cells. *See also* Erythrocytes
  magnesium content of
    assessment of, 83–84
  in cancer, 159–160
  free magnesium, 243
  protein-bound magnesium, 243
Blood transfusions, 57
Bone
  calcium content of, 94, 266
    in magnesium deficiency, 109, 110
  calcium-magnesium ratio of, 318–319
  calcium release from, 94, 101, 102
  effect of magnesium deficiency on,
      109–110, 268, 270–271
    bone strength effects, 268
    mineral/matrix analysis of, 270–271
  magnesium content of, 94, 266, 318
    bone cell-regulating functions of,
        263
    in cortical bone, 303
    in hemodialysis patients, 307, 319
    as percentage of total body
        magnesium, 69–70, 158, 263, 316
    in renal disease, 307, 309–310, 319
    in trabecular bone, 303
  magnesium-hydroxyapatite complex
      of, 303
  magnesium incorporation into, 264
  magnesium release from, 94, 101, 102,
      309
  mineral content of, in magnesium
      deficiency, 109–110
Bone disease, chronic renal failure-
    related, 308–310
Bone metabolism
  effect of exercise on, 17
  effect of magnesium on, in
      hemodialysis patients, 79, 266,
      321–322
Bone minerals, 101
Bone remodeling, effect of calcium
    intake on, 15, 17
Brain
  effect of magnesium deficiency on,
      110
  MAT1 gene expression in, 24
Brain injury
  as Alzheimer's disease risk factor, 344
  magnesium as treatment for, 345–348
  magnesium in, 333, 335, 339–342
Burn injuries, magnesium loss through,
    75

C

Caffeine, effect on magnesium excretion, 290
Calcification, vascular, in hemodialysis patients, 322–323
Calcilytics, 274, 281
Calcimimetics, 266–267, 274, 281
Calcitonin, effect on magnesium renal reabsorption, 74, 276–277, 290, 294
Calcium. *See also* Hypercalcemia; Hypocalcemia
  bone content of, 94
    in magnesium deficiency, 109, 110
    release of, 101, 102
  as bone mineral, 101
  bone strength effects of, 268, 269
  brain content of, in magnesium deficiency, 110
  cancer-related increase in, 159
  as carbamazepine antagonist, 371
  dietary intake of
    balance studies of, 95–99
    daily, 96
    effect on bone remodeling, 15, 17
    in prehistoric humans, 12, 15, 17–18, 157
  drinking-water content of. *See* Drinking-water calcium
  as essential element, 100
  extracellular, 101
    in vascular contractility, 244–245
  interactions with magnesium, 127, 174
    antagonistic, 127, 174, 233, 234, 346
    effect on magnesium absorption, 72, 158
    effect on magnesium excretion, 306
    effect on sodium/magnesium exchange transport, 299–300
    intestinal, 106
    in myocardial cells, 234
    in myocardial infarction, 233
  intracellular, 128
    in atherogenesis, 239
    glucose-related increase in, 218
    hypertensive effects of, 229
  muscle content of, in magnesium deficiency, 109
  neurological effects of, 333–334
  plasma content of
    adverse effects of, 101
    effect of urinary magnesium excretion on, 101
  urinary excretion of
    circadian variations in, 99, 100
    trans-tissue calcium transport and, 102
Calcium carbonate, 268, 269
Calcium channels, magnesium-related blockage of, 127, 128
Calcium/magnesium ratio, 101
  in hypertension, 229
  in urine, 94, 99
Calcium metabolism, effect of exercise on, 17
Calcium-sensing receptor (CaSR), 266–267, 272–285
  activation of, 264
    by magnesium, 264, 274–275
  G-protein-coupled superfamily and, 273–274
  in magnesium homeostasis, 272–285, 289
  parathyroid hormone-regulating activity of, 263, 264, 272, 273
  pharmacological modulation of, 281, 282
  phospholipase C activation by, 274
  renal functions of, 277–279
Calcium-sensing receptor (*CaSR*) gene mutations of
  Bartter-like phenotype associated with, 272, 280
  disorders associated with, 272, 279–281
  polymorphisms of, 281
Cancer. *See also specific types of cancer*
  magnesium and, 111, 159–167
Carbamazepine, 370, 371, 372
Carbohydrate metabolism, 214
Cardiovascular disease, 227–238. *See also* Atherogenesis/atherosclerosis
  French paradox of, 12
  magnesium as treatment for, 131–132
  magnesium deficiency-related, 11, 128
    hyperlipidemia associated with, 108

Cardiovascular disease (cont.)
    ionized and total magnesium levels in, 58
    reactive oxygen species in, 229–230
    role of exercise in, 176
    magnesium's cardioprotective effects in, 227, 241–242
        β-adrenergic desensitization inhibition, 234–235
        anti-arrhythmic activity, 131, 132, 236–237
        anti-oxidative activity, 235
        drinking-water magnesium-related, 12–17
        in ischemic-reperfusion injury, 233–234
        in myocardial infarction, 230–233
    risk factors for, 190
Catecholamines, in heart failure, 234–235
Cats, urolithiasis in, 381, 386–388
Central nervous system. *See also* Brain; Cerebrospinal fluid; Neurodegenerative disease; Neurological disease; Neuronal cells
    magnesium in, 338–355
        basal content of, 339–340
        in ischemia/stroke, 342–343
        in neurodegenerative diseases, 343–344
        neuroprotective effects of, 333, 335, 345–348
        in trauma, 128–129, 340–342
    trauma to
        magnesium as treatment for, 130, 344–345
        pathophysiology of, 130
Ceramide, 129, 254
Cerebral palsy, 81
Cerebrospinal fluid, magnesium content of
    in grass tetany, 385
    in major depression, 371–372
    in schizophrenia, 373
Chemotherapy, effect on ionized magnesium/total magnesium ratio, 60

Chloride, effect on sodium/magnesium exchange transport, 299–300
Cholesterol, 159
Chronic fatigue syndrome, 119
Chronic inflammatory syndrome, 143, 144
Chronic obstructive pulmonary disease, 133
Circadian rhythms
    of calcium and magnesium excretion, 99, 100
    of ionized and total magnesium levels, 61
Cisplatin, 60
Citrate. *See also* Magnesium citrate
    effect on ionized magnesium/total magnesium ratio, 57
Claudin, 16, 289
Colon
    magnesium transporter protein expression in, 25
    pseudo-obstruction of, 134
Colon cancer, relationship to drinking-water hardness levels, 161
Constipation, magnesium salts treatment for, 134
Cooking, dietary magnesium loss during, 84–85, 156–157
Copper, body content of, in magnesium deficiency, 109, 110
CorA protein, as magnesium transport protein, 22, 46, 47, 50, 52–53
    structure and function of, 48–49, 51
C-reactive protein, 144
Cycad seeds, 9
Cyclosporin A, 59, 60
Cystic fibrosis, 21, 60

D

Darkness therapy. *See* Scototherapy
Delayed sleep phase syndrome (DSPS), 118–119
Depression, magnesium deficiency-related, 105
Dermatologic disease, magnesium sulfate treatment for, 135
Diabetes mellitus
    in children, 195

ionized and total magnesium content in, 58, 215
magnesium deficiency in, 58, 108–109, 143–154, 192, 197–212, 215–216, 219
   clinical research studies of, 148–150
   effect of glycemic control on, 199
   epidemiological and clinical evidence of, 204–206
   as insulin resistance cause, 216
   intracellular magnesium levels in, 215, 216
   ionized magnesium levels in, 58, 215, 216
   low magnesium intake-related, 215–216
   possible mechanisms of, 198–199
   prospective studies of, 144–148
   total magnesium levels in, 58, 215, 216
   urinary magnesium excretion-related, 58, 215 216
magnesium depletion in, 290–291
   age-related, 217
magnesium metabolism in, 213
prevalence of, 143
treatment of
   effect on magnesium content, 219
   with magnesium supplementation, 78, 145, 148, 149, 150, 193, 213, 219–220
Dialysate, magnesium-containing, 306–307, 308, 309, 310, 318, 321
Dietary Approaches to Stop Hypertension (DASH), 230
Dietary intake, of nutrients, 94. See also Calcium, dietary intake of; Magnesium, dietary intake of
Digestive system, comparative anatomy of, 381–382
Digitalis intoxication, 236
Diuretics
   effect on magnesium renal handling, 290
   effect on serum magnesium levels, 306
   effect on TRPM6, 60
   as magnesium deficiency cause, 291

as magnesium depletion cause, 243
as magnesium renal wasting cause, 215–216
DNA synthesis, magnesium-based regulation of, 242
Docosahexaenoic acid, 108
Drinking-water calcium, 11–18
   geographic differences in, 11–12, 15–17
   relationship with cardiovascular disease, 11, 131
Drinking-water magnesium, 11–18
   absorption of, 165
      intestinal, 155–156
   bioavailability of, 156, 157
   effect on total body magnesium content, 155–156, 157
   geographic differences in, 11–12, 15–17
   recommended levels of, 157
   relationship with cardiovascular disease, 6, 11, 12–17
   relationship with hepatic cancer, 161–167
      age factors in, 163, 164, 165, 166
      effect of alcohol abuse on, 166
      genetic factors in, 163, 164, 165, 166
Drugs. See also names of specific drugs
   as hypomagnesemia cause, 59–60
   as magnesium depletion cause, 243
Dyslipidemia. See also Hyperlipidemia; Hypolipidemia
   in hemodialysis patients, 325–326
   magnesium deficiency-related, 76, 326

E
Eclampsia, 21, 133
Elderly persons
   delayed sleep phase syndrome (DSPS) in, 118–119
   diabetic, magnesium supplementation in, 199–200
   magnesium intake and absorption in, 158
Endocrine disorders, 75
Energy metabolism, magnesium in, 71, 143–144, 158, 347
Enterolithiasis, 381, 386, 387

Enzymes. *See also specific enzymes*
  magnesium as cofactor of, 69, 71, 104, 129, 144, 173, 241, 242
  in carbohydrate metabolism, 214
  in membrane-bound enzymes, 129
Epilepsy. *See also* Seizures
  photosensitive magnesium depletion-related, 119
Epinephrine, 56
Epsom salts, 4
Erythrocyte membranes, lipid metabolism in, 108
Erythrocytes
  magnesium content of, 70, 318
  in bipolar disorder, 369
  in diabetes mellitus, 197, 198
  during exercise, 175
  in hemodialysis patients, 319
  low levels of, 105
  in magnesium deficiency, 108
  in major depression, 372
  as nutritional status marker, 107
  sodium/magnesium exchange in, 25
Esophageal cancer, 161
Essential elements, speciation of, 100
Estimated average requirement (EAR), of magnesium intake, 94–103
Exercise
  effect on bone and calcium metabolism, 17
  magnesium and, 173–185
    compartmental magnesium shifts and, 174, 175
    immune system responses, 177–180
    oxidative stress and, 175–176

F
Fat, dietary, in magnesium excretion, 99
Fatigue, magnesium deficiency-related, 105
Fatty acids
  effect on magnesium uptake, 158
  free, 129, 130, 176
  in magnesium deficiency, 76
  volatile, in ruminants, 383
Ferrylmyoglobin radicals, 251–252
Fiber, dietary, effect on magnesium absorption, 72
Fibromyalgia, 119

Food, as magnesium source, 144, 156, 157, 198, 204, 206, 303–304
  cooking-related loss of, 84–85, 156–157
Foscarnet, 60
Free radicals, 235
  exercise-induced, 176–177
  magnesium deficiency-related, 247–249
  magnesium-related reduction in, 347–348
French paradox, of cardiovascular disease, 12
Fungi, magnesium transport proteins in, 49–50, 51–52
FXYD2 gene, 36

G
Gastric cancer, 161
Genetic disorders, 74, 75
Gitelman's syndrome, 75, 291
Glibenclamide, 206
Glucagon, effect on magnesium renal reabsorption, 73, 290, 294
Glucose
  effect on intracellular magnesium levels, 218
  hypermagnesemic effects of, 290–291
Glucose homeostasis, in magnesium deficiency, 108–109
Glucose intolerance, magnesium deficiency-related, 148, 150
Glucose metabolism, in magnesium deficiency, 213–214, 216
Glutamate, 345–346
  magnesium-mediated reduction in, 128–129
  magnesium-mediated release of, 345–346
Glutathione, 177, 179, 219, 347–348
Glutathione-stimulating hormone, in exercise, 177
Glycolytic cycle, magnesium in, 71
Grass tetany, 384–385
Grignard reagents, 5
Growth, effect of chronic magnesium deficiency on, 105–106
Growth factors, 229–230

H

Headaches
  magnesium deficiency-related, 334
  migraine, magnesium deficiency-
    related, 81, 105
  photosensitive magnesium depletion-
    related, 118, 119, 121, 122,
    123–124
Heart
  magnesium content of, 158
  magnesium transporter protein
    expression in, 25
  mineral content of, in magnesium
    deficiency, 109
Heart failure, 233–234
  catecholamines in, 234–235
  diuretics-related magnesium
    deficiency in, 291
  magnesium as treatment for, 243–244
Hemodialysis patients
  atherosclerosis n, 323–324
  hyperphosphatemia in, 320
  magnesium deficiency in
    as dyslipidemia cause, 325–326
    as hypertension cause, 325
    as vascular calcification cause,
      322–323
  magnesium in, 316–329
    effect on bone metabolism, 321–322
    effect on parathyroid hormone
      secretion, 310, 321
    as hypermagnesemia, 306
    ionized and total magnesium levels
      in, 306
    ionized magnesium/total
      magnesium ratio in, 56–57
  phosphate binder treatment for,
    320–321
Hemorrhage, subarachnoid, 344, 346
Hepatic cancer
  drinking-water magnesium content
    and, 155–169
  age factors in, 163, 164, 165, 166
  effect of alcohol abuse on, 166
  genetic factors in, 163, 164, 165, 166
  risk factors for, 165
  types of, 160–161
Hepatitis, 165
Homocysteine, 250–251

Hormones. *See also specific hormones*
  effect on magnesium and calcium
    excretion, 99
  effect on magnesium homeostasis,
    272–273
Human immunodeficiency virus (HIV)
  infection, 21
Hungry bone syndrome, 75
Hydroxyapatite, as tooth component,
  359, 360–362
p-Hydroxybenzenediazonium,
  genotoxicity of, 111
Hypercalcemia
  familial hypocalciuric, 273, 279, 280
  magnesium deficiency-related, 109
Hypercholesterolemia, as atherosclerosis
  risk factor, 249
Hyperglycemia, diabetes
  mellitus-related
  effect on magnesium excretion, 216
  as intracellular magnesium depletion
    cause, 218
Hyperhomocysteinemia, 250–251
Hyperinsulinemia
  diabetes mellitus-related
    effect on magnesium excretion, 216
    as intracellular magnesium
      depletion cause, 218
  low magnesium:calcium ratio-related,
    229
  magnesium deficiency-related, 148,
    150
Hyperlipidemia
  magnesium as treatment for, 193,
    323–324
  magnesium deficiency-related, 108,
    189, 193
  metabolic syndrome-related, 189, 190,
    191
Hypermagnesemia
  CaSr gene mutations-related, 279–281
  chronic renal failure-related, 307–308,
    317
  beneficial effects of, 308
  dysregulated. *See* Magnesium
    depletion
  effect on parathyroid gland function,
    309, 310
  exercise-related, 174

Hypermagnesemia (cont.)
    familial hypocalciuric hypercalcemia-related, 280
    in hemodialysis patients, 306
    neonatal severe hyperparathyroidism-related, 280–281
Hyperparathyroidism
    hereditary, 272
    neonatal severe, 279, 280–281
Hyperphosphatemia, in hemodialysis patients, 320
Hypertension
    age-dependent magnesium depletion in, 217
    calcium levels in, 128
    magnesium as treatment for, 174, 193, 241
    magnesium : calcium imbalance in, 229
    magnesium deficiency-related, 21, 77, 128, 189, 244–245
        in hemodialysis patients, 325
        reactive oxygen species and, 229–230
    metabolic syndrome-related, 189, 190, 191
    neonatal pulmonary, 134
Hypertriglyceridemia, magnesium deficiency-related, 108
Hypoalbuminemia, 57
Hypocalcemia
    autosomal-dominant, 272
    chronic renal failure-related, 309
    hypomagnesemia-related, 23, 36, 37, 39, 229, 291
    magnesium as treatment for, 158
    magnesium deficiency-related, 73, 75, 158
    neonatal, 58
    secondary, 295
Hypokalemia, magnesium deficiency-related, 73, 74
Hypomagnesemia. See also Magnesium deficiency; Magnesium depletion
    alcohol abuse-related, 157
    autosomal-dominant, 36, 291
    drug-related, 59–60
    exercise-related, 175
    familial renal, 294

    hereditary, genes involved in, 35–36
    with hypercalciuria, 291
    prevalence of, 104, 105
    with secondary hypocalcemia, 36, 229, 291
    renal magnesium reabsorption in, 38
    TRPM6 and, 23, 36, 37, 39
    skeletal magnesium decrease in, 268
Hypoparathyroidism, autosomal-dominant, 279–280, 291
Hypoxia-ischemia, 343, 344–345, 347

I
Imipramine, 28, 296, 372
Immune system, effect of magnesium on, 177–180
Immunosuppressive drugs
    effect on ionized and total magnesium levels, 58–59, 60
    as hypomagnesemia cause, 58–59, 60
Inflammation
    as atherosclerosis cause, 239, 245–246
    magnesium deficiency-related, 179–180, 324
Insulin
    effect on intracellular magnesium content, 218–219
    in magnesium metabolism, 291
Insulin receptors, effect of magnesium deficiency on, 213, 214
Insulin resistance
    cellular ion responsiveness in, 219
    low magnesium : calcium ratio-related, 229
    magnesium deficiency-related, 11, 77, 108, 144, 148, 149, 150, 189, 192, 213–214, 219
        in children, 195
        clinical evidence of, 202–203
    magnesium depletion-related, 213
Insulin secretion, effect of magnesium on, 199–200
Insulin sensitivity
    glucose clamp assessment of, 202–203
    magnesium deficiency-related decrease in, 148, 149

magnesium supplementation-related increase in, 150, 219–220
Intensive care unit patients, ionized and total magnesium levels in, 57
Interleukins, 178
International Symposia on Magnesium, 3, 7
Intestines, *MAT1* gene expression in, 24
Intrauterine fetal growth retardation, 81
Ionized magnesium/total magnesium ratio
  effect of chemotherapy on, 60
  inverse relationship with magnesium balance, 56
  in magnesium depletion, 57–61
Iron
  body content of, in magnesium deficiency, 109, 110
  as intracellular mineral, 101
Ischemia, cerebral, 335, 342–343
Ischemic heart disease, 230–231, 233, 239, 241
Isoproterenol, cardiotoxicity of, 234, 235

J
Jejunum, magnesium absorption in, 304–305

K
Ketoacidosis, diabetic, 59
Kidney. *See also* Hemodialysis patients; Magnesium, renal reabsorption and transport of; Magnesium, urinary excretion of; Renal failure
  calcium-sensing receptor functions in, 277–279
  magnesium transporter protein expression in, 24, 25
  mineral content of, in magnesium deficiency, 109
  role in magnesium homeostasis, 72, 275–277, 289–290
  TRPM6 localization in, 37–38

L
Lamotrigine, 370
Laxatives, 134
Learning, 334, 336

Lipid metabolism, magnesium in, 128, 129, 130
Lipid oxidation, in atherogenesis, 239
Lipid peroxidation, 177
  in magnesium deficiency, 247–249
  in magnesium depletion, 128, 130
  magnesium-related inhibition of, 177, 235, 347
Lipoprotein profile, in magnesium deficiency, 108
Lithium, 370, 371, 372
Liver, effect of magnesium deficiency on, 110, 159
Liver failure, acute, 134
Lung, *MAT1* gene expression in, 24
Lymphocytes, magnesium content of
  assessment of, 83–84
  in hemodialysis patients, 307, 319

M
Magnesia alba, 4
Magnesium
  absorption of
    in animals, 382–383
    disease-related decrease in, 76
    paracellular route of, 382
    transparacellular route of, 382–383
  in animals, 5, 381–391
  antioxidant activity of, 177, 230, 347–348, 373
  anti-peroxidant effect of, 177, 347
  balance studies of, 95–99
    correlation with calcium balance, 95
    correlation with phosphorus balance, 95
  biochemical functions of, 70
  as calcium antagonist, 127, 174, 241, 346
  cardioprotective effects of, 227, 241–242
  β-adrenergic desensitization inhibition, 234–235
  anti-arrhythmic activity, 131, 132, 236–237
  antioxidative activity, 235
  drinking-water magnesium-related, 12–17

Magnesium (*cont.*)
   in ischemic-reperfusion injury, 233–234
   in myocardial infarction, 230–233
  cellular extrusion of, 27
   regulation of, 27–28
   sodium/magnesium exchange in, 25–29
  cellular transport mechanisms of. *See also* Magnesium transport proteins
   of extrusion, 25–29
   of influx, 21–25
   sodium/magnesium exchange in, 25–29
  chemistry of, 69
  dialysate content of, 306–307, 308, 309, 310, 318, 321
  dietary intake of
   adequate, 102
   apparent absorption and, 96, 97
   average daily amount of, 304
   calcium balance and, 96, 97, 98, 99
   in chronic renal failure, 306
   critical amount of, 304
   effect on glucose metabolism, 216
   effect on renal magnesium conservation, 289
   estimated average requirement (EAR) of, 94–103
   high, cardioprotective effects of, 241–242
   inadequate, 104–105, 144, 145, 216, 240–241, 304
   low, magnesium intestinal absorption in, 304
   magnesium homeostasis and, 96, 97, 99
   phosphorus balance and, 96, 97, 98, 99
   in prehistoric humans, 12, 15, 17–18, 157
   Recommended Daily Allowance (RDA), 84, 85, 104–105, 144, 157
   relationship to diabetes mellitus development, 144–150
   relationship to fracture risk, 270
   sodium intake and, 98, 99
   urine output and, 96, 97
  dietary sources of, 84–85, 144, 156, 157, 198, 204, 303–304
  digestion-related loss of, 144
  distribution in the body, 69–70, 158
   in cancer, 159–160
  drinking water content of. *See* Drinking-water magnesium
  effect on sodium/magnesium exchange transport, 299–300
  as enzyme cofactor, 69, 71, 104, 144, 158, 173, 241, 242
   in carbohydrate metabolism, 214
   with membrane-bound enzymes, 129
  as essential element, 100
  extracellular, 21, 70
   in atherogenesis, 239
  extracellular fluid compartment content of, 316
  fecal excretion of, in animals, 383–384
  food preparation-related loss of, 84–85, 156–157
  free, in cardiovascular disease, 243–244
  free cystolic levels of, 21
  gastric absorption of, 304
  interactions with calcium, 127, 174
   as calcium antagonist, 127, 174, 346
   effect on magnesium absorption, 72, 158
   effect on magnesium excretion, 306
   effect on sodium/magnesium exchange transport, 299–300
   intestinal, 106
  intercellular effects of, 174
  intestinal absorption of, 71–72, 198, 293, 304, 317–318
   in animals, 382, 383
   in chronic renal failure, 303–315
   effect of protein intake on, 305
   effect of vitamin D on, 317–318
   factors affecting, 99–100
   relationship to magnesium urinary excretion, 306
   in uremia, 318
  intestinal content of
   interaction with calcium, 106

interaction with phosphorus, 106
interaction with zinc, 106
in magnesium deficiency, 106–107
intracellular, 55, 70, 83–84, 128, 158, 318–319
  aging-related decrease in, 217
  assessment of, 83–84
  in brain injury, 339–342
  in diabetes mellitus, 197–198
  effect of exercise on, 175
  glucose-related decrease in, 218
  insulin-induced stimulation of, 218–219
  as percentage of total body magnesium, 214, 303
  regulation of, 336
  relationship to serum magnesium levels, 318
  in stroke patients, 342–343
intracellular ionized
  in hemodialysis patients, 319
  negative correlation with hypertension, 325
as intracellular mineral, 101
ionized, 55, 83. *See also* Ionized magnesium/total magnesium ratio
  in acid-base imbalances, 59
  circadian variations in, 61
  concentration of, 55
  in diabetes mellitus, 59
  effect of chemotherapy on, 60
  in hemodialysis patients, 306
  as magnesium status clinical marker, 55, 56, 57
  measurement of, 213, 215
  in renal transplant patients, 60
  serum content of, 83, 289
measurement of
  with magnesium-specific ion-reductive electrodes, 213, 215
  spectroscopic, 6, 213, 215
membrane functions of, 128, 129, 173–174, 241, 242
metallic, 4
neuroprotective effects of, 333, 335, 344–348
pharmacodynamic studies of, 5

pharmacological functions of, 127–130
  differentiated from physiological functions, 7–8
  toxicity, 8
physiological functions of, 71, 127, 173, 241, 242
  differentiated from pharmacological functions, 7–8
  early research in, 5–6
in plants, 5
plasma content of
  chelation fraction of, 214
  in diabetes mellitus, 197–198
  effect of magnesium excretion in, 101
  ionized fraction of, 214
  normal, 303
  protein-bound fraction of, 214
  renal filtration of, 293, 305
renal handling of, 289
  effect of acid-base status on, 39
  effect of diuretics on, 290
renal reabsorption and transport of, 72–73, 272, 289, 293–302
  in animals, 382–383
  CaSr in, 276, 290
  cellular basis of, 293–297
  in distal convoluted tubules, 34–35, 36, 39–40, 272, 275–277, 289, 290, 293, 294–295
  effect of acid-base balance on, 305
  hormonal influences on, 272–273, 276–277, 294
  in hypomagnesemia with secondary hypocalcemia, 38
  in magnesium deficiency, 293
  paracellin/paracellular protein in, 276, 294
  paracellular pathways of, 293, 294
  in proximal tubules, 34, 289, 305
  sodium/magnesium exchange in, 295–300
  in thick ascending loop of Henle, 34, 275–276, 289, 290, 293, 305
  transient receptor potentials channels in, 294–295
  TRPM6-mediated, 38–40

Magnesium (*cont.*)
  renal wasting of
    in acute renal failure, 60
    aminoglycosides-related, 60, 281
    diuretics-related, 215–216
    TRPM6 and, 39
  serum content of, 81–82, 158
    assessment of, 81–82
    in chronic renal failure, 306–307
    ionized, 83, 289
    normal content, 289
    in peritoneal dialysis patients, 307
    protein-bound, 289, 293
    ultrafilterability of, 289, 293
  structural functions of, 241
  tissue content of
    in chronic renal failure, 306–307
    in uremia, 307
  total, 55, 69–70, 303, 316. *See also*
      Ionized magnesium/total
      magnesium ratio
    in brain injury, 342
    circadian variations in, 61
    components of, 55
    in cystic fibrosis, 60
    in diabetes mellitus, 59
    in diabetic ketoacidosis, 59
    effect of chemotherapy on, 60
    in hemodialysis patients, 306
    as magnesium status clinical
      marker, 55, 56, 57
    measurement of, 214–215, 243
    normal range of, 55
    physiological effects of, 158–159
    ranges of, 293
    in renal transplant patients, 60
  uptake of
    into bacteria, 46
    into eukaryotic cells, 46–53
    factors affecting, 158
    into mitochondria, 46, 47, 50, 51
    optimal, 158
  urinary excretion of, 72, 76, 198, 293
    in animals, 383
    circadian variations in, 99, 100
    in diabetes mellitus, 58, 198
    effect of antacids on, 316–317
    effect of calcium on, 306
    effect of diets on, 290
    effect of drugs on, 290
    effect of sodium on, 306
    effect of urine pH on, 305
    effect on plasma magnesium
      content, 101
    during exercise, 174
    factors affecting, 99–100
    in familial renal hypomagnesemia,
      294
    in hypomagnesemia with secondary
      hypocalcemia, 38
    in magnesium tolerance test, 84
    normal amounts of, 316
    in renal insufficiency, 305
    trans-tissue calcium transport and,
      102
  as vasodilator, 346
Magnesium-adenosine diphosphate
    (ADP) complex, 173, 303
Magnesium-adenosine triphosphate
    (ATP) complex, 69, 71, 173, 214,
    228, 303
  in glycolysis, 214
  stroke-related decrease in, 342, 343
Magnesium albumin, 55
Magnesium aspartate, 371
Magnesium aspartate-hydrochloride,
    249–250
Magnesium balance. *See* Magnesium
    homeostasis
Magnesium balance studies, 84
Magnesium bicarbonate, 55
Magnesium carbonate, 310
Magnesium chloride, 131
  lack of neuroprotective activity of, 345
  as magnesium depletion treatment,
    120
Magnesium citrate, 55, 134, 303
Magnesium complexes, 69
Magnesium deficiency, 69, 73–81
  acute, 104
  adverse effects of, 11
  alcohol abuse-related, 159
  in animal models, 73
  in animals, 381, 384–386
  anti-carcinogenic activity of, 160
  as atherogenesis cause, 244–245, 247
  cellular signaling mechanisms in,
    252–254

fatty acid saturation in, 253–254
ferrylmyoglobin radicals in, 251–252
free radicals in, 247–249
growth factors in, 246
leukocytes in, 254
lipid peroxidation in, 247–249
membrane phospholipids in, 252–254
pro-inflammatory agents in, 245–246, 254
sphingolipids in, 239, 254
as calcium metabolism impairment cause, 229
carcinogenic activity of, 111, 160
as cardiovascular disease cause, 11, 76–77, 128, 239–260, 243
    ionized magnesium/total magnesium ratio in, 58
chronic
    in animal models, 104–116
    nutrient metabolism in, 107–110
clinical, 74–81
as cystic fibrosis risk factor, 21
definition of, 8
in diabetes mellitus, 58, 79, 108–109, 143–154, 192, 197–212, 215–216, 219
    clinical research studies of, 148–150
    as insulin resistance cause, 216
    intracellular magnesium levels in, 215, 216
    ionized and total magnesium levels in, 215, 216
    low magnesium intake-related, 215–216
    magnesium excretion-related, 215–216
    prospective studies of, 144–148
diagnostic tests for, 8
differentiated from magnesium depletion, 3, 94, 117–118, 130
as disease cause, 157–158
diuretics-relatedr, 290
as eclampsia cause, 21
effect on bone, 109–110, 268, 270–271
    bone strength effects, 268
    mineral/matrix analysis of, 270–271
exercise-related, 175

experimental, 74
of free magnesium, in cardiovascular disease, 243–244
in hemodialysis patients
    as dyslipidemia cause, 325–326
    as hypertension cause, 325
    as vascular calcification cause, 322–323
as HIV infection risk factor, 21
hyperlipidemia-related, 193, 323–324
as hypertension cause, 21, 77, 128, 228–229
as hypocalcemia cause, 74–75, 158
as insulin resistance cause, 11, 77, 108, 148, 149, 150, 189, 192, 213–214, 219
    in children, 195
    clinical evidence of, 202–203
as insulin sensitivity cause, 148, 149, 150
intestinal magnesium levels in, 106–107
intracellular
    in diabetes mellitus, 197–198
    effect on insulin secretion, 199
intracellular ion content in, 173
ionized magnesium levels in, 55
as ischemic heart disease risk factor, 21
magnesium as treatment for, 55, 130–131
magnesium renal reabsorption in, 293
magnesium status measurement in, 214–215
as major depression cause, 371–372
marginal, 76–81
as neurological disease cause, 343–344
as osteoporosis cause, 79–81, 263–264, 268, 270
prevention of, 102
reactive oxygen species formation in, 229–230
renal effects of, mineral content and, 109
signs and symptoms of, 105, 176
as sterility cause, 5
total magnesium levels in, 55
urinary magnesium content in, 289

Magnesium deficit, 3
Magnesium depletion, 55, 56–57
  in animal models, 73
  cancer-related, 159–160
  as cardiovascular disease cause, 230–231
  differentiated from magnesium deficiency, 3, 94, 117–118, 130
  effect on lipid peroxidation, 128, 130
  homocysteine-related, 250–251
  insulin resistance in, 213
  ionized magnesium/total magnesium ratio in, 57–61
  magnesium as treatment for, 130–131
  photosensitization-related, 117–126
    clinical forms of, 117, 118–119
    treatment of, 120–124
  post-traumatic, 129
  renal magnesium handling in, 57
  serum magnesium content in, 82
  signs and symptoms of, 76
  total serum magnesium content in, 214
Magnesium-DNA complex, 303
Magnesium excess, 56
  in animals, 386–388
Magnesium homeostasis, 55, 71, 303–306
  factors in, 56
  in hemodialysis patients, 306–307
  hormonal regulation of, 272–273, 290
  ionized magnesium/total magnesium ratio in, 56
  kidney's role in, 272, 316–317
  mechanisms of, 198
  parathyroid hormone in, 290
  total magnesium levels in, 56
  transient receptor potential melastatin chanzymes in, 34–35
Magnesium hydroxide, 134
Magnesium hydroxyapatite, 303
Magnesium-inhibited cation channel (MIC), 40–41
Magnesium intoxication, antidote to, 5
Magnesium lactate, 55
Magnesium load tests, 8
Magnesium nutrition, overview of, 69–89
Magnesium nutritional status, assessment of, 82–83

Magnesium oxalate, 291
Magnesium phosphate, 55, 135
Magnesium-protein complex, 55
Magnesium research
  current trends in, 7–9
  history of, 3–6
Magnesium salts. *See also specific magnesium salts*
  as antacids, 134
  as arrhythmia treatment, 131, 132
  as laxatives, 134
  as magnesium depletion treatment, 120–121
Magnesium sulfate, 131, 132
  as arrhythmia treatment, 236
  as asthma treatment, 81
  as bipolar disorder treatment, 371
  dehydrated, 135
  as dermatologic disease treatment, 135
  as eclampsia treatment, 133
  as Epsom salts component, 4
  history of, 4
  intravenous, 120
  lack of neuroprotective activity of, 344–345
  as magnesium depletion treatment, 120, 121
  as neonatal pulmonary hypertension treatment, 134
  as neuromuscular disease treatment, 5
  overdose of, 133
  as pre-eclampsia prophylaxis, 133
  as preterm labor prophylaxis, 133
  side effects of, 133
Magnesium supplementation
  in athletes, 176
  comparison with magnesium-containing drinking water, 156
  effect on insulin secretion, 199–200
  effect on lipid metabolism, 193
  insulin sensitivity-modifying effects of, 150
  side effects of, 206, 208
  therapeutic indications for, 130–135
    in cardiology, 131–132
    diabetes mellitus, 79, 145, 148, 149, 150, 193, 206, 208, 213, 219–220

in gastroenterology, 134–135
in gynecology, 132–133
heart failure, 243–244
hyopcalcemia, 158
hyperlipidemia, 323–324
hypertension, 77, 174, 193, 241
liver failure, 134
magnesium deficiency, 8
magnesium deficits, 130–131
magnesium depletion, 118, 120
metabolic syndrome, 213
myocardial infarction, 131, 230–233, 243–244
osteoporosis, 268
physiological processes in, 127–130
in pneumology, 133–134
TRPM7 inactivation and, 38
use in scototherapy, 122–123
Magnesium tolerance test, 84
Magnesium transport proteins
ACDP2, 22, 23, 24, 25
Alr1, 49–50, 51, 52
Alrlp, 46
CorA, 22, 46, 47, 48–49, 50, 51, 52–53
eukaryotic, 46–53
MagT1, 22, 23, 24
mgtA, 47–48
mgtBC, 47–48
mgtE, 48
Mrs2, 46, 50–52
SLC41, 48
SLC4A1 member a2, 25
SLC41 member a1, 22, 23
SLC41 member a2, 22, 23
TRM6, 22–23, 36, 37–41, 295
TRM7, 22–23, 27, 34, 38–42
Magnium, 4
MagT1, 22, 23, 24
Major depression, 369, 371–373
Malabsorption syndromes, as magnesium depletion cause, 74, 75, 76
Malnutrition, protein-energy, magnesium deficiency-related, 107–108
Manganese, in magnesium deficiency, 109

Matrix metalloproteases, 245–246
Melatonin, 117, 121, 124
Membrane functions, of magnesium, 128, 129, 173–174, 241, 242
Memory, 334, 336
Menatetrenone, 270
Metabolic syndrome
definition of, 190
lifestyle factors in, 189
magnesium as treatment for, 213
magnesium deficiency-related, 189–196
magnesium-to-calcium ratio in, 229
Metabolic therapy, with magnesium. See Magnesium supplementation, therapeutic indications for
Metformin, 219
N-Methyl-D-aspartate channels, magnesium blockage of, 129, 174, 333, 336, 345–346
Migraine headache, magnesium deficiency-related, 81, 105
Milk, magnesium content of, 384
Milk tetany, 386
Mineral balance studies, 94–103
Minerals
absorption of, in magnesium deficiency on, 106–107
essential, 101
extracellular, 101
intracellular, 101
tissue content of, in magnesium deficiency, 109–110
Mood-stabilizing drugs, interaction with magnesium, 370, 371, 372
Muscle. See also Vascular smooth muscle cells
magnesium content of, 158, 316, 318
in diabetes mellitus, 198
in hemodialysis patients, 319
as percentage of total body magnesium, 303
magnesium-mediated contraction of, 174
magnesium transporter protein expression in, 25
mineral content of, in magnesium deficiency, 109

Myocardial infarction
  magnesium as treatment for, 131, 230–233, 243–244
  magnesium levels in, 76
Myosin II, as TRPM7 kinase substrate, 41–42

N
Natural killer cells, 178
Neonates, hypocalcemia in, 58
Nephrolithiasis, calcium oxalate, 291
Neurodegenerative diseases, magnesium depletion-related, 8–9
Neurological diseases, magnesium deficiency-related, 8, 343–344
Neuromuscular disorders, photosensitive magnesium depletion-related, 118
Neuronal cells, magnesium in, 174, 333–337
  neuroprotective effects of, 81, 333, 335–336
  therapeutic importance of, 334–335
Neurotransmitters, 336
  effect on intracellular calcium levels, 128
  magnesium-mediated release of, 128–129
  magnesium-related inhibition of, 346
Nifedipine, interaction with magnesium sulfate, 133
Nitric oxide synthase, 252
Nobel Prize, 5
No-reflow phenomenon, 233–234
Norepinephrine, 193–194
Nutrients
  absorption of, in magnesium deficiency, 106
  effect on magnesium and calcium excretion, 99
  metabolism of, in magnesium deficiency, 107–110
Nuts, as magnesium source, 204, 206, 303

O
Obesity
  in children, 195
  as diabetes mellitus risk factor, 143
  as metabolic syndrome risk factor, 190, 191
Oral contraceptives, effect on magnesium absorption, 158
Organomagnesium compounds, 5
Osteodystrophy, renal, 309, 310, 321–322
Osteoporosis, 195, 266–271
  magnesium as treatment, 268
  magnesium deficiency-related, 79–81, 263–264, 268, 270
Osteoprotegerin, 264, 268, 270
Ovarian cancer, 161
Oxidative stress, 176–177, 235
  glutathione depletion in, 179

P
Pancreatic cancer, 161
Panic attacks, 119
Paracellin protein, 276, 294
  encoding gene for, 35–36
Parathyroid hormone, 74, 109, 110
  CaSr-related regulation of, 263
  in chronic renal failure, 309–310
  in hemodialysis patients, 321
    effect on bone metabolism, 322
    as hypertension cause, 325
  in hypermagnesemia, 308
  in magnesium deficiency, 79
  in magnesium renal reabsorption, 276–277, 289, 294
  secretion of
    calcilytics-related increase in, 281
    calcimimetics-related decrease in, 281
    magnesium-mediated, 266, 321
Parathyroid hormone-calcitonin couple, 56
Parkinson's disease, 343–344
Peanut butter, as magnesium source, 204, 206
Peritoneal dialysis patients, serum magnesium levels in, 307
Phosphate, effect on magnesium absorption, 72
Phosphate binders, 320–321
Phospholipase C, magnesium-activated, 40
Phospholipases, 128, 129, 130

Phosphorus
  balance studies of, 95–99
  bone content of, 266
  brain content of, in magnesium
      deficiency, 110
  daily intake of, 96
  intestinal content of, interaction with
      magnesium, 106
  as intracellular mineral, 101
Pioglitazone, 219
Plants
  magnesium content of, 157
  magnesium transport proteins in,
      49–50
Platelet aggregation, 108, 229
Polycystins, 37
Polyphenols, red wine as source of, 12
Postoperative patients, ionized and total
      magnesium levels in, 57
Potassium
  cancer-related deficiency in, 159
  effect on renal magnesium
      reabsorption, 290
  effect on sodium/magnesium
      exchange transport, 299–300
  as intracellular mineral, 101
  serum levels of, in arrhythmia, 132
Potassium channels, magnesium-related
      regulation of, 346
Pre-eclampsia, 132
Pregnancy
  in diabetic patients, 79
    ionized and total magnesium levels
      in, 58
  magnesium depletion during, 58
Prehistoric humans, calcium and
      magnesium intake in, 12, 15,
      17–18, 157
Preterm labor, 133
Prostanoids, 77
Protein, effect on magnesium uptake,
      158
Protein kinase C, 129
Protein metabolism, in magnesium
      deficiency, 107–108
Psychoses, magnesium in, 369–377
  in bipolar disorder, 369–371
  in major depression, 369, 371–373
  in schizophrenia, 369, 373–374

Pyridoxine metabolism, in magnesium
      deficiency, 110
Pyrophosphates, 242

Q
Quinidine, 28

R
Reactive oxygen species (ROS), 130,
      176–177, 229–230, 235
Rectal cancer, 161
Red wine, as polyphenol source, 12
Renal disease, ionized and total
      magnesium levels in, 58–59
Renal failure
  acute, renal magnesium wasting in, 60
  chronic
    bone disease associated with,
      308–310
    bone magnesium content in,
      309–310
    calcium excretion in, 306
    hypermagnesemia in, 307–308
    intestinal magnesium absorption
      in, 303–315
    ionized and total magnesium levels
      in, 56–57, 58
    sodium excretion in, 306
    as magnesium deficiency cause, 316
Renal insufficiency, magnesium
      excretion in, 305
Renal transplant patients
  intestinal magnesium absorption in,
      305
  ionized and total magnesium levels
      in, 58–59, 60
Ruminants
  hypomagnesemia in, 384–386
  magnesium absorption in, 382, 383
  spontaneous atherosclerosis in, 386
  urolithiasis in, 388

S
Salbutamol, 133
Schizophrenia, 344, 369, 373–374
Scotothuerapy, 121–124
Seizures, 74, 76, 176
Selenium, 106
Sertraline (Zoloft), 373

SLC4A1 member a2, 25
SLC41 member a1, 22, 23
SLC41 member a2, 22, 23
Sleep disorders, magnesium deficiency/depletion-related, 105, 118–119
Society for the Development of Research on Magnesium, 7
Sodium
  cancer-related increase in, 159
  effect on magnesium excretion, 306
  effect on sodium/magnesium exchange transport, 299–300
Sodium-calcium exchange system, 336
Sodium chloride, as extracellular mineral, 101
Sodium-magnesium exchange system, 336
  in basolateral membrane, 295–300
  characteristics of, 297–300
  in magnesium extrusion, 25–29
Sodium-potassium-adenosine triphosphatase pump, γ-subunit of, 36
Sodium-potassium pump, magnesium-ATP-driven, in hypertension, 228, 229
Sodium-potassium ratio, in hypertension, 228–229
Sodium valproate, 370, 371, 372
Soft tissue, magnesium content of, 69–70, 303
Sphingomyelinases, 129
Spleen, *MAT1* gene expression in, 24
Stomach, magnesium absorption in, 304
Stomach cancer, 161
Stress
  exercise-related, 178
  as magnesium depletion cause, 56, 118, 120
  oxidative, 179, 235
Stroke, 192, 239, 250–251, 342–343
Substance P, 179–180, 270
Sudden infant death syndrome, 117, 119
Suicide, 371–372
Surgery, as magnesium depletion cause, 75
Sweat, magnesium excretion in, 174
Syndrome X. *See* Metabolic syndrome

T
Taurine, 56
  as scototherapy agent, 122, 123
Teeth
  chemical composition of, 359–360
  magnesium in
    as apatite component, 361–364
    distribution of, 359, 362
    effect on cell adhesion, 359, 362–364
Testes, magnesium transporter protein expression in, 25
Tetany, 76, 175, 381
  grass, 384–385
  milk, 386
  winter, 385–386
Thromboxane A_2, 108
Thyroid, magnesium transporter protein expression in, 25
Tissue, magnesium content of, 158
Torsades de pointes, 236
Tramterene, 290
Transient receptor potential ion channels
  in magnesium renal reabsorption, 294–295
  superfamily of, 36–37
Transient receptor potential melastatin (TRPM) 6 chanzyme, 22–23, 34
  assembly with TRMP7, 41
  functional characteristics of, 39–41
  gene mutations of, 23, 36, 37, 39, 295
  homology with TRPM7, 36, 37
  α-kinase domain of, 22–23, 41, 42
  localization of
    along the distal convoluted tubules, 37–38
    in the intestines, 38
  regulation of, 38–39
  renal expression of
    effect on magnesium renal reabsorption, 38–39
    effect on metabolic acidosis, 39
  structural organization of, 22–23, 37
Transient receptor potential melastatin (TRPM) 7 chanzyme, 34
  amyotrophic lateral sclerosis/parkinsonism/dementia and, 23
  assembly with TRPM6, 23
  cellular functions of, 38

functional characteristics of, 39–41
inactivation of, 40
α-kinase domain of, 22–23, 41–42
magnesium binding with, 40–41
in magnesium homeostasis, 38
magnesium-nucleotide-regulated metal ion current of, 22
potassium-mediating activity of, 22
regulation of, 38–39
renal expression of, 38
in smooth vascular muscle cell growth, 27
sodium-mediating activity of, 22
structural organization of, 22–23
Tricyclic antidepressants, 372
Troponin, 174
L-Tryptophan, as scototherapy agent, 122, 123, 124
Tumor necrosis factor-α, 144, 270

U

Uremia
 intestinal magnesium absorption in, 318
 tissue magnesium content in, 307
Urine, calcium/magnesium molar ratio in, 94
Urolithiasis, 381, 386–388

V

Vascular smooth muscle cells
 effect of hyperglycemia on, 218
 effect of magnesium deficiency on, 242
  fatty acid unsaturation and, 253
 magnesium:1 calcium imbalance in, 229
 magnesium efflux from, 27
 magnesium-mediated contractility of, 227–228, 244–245
 signaling pathways in, 252–254
Vasoconstriction
 diabetes mellitus-related, 218
 magnesium:calcium imbalance-related, 229

Vasodilatation
 magnesium-related, 346
 magnesium-related reduction in, 241–242
Vasopressin, 74, 291, 294
Vasospasm, magnesium deficiency-related, 241–242, 244–245
Verapamil, interaction with magnesium, 371
Veterinary medicine, magnesium in, 6, 381–391
Vitamin B_6, 250
Vitamin B_{12}, 250
Vitamin C deficiency, magnesium deficiency-related, 110
Vitamin D, 72
 in intestinal magnesium absorption, 304–305
 in magnesium absorption, 317–318
 in magnesium depletion, 79–80
 in magnesium renal absorption, 291
 in magnesium uptake, 158
Vitamin D_3
 effect on magnesium renal re-absorption, 276–277
 effect on serum magnesium levels, 306
Vitamin E, 219
Vitamin K, 270–271
 in bone abnormalities, 264
Vitamin metabolism, in magnesium deficiency, 110

W

Winter tetany, 385–386
Wolff-Parkinson-White syndrome, 236–237

Z

Zinc
 body content of, in magnesium deficiency, 109, 110
 intestinal content of, interaction with magnesium, 106
 as intracellular mineral, 101

Printed in Singapore